Lecture Notes in Physics

T0188937

Volume 838

For further volumes:
http://www.springer.com/series/5304

The Lecture Notes in Physics

The series Lecture Notes in Physics (LNP), founded in 1969, reports new developments in physics research and teaching—quickly and informally, but with a high quality and the explicit aim to summarize and communicate current knowledge in an accessible way. Books published in this series are conceived as bridging material between advanced graduate textbooks and the forefront of research and to serve three purposes:

- to be a compact and modern up-to-date source of reference on a well-defined topic
- to serve as an accessible introduction to the field to postgraduate students and nonspecialist researchers from related areas
- to be a source of advanced teaching material for specialized seminars, courses and schools

Both monographs and multi-author volumes will be considered for publication. Edited volumes should, however, consist of a very limited number of contributions only. Proceedings will not be considered for LNP.

Volumes published in LNP are disseminated both in print and in electronic formats, the electronic archive being available at springerlink.com. The series content is indexed, abstracted and referenced by many abstracting and information services, bibliographic networks, subscription agencies, library networks, and consortia.

Proposals should be sent to a member of the Editorial Board, or directly to the managing editor at Springer:

Christian Caron
Springer Heidelberg
Physics Editorial Department I
Tiergartenstrasse 17
69121 Heidelberg/Germany
christian.caron@springer.com

Hans J. Krappe · Krzysztof Pomorski

Theory of Nuclear Fission

A Textbook

 Springer

Dr. Hans J. Krappe
Berlin
Germany

Prof. Krzysztof Pomorski
Theoretical Physics Division
Marie Curie-Sklodowska University
Lublin
Poland

ISSN 0075-8450
ISBN 978-3-642-23514-6
DOI 10.1007/978-3-642-23515-3
Springer Heidelberg Dordrecht London New York

e-ISSN 1616-6361
e-ISBN 978-3-642-23515-3

Library of Congress Control Number: 2011940203

Springer is part of Springer Science+Business Media (www.springer.com)

Preface

In his book "Theories of Nuclear Fission" of 1964, L. Wilets summarized the state of understanding nuclear fission at that time. Since then the field has considerably expanded owing to the vast increase in the amount and quality of experimental data, in particular those obtained in heavy-ion fusion-fission reactions or made possible by the development of mass separators and high-flux neutron sources. Motivated in many cases by the new data and greatly aided by the phenomenal increase in computing power over the past decades, considerable progress has also been made in our theoretical understanding of nuclear fission. Notably the construction of the potential energy surface profited from Strutinsky's shell correction and the introduction of energy-density functionals, which made realistic calculations possible in both, the Hartree-Fock and the Hartree-Bogolyubov frame. The progress has been less convincing for inertial and especially friction coefficients, needed to model fission dynamics. A variety of models have been proposed for the friction tensor with widely differing results.

Our aim in this book is to give an account of the development of the theory of fission in recent years. In order to keep the size of the book within reasonable bounds, we focus on low and medium energy fission. Thus we shall not discuss the decay of highly excited nuclei as in spallation and multifragmentation reactions or processes involving antiprotons or hyperons. Heavy-ion fusion reactions will be treated only marginally.

The reader is expected to be familiar with the basic elements of theoretical physics. We will frequently make reference to the first 6 volumes of the "Course of Theoretical Physics" by Landau and Lifshitz and to the textbook "The Nuclear Many-Body Problem" by P. Ring and P. Schuck for more detailed derivations and discussions of results. To represent multidimensional dynamics in the collective coordinates we will use tensor notation in Chap. 5. The reader may consult A. Lichnerowicz's "Elements of Tensor Calculus" for some elementary results from tensor analysis.

We thank the staff of the library in the Helmholtz-Zentrum Berlin für Materialien und Energie (formerly Hahn-Meitner-Institut für Kernforschung

Berlin) for providing access to literature online, to printed material and to microfiche documents. We are also indebted to the Maria Curie-Skłodowska University for continuous financial support, which covered the cost of travels between Berlin and Lublin.

We thank P. Möller, T. Leisner, F. Gönnenwein, and L. M. Robledo for allowing us to use unpublished figures. We appreciate the comments of D. Fick and F. Gönnenwein on sections of the forth chapter. Our particular thanks go to P. Fröbrich for a critical reading of the whole manuscript and valuable advice.

Over many years we have profited from discussions with colleagues at our home institutions and abroad. It would not be possible to name all of them. One of us (HJK), however, would like to mention in particular a short, but very fruitful collaboration with R. Nix and the lasting impression made on him by S. Großmann and W. Swiatecki and their style of doing physics. The other (KP) is particularly grateful to K. Dietrich and A. Sobiczewski for long years of collaboration and discussions.

Last, but not least, we want to thank our wives, E. Holub-Krappe and B. Nerlo-Pomorska, for their support and for their patience.

Berlin   H. J. Krappe
Lublin   K. Pomorski
July 2011

Contents

1

Introduction

After the unexpected discovery of nuclear fission in 1938 by Hahn and Straß-mann in Berlin [1] the basic concepts for a theoretical modeling of this reaction had been established very quickly. Many concepts were taken from other fields of physics and adapted for the specific use in nuclear fission theory. They came from as diverse disciplines as hydrodynamics, aerosol physics, statistical mechanics, atomic physics, molecular spectroscopy, and chemical reaction theory. Even the name "fission" was borrowed by Frisch from cell biology [2]. To introduce these concepts and discuss the similarities and differences of nuclear fission with other physical phenomena, we shall first present the evolution of fission theory along historical lines before using a more systematic approach in the following chapters.

1.1 The discovery of nuclear fission

In 1934 E. Fermi and his collaborators irradiated heavy elements, in particular natural uranium and thorium, with neutrons, moderated in paraffin to thermal energies [3]. Several β activities were identified in the irradiated probe and interpreted as decay chains following neutron absorption in uranium [4]. Radiochemists in Paris (I. Curie and P. Savitch [5]) and in Berlin (O. Hahn, L. Meitner, and F. Straßmann [6]) confirmed and extended these findings. Fermi's interpretation of the radiochemical results in terms of β and α decay chains of uranium and transuranium isotopes did not go without any objection [7]. It is however the merit of the Berlin group to have finally shown that the irradiation of uranium as well as thorium yields decay products of medium charge, specifically barium, strontium, yttrium, and krypton, i.e. fission products [1]. We know today that most of the early "transuraniums" had actually been fission products [8].

The observation of nuclear fission came completely unexpected. So much that Hahn hesitated to publish his sensational findings without having a plausible physical explanation. He communicated his scruples in a letter in

H. J. Krappe and K. Pomorski, *Theory of Nuclear Fission*,
Lecture Notes in Physics 838, DOI: 10.1007/978-3-642-23515-3_1,
© Springer-Verlag Berlin Heidelberg 2012

mid-December 1938 to his collaborator for decades, Lise Meitner, who was forced to emigrate from Nazi Germany to Stockholm only a few months earlier [9–13]. Lise Meitner discussed the findings of Hahn and Straßmann with her nephew Otto Robert Frisch during a Christmas holiday in Sweden. Invoking the concept of a volume-charged, incompressible liquid-drop to describe a heavy nucleus, developed earlier by v. Weizsäcker [14], and Niels Bohr and his collaborators [15, 16], they considered the binding energy of uranium as a function of global changes of the shape from spherical to elongated configurations. They concluded that uranium should become unstable against a split into equal fragments once it is sufficiently deformed from its ground state to overcome an energy barrier for which the modest amount of energy, gained by the absorption of a neutron, should be sufficient [17].

Within months after Hahn and Straßmann's discovery and Meitner and Frisch's explanation became known, the most important features of low-energy fission were established: the energy release or Q value of almost 200 MeV, resulting from the Coulomb repulsion of the fission fragments [18] and the neutron yield per fission event, $\nu > 1$, [19], thus opening the possibility for a chain reaction. Gamow had explained the α decay and its sometimes rather long half-lives by a quantum tunneling process of a preformed α particle through a Coulomb barrier. If the concepts of Meitner and Frisch were correct, one could expect spontaneous fission of uranium also from the ground state, but with a considerably longer half-life than for the α decay because of the larger reduced mass for (almost) symmetric fission. This was in fact observed by Flerov and Petrzhak in Leningrad [20] already one year and a half after Hahn and Straßmann's discovery.

The basic theoretical concepts on which our understanding of nuclear fission is still based today were also developed very quickly after Meitner and Frisch's paper [17]. Feenberg [21], Frenkel [22] and v. Weizsäcker [23] showed independently that a classical, spherical, homogeneously charged drop looses its stability against spontaneous deformation when the ratio of its electrostatic energy E_{Coul} to twice the surface energy E_{surf} becomes larger than 1. This ratio, $x = E_{Coul}/2E_{surf}$, became later known as the fissility. In a seminal paper [24] Niels Bohr and John A. Wheeler fully developed the concept of the energy surface of a nucleus as a function of a set of deformation parameters. Considering the systematics of nuclear binding energies, it is pointed out that decays of uranium into more than two fragments would be connected with an even larger Q value than binary fission. However, it is not the Q value that determines which decay mode actually occurs, but the height of the minimal energy barrier that has to be overcome in the multidimensional energy hypersurface as function of deformation parameters on a decay path leading from the compact shape of the uranium nucleus to the configuration of its decay fragments at infinity. In two dimensions one may visualize the situation in terms of a mountain ridge surrounding the compact shape of the initial state. A saddle point on that ridge opens into the fission valley leading eventually to two separated, in general elongated, fragments. A cloverleaf ternary fission process e.g. would, in this picture, be connected with a considerably higher saddle point, corresponding to a fission valley ending with three equally sized outgoing fragments. The deformation of the newly born fission fragments yields an explanation for the observed emission of neutrons and γ's from the fission fragments since the release of their deformation energy provides them with sufficient excitation energy to allow these emissions.

Another very important concept introduced in this early paper by Bohr and Wheeler is the treatment of the excited uranium nucleus in terms of a microcanonical ensemble. It allows to introduce the phase-space volume of the excited nucleus, its logarithm, which is the entropy, and its derivative with respect to the energy, which is the level density. Using these quantities, the fission decay-rate is expressed in the transition-state theory in terms of the phase-space volume at the fission saddle-point. One also obtains from the ratio of the phase-space volumes of fission and of the other decay channels the ratio of the fission rate to the rates of the other decays. The transition-state theory had been developed earlier by Pelzer and Wigner [25, 26] to calculate chemical reaction rates. It is the great advantage of this thermodynamic approach to the decay rates that only equilibrium properties, like energy barriers and level densities are needed without having to address details of the dynamics of the decay process. A detailed presentation of the statistical theory will be given in the third and fourth chapters.

Immediately after the discovery of fission technical applications of the chain reaction were considered. As an example we mention a detailed paper by Flügge [27], which appeared already in June 1939. With the beginning of the Second World War in September 1939 nuclear research was considered in most countries involved as of military importance with the consequence that style of research and research objectives were determined by the military for the next decade. An account of the immediate impact of the discovery of fission on nuclear science in various countries was presented in the first two sessions of the conference "50 Years with Fission" by early participants and witnesses of the development [28].

1.2 The classical liquid-drop model

1.2.1 Volume-charged drops

In the model of a homogeneously charged drop with a sharp surface the energy can be represented by the three-term expression [29, 30]

$$E(A, I, \beta) = -c_1(I)A + c_2(I)S(\beta) + \frac{\rho_c^2}{2} \int d\mathbf{r} \int \frac{d\mathbf{r}'}{|\mathbf{r} - \mathbf{r}'|} , \qquad (1.1)$$

where $I = (N - Z)/(N + Z)$ is the neutron excess, β is a set of deformation parameters, characterizing the nuclear shape, and $c_1(I)$ is the volume energy per nucleon. It turns out that a fairly good average description of empirical binding energies can be achieved with the ansatz $c_1 = a_V(1 - \kappa_V I^2)$ and universal nuclear parameters a_V and κ_V. The second term of the liquid-drop formula (1.1) is the surface energy. The I dependence of the surface tension c_2 is parametrized analogously: $c_2 = a_S(1 - \kappa_S I^2)$. $S(\beta)$ is the surface area. Assuming that the radius of a spherical nucleus is proportional to $A^{1/3}$, $R(A) = r_0 A^{1/3}$, one has for the surface area of the sphere $4\pi r_0^2 A^{2/3}$ and for its surface energy $E_{\text{surf}}^{(\text{spher})} = 4\pi c_2 r_0^2 A^{2/3}$. In the third, the Coulomb term of the liquid-drop formula (1.1), ρ_c is the charge density. The integral is to be extended in $\mathbf{r}$ and $\mathbf{r}'$ over the volume of the nucleus. One obtains for the Coulomb energy of the sphere $E_{\text{Coul}}^{(\text{spher})} = 3Z^2 e^2/(5r_0 A^{1/3})$.

To describe deformed nuclei it is useful to introduce the ratio of the surface of the deformed nucleus to that of the spherical nucleus with the same volume. This function, $B_{\text{surf}}(\beta) = S(\beta)/S(0)$, depends only on the shape parameters and is independent of the size of the nucleus. Similarly the ratio of the electrostatic energy of the deformed nucleus and that of a sphere of equal volume and charge, $B_{\text{Coul}}(\beta) = E_{\text{Coul}}(\beta)/E_{\text{Coul}}(0)$, can be defined. In terms of these shape functions and the dimensionless ratio $x = E_{\text{Coul}}^{(\text{spher})}/(2E_{\text{surf}}^{(\text{spher})})$ the deformation energy with respect to the energy of the sphere has a particularly transparent form when measured in units of the surface energy of the sphere of equal volume

$$\frac{E_{\text{def}}}{E_{\text{surf}}^{(\text{spher})}} = \xi = (B_{\text{surf}}(\beta) - 1) + 2x(B_{\text{Coul}}(\beta) - 1). \tag{1.2}$$

In this expression the first term on the right hand side tends to make the shape of the nucleus compact, whereas the second represents the disruptive effect of the Coulomb forces. All size and material parameters are lumped together in the fissility

$$x = \frac{Z^2}{A} \frac{3e^2}{40\pi r_0^3 c_2},$$

while the shape dependence is contained in the rather complicated dependence of the shape functions B_{surf} and B_{Coul} on the deformation parameters.

If the surface of the deformed nucleus is described in spherical coordinates by the expansion

$$R(\Omega) = r_0 A^{1/3}\left(1 + \sum_{l=0}^{\infty}\sum_{m=-l}^{l} \beta_{lm} Y_{lm}(\Omega)\right) \tag{1.3}$$

in spherical harmonics $Y_{lm}(\Omega)$, one obtains up to second order in the deformation parameters β_{lm} [30]

$$B_{\text{surf}} = 1 + \frac{1}{8\pi}\sum_{l,m}(l-1)(l+2)|\beta_{lm}|^2,$$

$$B_{\text{Coul}} = 1 - \frac{5}{4\pi}\sum_{l,m}\frac{l-1}{2l+1}|\beta_{lm}|^2. \tag{1.4}$$

Insertion into Eq.(1.2) shows that the first multipole deformation for which the restoring force

$$\frac{\partial \xi}{\partial \beta_{lm}} = \left\{ \frac{1}{4\pi}(l-1)(l+2) - x\,\frac{5}{\pi}\,\frac{l-1}{2l+1} \right\} \beta_{lm} \tag{1.5}$$

changes sign is the quadrupole $l = 2$ when $x = 1$, which indicates that $x = 1$ is the stability limit for a sphere. Analytic expressions for the shape functions have been derived for higher orders in the β_{lm} than the second [31,32]. They become quickly rather complicated so that the integrals in the surface and the Coulomb energies are better evaluated numerically in order to map the whole hypersurface $\xi(\beta; x)$. It is quite helpful in this context that the six-dimensional Coulomb integral can be reduced to a double surface integral [33].

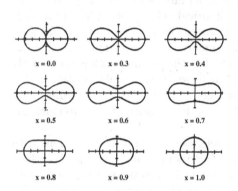

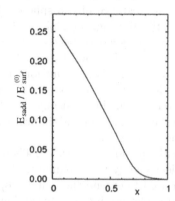

Fig. 1.1. Symmetric saddle-point shapes in the liquid-drop model (after Ref. [34]).

Fig. 1.2. Relative barrier heights as function of the fissility x (after Ref. [35]).

The most important features of the deformation-energy hypersurface $\xi(\beta, x)$ are the extrema for fixed x. When the multipole coefficients β_{lm} are chosen as deformation parameters in the vicinity of the sphere, the extrema are located at the zeros of the curly bracket in Eq. (1.5). In the relevant range of deformation parameters there is a minimum at the spherical shape, $\beta_{lm} = 0$ for all l, m, if $x < 1$, which becomes a maximum for $x > 1$ and $l = 2$, i.e. the matrix of the second derivatives $M_{2m;2m'} = \partial^2 \xi / \partial \beta_{2m} \partial \beta_{2m'}$ vanishes for $x = 1$. Then there is the fission saddle point, at which this matrix has one negative eigenvalue, all other eigenvalues being positive. The sequence of shapes corresponding to the fission saddle is plotted in Fig. 1.1 as a function of the fissility x. The corresponding barrier heights are shown in Fig. 1.2. For fissilities $x < 0.394$ one more eigenvalue of $M_{lm;l'm'}$ becomes negative. The point $(\beta^{(\text{saddle})}, x_{\text{Bus}} = 0.394)$ is called the Businaro-Gallone point. The eigenvector connected with this additional negative eigenvalue involves only multipole modes with odd l's, i.e. the saddle loses stability with respect to deformations without mirror symmetry [34,36]. The advantage of plots like the ones in Figs

1.1 and 1.2 is their independence of the specific values of the liquid-drop parameters c_1, c_2, and r_0. However, if one wants to know the barrier height of a specific nucleus in absolute units, one has to refer to a specific set of these parameters. Often used is the set $a_V = 16.99$ MeV, $\kappa_V = \kappa_S = 1.79$, $a_S = 18.56$ MeV, and $r_0 = 1.205$ fm, fitted by Myers and Swiatecki [37] to empirical ground-state masses and some fission-barrier heights. With these parameters one obtains for instance for the liquid-drop barriers of ^{236}U ($x = 0.7587$) and ^{210}Po ($x = 0.6991$) 6.957 MeV and 13.763 MeV, respectively.

In the liquid-drop model passage over the saddle does not automatically mean eventual separation into two disconnected fragments. In fact, two spheres with a thin thread connecting them would be a configuration which has asymptotically the same energy as two independent spheres since the energy of the thread goes to zero when the radius shrinks sufficiently fast with increasing distance of the spheres. What actually leads to a rupture of the neck is the Rayleigh instability of a cylindrical column of an incompressible liquid when its length becomes larger than its circumference [38], a phenomenon described first by Plateau [39]. Some modifications of Rayleigh's classical result are needed to account for the fact that the forming neck is not exactly a cylinder and that one is not dealing in fission with a static situation [40]. However, it remains basically the Rayleigh instability which determines the rupture configuration and, in particular, the deformation of the two emerging fragments at scission.

With the advent in the 1960's of accelerators generating heavy ions with sufficient energy to overcome the fusion barrier of heavy target nuclei, there was a need to extend the liquid-drop model (1.1) in two respects. First, compound nuclei produced in heavy-ion fusion reactions can carry large angular momenta. Therefore a rotational energy term was added to the liquid-drop formula. Second, since nuclear densities have exponential tails rather than a sharp surface, there is a short-ranged, attractive nuclear force before two nuclei approaching each other actually touch. It was called proximity force in Ref. [41].

Classical droplets rotate rigidly in equilibrium – because they are viscous, a superfluid drop of helium does not. We shall see in the next section that nuclei behave for moderate angular momenta rather like a superfluid drop. At higher excitation energy or for large angular momenta they tend to rotate rigidly. The investigation of rigidly rotating liquid bodies has a long history. Volume-charged drops have been considered in detail by Pik-Pichak [42, 43] and by Cohen, Plasil and Swiatecki [44]. In analogy to the shape functions in Eq. (1.2) the ratio $B_{rot}(\beta)$ of the rotational energy of a deformed drop to the rotational energy of a sphere with the same volume and the same angular momentum

$$B_{\text{rot}} = \frac{\mathcal{J}_{\text{spher}}}{\mathcal{J}_{\text{def}}} = \frac{(2/5)M_{\text{nucl}}AR_0^2}{M_{\text{nucl}}A/(4\pi R_0^3/3)\int \rho^3 d\rho d\phi dz} = \frac{8\pi R_0^5}{15\int \rho^3 d\rho\, d\phi\, dz}$$

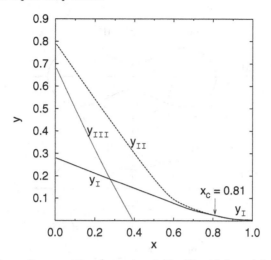

Fig. 1.3. Phase diagram of a charged, rotating liquid drop (after Ref. [44]).

is introduced, where $\mathcal{J}_{\text{spher}}$ and $\mathcal{J}_{\text{def}}$ are the moments of inertia for rigid rotation of the sphere and the deformed nucleus, respectively, and the integrand is written in cylindrical coordinates, the z axis being the rotation axis and the integral is to be extended over the nuclear volume. Measuring the deformation as before in units of the surface energy of the sphere, one obtains instead of Eq. (1.1)

$$\xi = (B_{\text{surf}} - 1) + 2x(B_{\text{Coul}} - 1) + y(B_{\text{rot}} - 1), \qquad (1.6)$$

where

$$y = \frac{E_{\text{rot}}^{(\text{spher})}}{E_{\text{surf}}^{(\text{spher})}} = \frac{5}{16\pi} \frac{\hbar^2 J^2}{M_{\text{nucl}} r_0^4 c_2 A^{7/3}} \qquad (1.7)$$

is a measure of the angular momentum $\hbar J$ and M_{nucl} is the mass of a nucleon. Inserting nuclear parameters, one obtains $y \approx 2J^2 A^{-7/3}$. The centrifugal force leads to a spheroidal deformation of the liquid-drop ground state. The stability properties of the extended liquid-drop model (1.6) are conveniently discussed in terms of a phase diagram in the x, y plane. In the area below the curve y_I in Fig. 1.3 the ground-state shapes are close to oblate spheroids rotating around the symmetry axis. They have been called Hiskes shapes [45]. There is a shape phase-transition at the line y_I to triaxial shapes, called Beringer-Knox [46] shapes in Ref. [44]. They rotate around the smallest axis. The line y_{II} branches off the line y_I at $x \approx 0.81$ [43]. Beyond the curve y_{II} no stable ground state exists any more.

The saddle-point shapes are more compact for finite y than without rotation. They are reflection symmetric between curves y_{II} and y_{III} and reflection asymmetric below curve y_{III} in Fig. 1.3. They are axially symmetric for $y = 0$ and then resemble cylinders with rounded ends for $0.7 < x < 1$ and are dumbbell-shaped for smaller fissility. For $y > 0$ the saddle points are

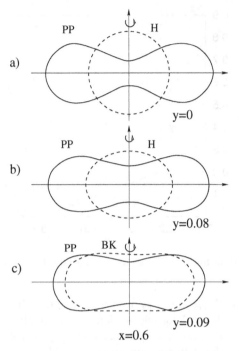

Fig. 1.4. Mean transverse cross sections of ground-state shapes for a sequence of drops with fissility $x = 0.6$ and increasing angular momentum (dashed lines). They have Hiskes shapes (H) for the two smaller y parameter values, and triaxial Beringer-Knox shapes (BK) for $y = 0.09$. The corresponding saddlepoint shapes are shown as full lines (Pik-Pichak shapes, PP) (after Ref. [44]).

in general triaxial and were called Pik-Pichak shapes. In Fig. 1.4 a sequence of ground-state and saddle-point shapes is shown for $x = 0.6$. Approximate analytical expressions for the phase-transition lines in Fig. 1.3 and for the corresponding deformation energies are given in the appendix of Ref. [47].

Long before interest was directed towards the study of charged, rotating, liquid drops in nuclear physics, equilibrium shapes of rotating, gravitating drops have been investigated by several mathematicians and astronomers, an account of which can be found in the textbook by Appell [48] or in the monographs by Lichtenstein [49] and Chandrasekhar [50]. Eq. (1.6) can be used for this case also by allowing the fissility parameter x to become negative to account for the attractive character of the gravitational force in contrast to the repulsive Coulomb force. The stability diagram, Fig. 1.3, has therefore an extension for $x < 0$ with more than just a formal meaning [52]. In astrophysics the most interesting case is a rotating body with negligible surface tension $(x \rightarrow -\infty, y \rightarrow \infty)$. On its surface the centrifugal force must just balance the gravitational force. One can show that the resulting equilibrium shapes are exactly ellipsoids [50, 51]. Their moments of inertia are plotted in units

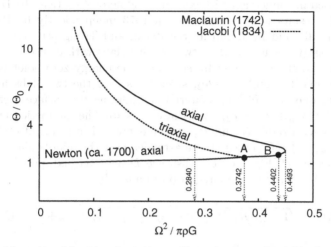

Fig. 1.5. Moments of inertia of rotating, self-gravitating celestial bodies at which the total energy is stationary as function of the angular velocity (after Ref. [51]).

of the moments of inertia of the sphere with equal volume in Fig. 1.5 against the dimensionless variable $2h = \Omega^2/(\pi G\rho)$, where Ω and ρ are the angular velocity and mass density of the celestial body, respectively, and G is the constant of gravitation. If one replaces $e^2 Z^2$ in the definition of the fissility x by $-GM_{\text{tot}}^2$, one obtains

$$h = - \lim_{E_{\text{surf}} \to 0} y/x = \Omega^2/(2\pi\rho G) \ .$$

There is no stationary solution for $2h > 0.4493$; two solutions exist for $0.3742 < 2h < 0.4493$ and three solutions for $2h < 0.3742$. The full line corresponds to oblate spheroids, rotating around their smaller axis. They are called Maclaurin spheroids. Along the lower branch the eccentricity and the moment of inertia increase with increasing Ω. On the upper branch they continue to increase, now with decreasing Ω. The dashed line in Fig. 1.5 corresponds to ellipsoids with three different axes. They are called Jacobi ellipsoids. Their eccentricity and moment of inertia increase with decreasing Ω. For $\Omega \to 0$ the Maclaurin spheroids become spheres and infinitely thin disks with infinite radius for the lower and upper branch, respectively. The Jacobi ellipsoids become infinitely long needles in this limit. They become unstable against pear-shaped deformations for $2h \leq 0.284$, the Poincaré point. The Maclaurin spheroids become unstable at the point B in Fig. 1.5, the Riemann point with $2h = 0.4402$, and remain unstable for larger eccentricity. Continuing the fissility from $x = -\infty$ to positive values, the Jacobi ellipsoids pass over to Beringer-Knox shapes [52].

Another feature added to the liquid-drop model when heavy-ion scattering experiments become possible were the proximity forces. They were first discussed in connection with colloidal and aerosol particles by Derjaguin [53, 54]. His results were rediscovered in 1973 independently by Bass [55], Wilczyński [56], and by Randrup, Swiatecki, and Tsang [41, 57]. If the interaction energy per unit area of two parallel slabs with distance s between half density points is denoted by $e(s)$, and the energy zero-point is chosen at $s = \infty$, one has $e(0) = -2c_2$, since for $s = 0$ the two slabs have just merged into one piece of bulk material annihilating the surface-energy density $c_2 A^{2/3}/(4\pi r_0^2 A^{2/3}) = c_2/(4\pi r_0^2)$ of each slab. The interaction energy of two spheres with radii R_1 and R_2 can be expressed in terms of $e(s)$ by the integral $V(s) = \int e(D)\rho \, d\phi \, d\rho$ to be extended over the middle plane between the two spheres, cf. Fig. 1.6. The distance D between juxtaposed points on the surfaces is related to the integration radius ρ by

$$D = s + R_1(1 - \sqrt{1 - (\rho/R_1)^2}) + R_2(1 - \sqrt{1 - (\rho/R_2)^2}) = s + \frac{\rho^2}{2\overline{R}} + \mathcal{O}(\rho^4/\overline{R}^3),$$

where $\overline{R} = R_1 R_2/(R_1 + R_2)$ is half of the harmonic mean of R_1 and R_2. Changing integration variables from ρ to D, and neglecting terms of order $\rho^3/\overline{R}^3$, one obtains

$$V(s) = 2\pi\overline{R} \int_s^\infty e(D)dD. \qquad (1.8)$$

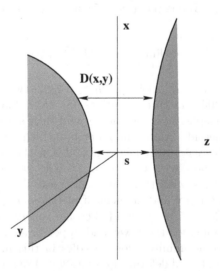

Fig. 1.6. Proximity interaction between two spheres with central radii R_1 and R_2 and shortest distance s between their surfaces.

Use has been made of the fact that $e(s)$ decreases exponentially with increasing s so that the upper limit of the integral in Eq. (1.8) could be shifted

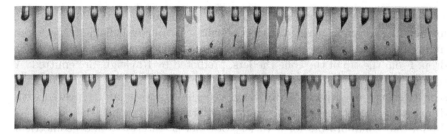

Fig. 1.7. Jet formation on a pending drop in a supercritical electric field (from Ref. [61]).

to infinity. Corrections to this lowest order expression have been considered by Błocki and Swiatecki [58]. The function $e(s)$ was calculated in Thomas-Fermi approximation in Ref. [41]. Implicitly the assumption was made that the density profiles of the two approaching drops remain "frozen" when they overlap.

The proximity formula (1.8) has the important consequence that $V'(s = 0) = (\overline{R}/r_0^2)c_2$, relating the proximity potential to the surface tension constant c_2. This formula was first derived by Bradley [59]. In addition, the saturation properties of nuclear matter imply $e'(s = 0) = 0$, i.e. $V''(s = 0) = 0$.

1.2.2 Surface-charged drops

Volume-charged drops are the exception, rather than the rule in nature since it takes the strong nuclear forces to keep the charges from assembling on the surface to minimize the electrostatic energy. Charged classical drops, like aerosol droplets, and liquid clusters of metal atoms carry surface charges in equilibrium. For such drops Lord Rayleigh derived a stability estimate [60]. He found that the restoring force of small multipole vibrations around the spherical shape vanishes first for the quadrupole mode and this happens when the fissility x becomes one. For the conducting sphere with radius R, surface tension σ, and charge Q the surface and Coulomb energies are $4\pi\sigma R^2$ and $Q^2/(2R)$, respectively. Therefore the fissility is in this case $x = Q^2/(16\pi\sigma R^3)$. One sees immediately that the outward directed Maxwellian tension, produced by the surface charge $Q^2/(8\pi R^4)$, just equals the inwards directed surface tension $2\sigma/R$ when $x = 1$, i.e. the surface looses locally its mechanical stability at this fissility. Only for the sphere do the two types of instability, the local instability, called Rayleigh-Taylor instability, and the instability against global changes of the shape, sometimes called secular instability, happen to coincide. Rayleigh concludes this section of his paper [60] on the stability of the charged, conducting sphere with the remark that for supercritically charged drops "the liquid is thrown out in fine jets, whose fineness, however has a limit."

The jets were first observed by Zeleny [63], not for free charged drops, but for pending drops in a vertical electric field. At a critical field strength the induced electric surface-charge density leads to the formation of a cone on the drop. From the tip of the cone the jets, predicted by Rayleigh, are emitted. A sequence of pictures was taken [61], which is shown in Fig. 1.7. The process is seen to repeat itself periodically. Such experiments were later repeated with drops falling freely through a horizontal, polarizing electric field [64,65] and by Taylor and collaborators for a variety of different interfaces and geometrical arrangements [66].

Only more recently have free, charged drops of some tens of μm radius been levitated in electromagnetic traps and their dynamical behaviour has been observed close to the critical fissility, see Ref. [68] and further references given there. In these experiments the fissility was directly determined by measuring the amplitude and slippage phase of a quadrupole oscillation induced by a small electric AC field as a function of the surface-charge density. Close to $x = 1$ the amplitude becomes very large and the motion anharmonic. Fig. 1.8 shows a sequence of snapshots obtained by Leisner and his collaborators [62]. The originally spherical drop has deformed to a prolate spheroid in frame a, develops a spindle-like shape in b, and ejects jets of charged particles in c. Thereafter it returns to a spherical shape. Although the loss of charge during the Coulomb explosion amounts to more than 30% of the total charge, there is only a tiny loss of mass connected with it.

The calculation of the Coulomb energy is more complicated in the case of surface-charged drops than for volume-charged drops. As the given surface is required to be an equipotential surface, one has to solve an electrostatic boundary-value problem. Writing the Coulomb energy as a functional of the surface-charge density σ, one finds the equilibrium charge-distribution from minimizing the energy with respect to σ with the constraint of a given total

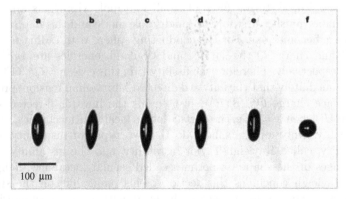

Fig. 1.8. Coulomb explosion observed on a levitated drop of glycol of 48μm diameter (after Ref. [62])

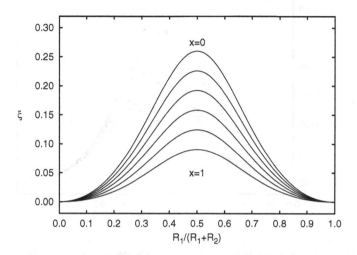

Fig. 1.9. Energy of two touching conducting spheres as function of the asymmetry of the system for fissilities x between 0 and 1 (after Ref. [67]).

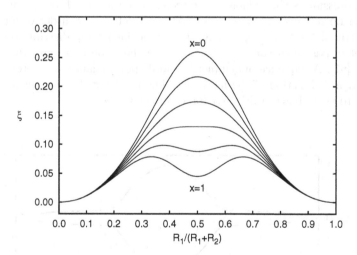

Fig. 1.10. Energy of two touching volume-charged spheres as function of the asymmetry of the system for fissilities x between 0 and 1 (after Ref. [67]).

charge [69]. In this way a formula for small multipole deformations β_{lm} can be derived

$$B_{\mathrm{Coul}} = 1 - \frac{1}{4\pi}\sum_{lm}(l-1)|\beta_{lm}|^2.$$

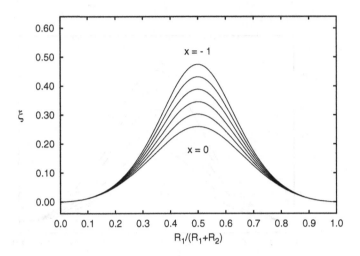

Fig. 1.11. Energy of two touching gravitating spheres as function of the asymmetry of the system for fissilities x between 0 and -1.

This expression is the analog of the second equation in (1.4). It was already given in Lord Rayleigh's paper [60]. Alternatively, one can derive for axially symmetric shapes an integral equation for a line-charge distribution on the symmetry axis, such that the given surface shape becomes an equipotential surface [67]. A sequence of symmetric saddle-point shapes was determined in this way as a function of the fissility x. The resulting shape sequence is very similar to that shown in Fig. 1.1 for volume-charged drops.

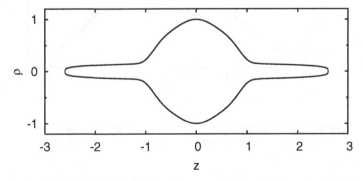

Fig. 1.12. Example of a Rayleigh-Taylor instability. The fissility $x = 0.9$ and the elongation variable $\alpha = 0.72$ were fixed. The next 19 symmetric deformation parameters α_{2n} in the generalized Cassinian-oval parametrization [70] are varied until a minimum in the deformation energy ξ was reached (after Ref. [67]).

All of these saddle points are unstable against asymmetric deformations, which was already noticed by Tsang [71] using a similar method. This finding is confirmed by calculating the deformation energy of a configuration consisting of two spheres with radii R_1 and R_2 in contact as a function of the asymmetry variable $R_1/(R_1+R_2)$ with $R_1^3+R_2^3$ =const. The result is shown in Fig. 1.9 for fissilities from 0 to 1 and compared with the corresponding deformation energy for volume-charged drops in Fig. 1.10. Energies are in units of the surface energy of the sphere with the volume $4\pi(R_1^3 + R_2^3)/3$. Formally Fig. 1.10 has an extension to negative fissilities x representing gravitating drops. This is shown in Fig. 1.11.

Apart from this global instability there is also an indication of the Rayleigh-Taylor instability or rather its precursor, the cone formation. When $x \geq 0.9$, numerical search for the spheroidal saddle leads always into energy "valleys" corresponding to shapes as given in Fig. 1.12. Note that the chosen parametrization was not flexible enough to represent shapes as in frame c of Fig. 1.8. Although this phenomenon of Coulomb spray is widely used in a variety of industrial applications [72], there is so far no dynamical theory for this decay process.

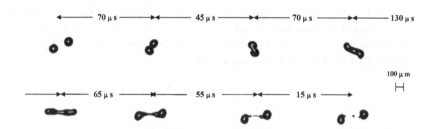

Fig. 1.13. Eight snapshots of a collision of two droplets of glycol with finite impact parameter. The time between consecutive pictures is indicated above in μs. The compound system remains approximately mirror symmetric and reseparates leaving a very small droplet from the neck between the two large fission fragments (after Ref. [73]).

In another experiment involving levitated micro-particles it was nicely shown that also drops with macroscopic dimensions can decay by fission into two fragments of comparable size, rather than emitting jets. In this experiment one drop, kept in the trap, was bombarded with another drop and the fusion and subsequent fission process was recorded. The total charge of the system was now kept below the critical fissility, however, in analogy to a heavy-ion fusion-fission reaction, a large amount of angular momentum was brought into the system. In the sequence of snapshots of this fusion-fission reaction, shown in Fig. 1.13, the highly viscous system can be seen to rotate until an elongated compound state is reached, which subsequently fissions into two fragments of comparable size [73, 74]. This time it is the interplay between the cohesive surface tension and the disruptive centrifugal force, i.e. a volume force, which controls the reaction, rather than the Maxwellian tension at the surface.

1.2.3 Hydrodynamics

The only use made so far of the liquid-drop model concerned the formula for the binding energy. Already very early, however, a much wider use of the concept of an incompressible nuclear liquid was made by invoking the Euler equations [24, 35, 75–78] or even the (linearized) Navier-Stokes equations [79–82] to describe nuclear dynamics. A microscopic justification of this wider use of Newtonian fluid dynamics is considerably more complicated and in fact still controversial as we shall see in the fifth chapter. Nevertheless, conceptually hydrodynamics has been very fruitful to understand nuclear collective motion, in particular low-lying vibrational states and the fission process as an evolution of nuclear shapes.

For an irrotational, incompressible flow the velocity field $\mathbf{v}(\mathbf{r})$ can be derived from a velocity potential $\phi(\mathbf{r})$, which satisfies the Laplace equation $\Delta\phi = 0$ inside the fluid. For the surface oscillations of a spherical drop, sometimes called capillary modes, one may represent the general solution of the Laplace equation in spherical coordinates by

$$\phi(r, \theta, \phi) = \sum_{lm} A_{lm}(t) r^l Y_{lm}(\theta, \phi),$$

where the coefficients A_{lm} have to be determined from the boundary condition $\hat{\mathbf{n}} \cdot \nabla \phi = \hat{\mathbf{n}} \cdot \dot{\mathbf{R}}$ on the free surface, where $\hat{\mathbf{n}}$ is the unit normal vector on the surface. Representing the surface modes by Eq. (1.3), one has $\hat{\mathbf{n}} = \mathbf{R}/r$ in lowest order in the parameters β and obtains

$$A_{lm}(t) = \dot{\beta}_{lm}/(l R_0^{l-2}).$$

For constant mass density ρ_m the kinetic energy of the fluid is

$$T = \frac{1}{2}\rho_m \int \mathbf{v}^2 d\tau = \frac{1}{2}\sum_{lm} B_l |\dot{\beta}_{lm}|^2 \qquad (1.9)$$

with the inertial parameter $B_l = \rho_m R_0^5/l$. Together with the potential energy from Eqs. (1.2) and (1.5) the classical Lagrangian for capillary modes becomes in terms of the generalized coordinates β_{lm}

$$\mathcal{L}(\dot{\beta}_{lm}, \beta_{lm}) = \frac{1}{2} \sum_{lm} (B_l |\dot{\beta}_{lm}|^2 - C_l |\beta_{lm}|^2) \tag{1.10}$$

with the restoring force constant [30]

$$C_l = R_0^2 c_2 (l-1) \left(l + 2 - \frac{20\,x}{2l+1} \right)$$

of the lth multipole mode. Eq. (1.10) represents a system of harmonic oscillators with frequencies $\omega_l = \sqrt{C_l/B_l}$.

For ellipsoidal drops it is natural to introduce body-fixed coordinates oriented along the axes of the inertial tensor. When the orientation of the body-fixed coordinates with respect to space-fixed coordinates is given by Euler angles $\theta_1, \theta_2, \theta_3$ the transformation from deformation parameters β_{lm} to body-fixed parameters a_{lm} is given by $\beta_{lm} = \sum_{m'} \mathcal{D}^{l*}_{mm'}(\theta_1, \theta_2, \theta_3) a_{lm'}$, where the $\mathcal{D}$ functions shall be used in the following as defined by Edmonds [83]. For the quadrupole mode $l = 2$ one can choose the Euler angles such that

$$a_{20} = \beta \cos\gamma, \qquad a_{2\pm1} = 0, \qquad a_{2\pm2} = \frac{\beta}{\sqrt{2}} \sin\gamma. \tag{1.11}$$

Instead of the five variables β_{2m} ($m = 0, \pm1, \pm2$) the two intrinsic variables β and γ and the three Euler angles θ_i are introduced as the new collective variables. A. Bohr has shown that the kinetic energy (1.9) expressed in the new variables has the form [84, 85] $T = T_{\text{vib}} + T_{\text{rot}}$ with

$$T_{\text{vib}} = \frac{1}{2} B_2 (\dot{\beta}^2 + \beta^2 \dot{\gamma}^2)$$

and

$$T_{\text{rot}} = \sum_{\kappa=1}^{3} \frac{M_\kappa^2}{2\mathcal{J}_\kappa},$$

where $\mathcal{J}_\kappa = 4B_2\beta^2 \sin^2(\gamma - 2\pi\kappa/3)$ are the irrotational moments of inertia in the three main inertial directions and the M_κ are the components of the angular momentum in these body-fixed directions. The parameter γ characterizes the deviation from axial symmetry: Prolate spheroids have $\gamma = 0$, oblate spheroids $\gamma = \pi/3$ and ellipsoids are characterized by $0 < \gamma < \pi/3$. The irrotational moment of inertia of a spheroid, rotating around an axis perpendicular to the symmetry axis, is $\mathcal{J}_{\text{irrot}} = 3B_2\beta^2 \approx 0.9\beta^2 \mathcal{J}_{\text{rigid}}$.

To account for the discrete nature of rotational and vibrational states and the tunneling process in spontaneous fission, a "quantization" of the hydrodynamical equations of motion is obviously needed. The theory is constructed in rather close analogy to the methods used in the spectroscopy of binary molecules [85, 86].

Neither the irrotational moments of inertia, nor the rigid moments describe experimental rotational spectra quantitatively. The same is true for the vibrational spectra. Already Hill and Wheeler [87] remarked that the nuclear fluid must be quite different from an ordinary liquid: "It (the nuclear fluid) is considered to be completely transparent internally with respect to motion of the constituent particles, and to receive disturbances solely by way of surface deformations. Its near incompressibility comes about, not by particle to particle push, as in an ordinary liquid, but by more subtle means. It is capable of collective oscillations, but it is the wall which organizes these disturbances, not nucleon to nucleon interactions. Oscillations experience a damping, but the mechanism of the damping is unlike that encountered in ordinary liquids." This mechanism was later called one-body dissipation.

1.3 Shell model and pairing correlations

The nuclear binding energy per particle shows maxima as function of the nucleon numbers when either the neutron or the proton number is one of the magic numbers 2, 8, 20, 28, 50, 82, or 126. An analogy was early noticed with the especially strong binding of the electrons in atomic physics at the shell closure for the nobel gas atoms. The physical concept of individual electrons moving without collisions in a common spherical potential well, used in atomic physics, was adapted to the nuclear case by Haxel, Jensen, and Suess [88] and Goeppert-Mayer [89] by adding a strong spin-orbit coupling term to the central potential in order to reproduce the nuclear magic numbers. The nuclear shell model became highly successful in nuclear spectroscopy, in particular after Rainwater [90] and A. Bohr [91] pointed out that the Jahn-Teller effect would remove the high degeneracy of the shell-model ground-state energies of non-magic nuclei by breaking the spherical symmetry of the single-particle potential. In particular, the quadrupole-deformed oscillator potential was used extensively by Nilsson [92] and his collaborators to calculate single-particle states, magnetic moments, spectroscopic factors, and transition matrix elements of various multipolarities of midshell nuclei [93].

In the first 1-2 MeV above the Fermi energy the level scheme of even-even nuclei has only a few collective states of vibrational or rotational character, in contrast to neighboring odd-even nuclei which show already a high level density just above the Fermi surface. The existence of an energy gap for even-even systems is not understandable in the framework of the independent particle model, but is reminiscent of the electronic energy gap in superconducting metals. Note that the latter systems are understood as being infinitely extended in space and would therefore not have collective rotational or vibrational states in the gap. Bohr, Mottelson, and Pines therefore proposed to adopt the BCS theory [94] developed for an infinite electron gas to the "gas" of nucleons in a finite potential well [95]. Making reasonable assumptions on the strength of short-ranged effective nucleon-nucleon forces it was in fact possible to understand not only the spectral density of even and odd nuclei above the Fermi surface, but also to account for a reduction of the moment of inertia of deformed nuclei compared with the rigid moment. Belyaev showed that in the BCS model for superfluidity one can reconcile the measured electric quadru-pole moments of these nuclei with the moments of inertia extracted from ground-state rotational bands [97].

One does not get the total energy of a nucleus by just summing the shell-model single-particle energies of all occupied levels; first of all, because one would count the potential energy twice. Now let us assume that the oscillator potential of the Nilsson model were the result of a self-consistent calcula-tion of the single-particle potential. By virtue of the virial theorem for the oscillator the kinetic and the potential energies contribute $1/2$ to the total single-particle energy. Reducing the potential energy by $1/2$ to avoid double counting therefore results in taking just $3/4$ of the total energy. This recipe however, does not lead to satisfactory results for the total binding energy. But, if one calculates in this way the energy as function of Nilsson's quadrupole deformation parameter δ, one finds energy minima for the rare-earth nuclei at the empirically determined deformations, in particular when pairing corre-lations are taken into account in the BCS scheme and Coulomb energies are calculated more carefully [98].

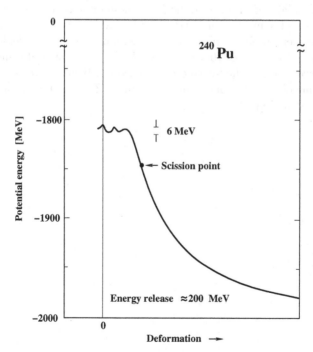

Fig. 1.14. Schematic plot of the potential energy along the fission path for ^{240}Pu (from Ref. [96]).

The tremendous success of the shell model plus pairing correlations in the spectroscopy of low-lying excited states called for some kind of unification with the seemingly completely opposite concepts of the liquid-drop model which had been as successful in understanding low-energy nuclear fission. Myers and Swiatecki [37] and Strutinsky [99] introduced in 1966 independently the concept of shell corrections to the liquid-drop energy which meant a considerable break-through in this respect. In particular with Strutinsky's shell-correction it is not only possible to account for the extra binding of nuclei with magic proton or neutron numbers, but one could obtain in addition the ground-state deformations of midshell nuclei in good agreement with experiments [100,101], and above all, understand the occurrence of secondary minima in the deformation energy surface of fissile nuclei, leading to fission isomeric states [102,103], which had been discovered some years earlier by Polikanov and his collaborators [104] in the framework of a program to produce superheavy fissile nuclei in heavy-ion fusion reactions. Although the shell effects on the potential energy of fissile nuclei are only of the order of a few MeV and therefore small compared to the total energy release in fission of the order of 200 MeV, as shown schematically for ^{240}Pu in Fig. 1.14, they have a decisive effect on spontaneous fission and the fission decay of heavy compound nuclei with moderate excitation energy.

Also finer details of the two-saddle ("double-humped") energy surface, like the pear shape of some fission barriers [105] and the near constancy of the fission-barrier heights from thorium to berkelium (see e.g. Vandenbosch and Huizenga: *Nuclear Fission* [106] Chap. VIII) could be naturally understood in the framework of the shell-correction procedure. It also gave rise to speculations that nuclei near the magic proton number $Z = 114$ might have a finite lifetime against fission [93,107].

2

Potential-Energy Surfaces

In this chapter we present a microscopic justification for the potential-energy surface, used in the introduction as an important ingredient in the description of nuclear fission. It will be shown in particular that the liquid-drop model and its shell corrections can be systematically derived from a mean-field approach to the quantal nuclear many-body problem. A number of variants of the liquid-drop model and for the shell corrections have been proposed. Their generic relations will be discussed in the following.

Phenomenological parameters in the liquid-drop energy and in the shell correction are usually fitted to describe empirically known ground-state properties of nuclei, in particular binding energies from the nuclear mass table and fission barrier heights. The same holds for parameters in the effective two-body interactions, used in the selfconsistent mean-field approach. The first sections of this chapter are therefore devoted to the calculation of nuclear ground-state properties. With parameters fitted to these properties, the construction of energy surfaces as functions of shape parameters is presented in the last section with a particular emphasis on secondary minima and saddle points.

2.1 Mean field theories

Many attempts have been made to obtain ground-state binding energies and radii on an a priori basis from empirical nucleon-nucleon potentials by solving the many-body Schrödinger equation in the self-consistent Brueckner-Hartree-Fock approximation. Even with the inclusion of three-body correlations the results were not very encouraging. For a discussion see e.g. Ref. [108] Sec. 5.6.1.1. The success of the local energy-density functional method in correlated many-electron systems [109–112], on the other hand, motivates the more modest approach of using phenomenological nuclear energy-density functionals to obtain nuclear ground-state properties. For historical reasons they are formulated in terms of short-range or zero-range two-body and zero-range

H. J. Krappe and K. Pomorski, *Theory of Nuclear Fission*,
Lecture Notes in Physics 838, DOI: 10.1007/978-3-642-23515-3_2,
© Springer-Verlag Berlin Heidelberg 2012

three-body potentials to be used in mean-field approximation. A large number of such effective potentials has been proposed and fitted to describe either binding energies of nuclei more or less close to the beta stability line of the mass table or extremely neutron rich nuclei or neutron matter. In addition, certain spectroscopic data, the nuclear level density or fission barriers were used in the fits. In an alternative approach the phenomenological relativistic field theory of Serot and Walecka [113] was used in mean-field approximation to describe a variety of nuclear data. In the following effective potentials and corresponding functionals of relevance in the theory of fission will be described.

We use here the term mean-field theory to include the non-relativistic approach specifically the Hartree-Fock approximation and its semiclassical analogues, the Thomas-Fermi (TF) and extended Thomas-Fermi (ETF) approximations. The term is also used for the Hartree-Fock-Bogolyubov approximation (HFB) when pairing correlations are to be included in the theory. Since these approaches to the many-body problem have been described in many text books (cf. e.g. Ring and Schuck, *The Nuclear Many-Body Problem* [108], Brack and Bhaduri, *Semiclassical Physics* [114], and Petkov and Stoitsov, *Nuclear Density Functional Theory* [112] or the review by Bender, Heenen, and Reinhard [115]), we do not give complete, formal derivations of these theories, but for later references the basic formulae are presented in the following.

2.1.1 The Hartree-Fock theory

The starting point is the Hamiltonian for the A nucleon system

$$H = -\frac{\hbar^2}{2M_{\text{nucl}}} \sum_{n=1}^{A} \Delta_n + \sum_{n,n';n<n'}^{A} V(\mathbf{r}_n - \mathbf{r}_{n'}; \sigma_n \sigma_{n'}, \tau_n \tau_{n'}) \,, \qquad (2.1)$$

where the two-particle interaction $V(\mathbf{r}_1 - \mathbf{r}_2; \sigma_1\sigma_2, \tau_1\tau_2)$ consists of the long-range Coulomb interaction between the protons and a spin and isospin-dependent, short-range, effective nuclear interaction $V_{\text{eff}} = V(\mathbf{r}_1 - \mathbf{r}_2)\hat{A}(\hat{P}_\sigma, \hat{P}_\tau)$, where the operator

$$\hat{A} = W + B\hat{P}_\sigma - H\hat{P}_\tau - M\hat{P}_\sigma\hat{P}_\tau \qquad (2.2)$$

represents the standard dependence of the central part of the effective two-body potential on the spin and isospin exchange-operators $\hat{P}_\sigma = \frac{1}{2}(1+\boldsymbol{\sigma}_1\boldsymbol{\sigma}_2)$ and $\hat{P}_\tau = \frac{1}{2}(1+\boldsymbol{\tau}_1\boldsymbol{\tau}_2)$ with empirical parameters W, B, H, and M, which are the strength parameters of the Wigner, Bartlett, Heisenberg, and Majorana interactions, respectively.

Let the functions $\chi_i(\mathbf{r}, \sigma, \tau)$ form a complete set of orthonormal single-particle states. They can, for instance, be the eigenstates of a shell-model single-particle Hamiltonian. With respect to this set of states nucleon creation and annihilation operators $\hat{a}_i^+$ and $\hat{a}_i$ are defined in terms of which the Hamiltonian (2.1) may be written in Fock-space representation as

$$\hat{H} = \sum_{l,l'} t_{ll'} \hat{a}_l^+ \hat{a}_{l'} + \frac{1}{2} \sum_{l_1 l_2 l_1' l_2'} V_{l_1 l_2 l_1' l_2'} \hat{a}_{l_1}^+ \hat{a}_{l_2}^+ \hat{a}_{l_2'} \hat{a}_{l_1'} \tag{2.3}$$

with

$$t_{ll'} = -\frac{\hbar^2}{2M_{\text{nucl}}} \int \chi_l^*(\mathbf{r}) \Delta \chi_{l'}(\mathbf{r}) d^3r,$$

and

$$V_{l_1 l_2 l_1' l_2'} = \int d^3r \int d^3r' \, \chi_{l_1}^*(\mathbf{r}_1 \sigma_1 \tau_1) \chi_{l_2}^*(\mathbf{r}_2 \sigma_2 \tau_2)$$
$$V(\mathbf{r}_1 - \mathbf{r}_2) \hat{A}(\hat{P}_\sigma, \hat{P}_\tau) \chi_{l_1'}(\mathbf{r}_1 \sigma_1 \tau_1) \chi_{l_2'}(\mathbf{r}_2 \sigma_2 \tau_2) \,.$$

In the Hartree-Fock approximation one seeks an approximate solution of the N-body problem in the set of Slater determinants $\det[\chi_i(\mathbf{r}_j \sigma_j \tau_j)]$, $i,j = 1 \ldots A$ built from A wave functions χ_i. In Fock space these A-particle states are represented by

$$|\text{HF}\rangle = \prod_{i=1}^A \hat{a}_i^+ |0\rangle \,.$$

It is the aim of the HF approximation to find the single-particle basis

$$\varphi_i(\mathbf{r}, \sigma, \tau) = \sum_{i'} U_{ii'} \chi_{i'}(\mathbf{r}, \sigma, \tau) \tag{2.4}$$

connected with the original basis of the functions χ_i by a unitary transformation $U_{ii'}$ and with creation and destruction operators

$$\hat{c}_i^+ = \sum_{i'} U_{ii'} \hat{a}_{i'}^+ \quad \text{and} \quad \hat{c}_i = \sum_{i'} U_{ii'} \hat{a}_{i'} \,,$$

for which the Slater determinant

$$|\text{HF}\rangle^{(0)} = \prod_{k=1}^A \hat{c}_k^+ |0\rangle \tag{2.5}$$

minimizes the expectation value $E^{\text{HF}} = \langle \text{HF} | H | \text{HF} \rangle$, while conserving the proton and neutron numbers. The density matrix is given by

$$\rho_{kl} = \langle \text{HF} | \hat{a}_l^+ \hat{a}_k | \text{HF} \rangle$$

with

$$\text{tr} \, \rho = A \,. \tag{2.6}$$

In terms of the density matrix and the antisymmetrized matrix element of the two-particle interaction

$$\overline{V}_{l_1' l_2' l_1 l_2} = V_{l_1' l_2' l_1 l_2} - V_{l_1' l_2' l_2 l_1} \,,$$

the expectation value E^{HF} is

$$E^{\mathrm{HF}} = \langle \mathrm{HF}|H|\mathrm{HF}\rangle = \sum_{l_1 l_2} t_{l_1 l_2}\rho_{l_2 l_1} + \frac{1}{2} \sum_{l_1 l_1' l_2 l_2'} \rho_{l_1 l_1'} \overline{V}_{l_1' l_2' l_1 l_2} \rho_{l_2' l_2} \,. \qquad (2.7)$$

Effective two-particle nuclear interactions are often density dependent: $V(\mathbf{r}_1 - \mathbf{r}_2; \rho(\mathbf{R}))$ with $\mathbf{R} = (\mathbf{r}_1 + \mathbf{r}_2)/2$. To calculate the variation of such a potential with respect to the density matrix, we consider

$$\rho(\mathbf{R}) = \sum_{ij} \rho_{ij}\chi_i^*(\mathbf{R})\chi_j(\mathbf{R})$$

and get

$$\frac{\partial \overline{V}}{\partial \rho_{ij}} = \frac{\partial \overline{V}}{\partial \rho}\frac{\partial \rho}{\partial \rho_{ij}} = \frac{\partial \overline{V}(\rho(\mathbf{R}))}{\partial \rho}\chi_i^*(\mathbf{R})\,\chi_j(\mathbf{R})\,. \qquad (2.8)$$

To obtain the minimum of E^{HF} with respect to variations of the density matrix with the constraint of preserving the trace (2.6) we take the derivative

$$\frac{\partial(E^{\mathrm{HF}} - \epsilon\,\mathrm{tr}\,\rho)}{\partial \rho_{kl}} = t_{kl} + \Gamma_{kl} - \epsilon\delta_{kl} = 0\,, \qquad (2.9)$$

where ϵ is a Lagrange multiplier and Γ_{kl} is the mean field, defined as

$$\begin{aligned}
\Gamma_{kl} &= \frac{\partial\,\mathrm{tr}[\rho\overline{V}\rho]}{\partial \rho_{kl}} \\
&= \sum_{l_2 l_2'} \overline{V}_{kl_2' ll_2}\rho_{l_2' l_2} + \frac{1}{2} \sum_{l_1 l_2 l_1' l_2'} \rho_{l_1 l_1'}\left[\frac{\partial \overline{V}}{\partial \rho(\mathbf{r})}\chi_k^*\chi_l\right]_{l_1' l_2' l_1 l_2}\rho_{l_2' l_2}\,. \quad (2.10)
\end{aligned}$$

The second term on the right-hand-side of the last equation appears when the effective two-particle potential is explicitly density dependent. One sees that the unitary transformation, introduced in Eq. (2.4) diagonalizes the eigenvalue Eqs. (2.9). In space representation the Hartree-Fock equation (2.9) becomes

$$-\frac{\hbar^2}{2M_{\mathrm{nucl}}}\Delta\varphi_k(\mathbf{r},\sigma,\tau) + \sum_{j=1}^{A}\sum_{\sigma'\tau'}\int d\mathbf{r}'\varphi_j^*(\mathbf{r}',\sigma',\tau')V(\mathbf{r}-\mathbf{r}')\hat{A}(\hat{P}_\sigma,\hat{P}_\tau)$$

$$\times[\varphi_j(\mathbf{r}',\sigma',\tau')\varphi_k(\mathbf{r},\sigma,\tau) - \varphi_j(\mathbf{r},\sigma,\tau)\varphi_k(\mathbf{r}',\sigma',\tau')] = \epsilon_k\varphi_k(\mathbf{r},\sigma,\tau)\,, \quad (2.11)$$

where the sum is to be extended over the A eigenstates with the smallest eigenvalues ϵ_k.

In the HF basis of the $\varphi_i(\mathbf{r},\sigma,\tau)$ the density matrix is diagonal with eigenvalues 1 for occupied and 0 for unoccupied states $\rho_{ij} = n_i\delta_{ij}$

$$n_i = \begin{cases} 1 & \text{if} \quad i \le i_{\mathrm{Fermi}} \\ 0 & \text{if} \quad i > i_{\mathrm{Fermi}}\,. \end{cases} \qquad (2.12)$$

The HF energy becomes

$$E^{\mathrm{HF}} = \sum_{i=1}^{A} t_{ii} + \frac{1}{2} \sum_{i,j=1}^{A} \overline{V}_{ij,ij} \tag{2.13}$$

and the eigenvalues ϵ_i of Eq. (2.11) are given by

$$\epsilon_i = t_{ii} + \Gamma_{ii} \; ;$$

therefore

$$E^{\mathrm{HF}} = \sum_{i=1}^{A} \left(\epsilon_i - \frac{1}{2} \Gamma_{ii} \right) . \tag{2.14}$$

One can show [108] that

$$[H, \rho] = 0 \tag{2.15}$$

in the HF-basis and admitted variations $\delta\rho_{k'k}$ must be between particle states (i.e. states above the Fermi energy) k' and hole states (i.e. states below the Fermi energy) k or vice versa. Therefore the behavior of the interaction potential in the particle-particle channel does not matter in the HF-approximation.

The HF equation (2.11) is a nonlinear eigenvalue equation. It is solved iteratively, starting e.g. with a Wood-Saxon mean field $\Gamma^{(0)}$. The eigenfunctions $\varphi_i^{(1)}$ obtained from solving the HF equations with this mean field were used to calculate a new mean field $\Gamma^{(1)}$ according to Eq. (2.10) and so forth until selfconsistency is reached. Two points require special care when using this iteration scheme.

First, the HF equation is valid for any stationary point, not only the ground state. It depends on the closeness of the shape of the initial shell-model potential to the final local minimum in the energy surface to which minimum the iteration converges if there are several local minima.

Second, the iteration preserves the symmetry of the initial field. If the desired solution has axial, but not spherical symmetry, the initial field should be given that reduced symmetry. If the solution is expected to be triaxial, the initial field should be triaxial. The same holds for violation of reflection symmetry.

Saddle points cannot be obtained in this way. Instead, one has to add a suitable constraining field $q Q_{kl}$ to Γ_{kl}. Usually $\hat{Q}$ will be the operator of the quadrupole moment and q is a Lagrange parameter. The latter has the meaning of a generalized force, necessary to keep the nucleus in equilibrium at a given value of the quadrupole deformation $Q = \langle \mathrm{HF}|\hat{Q}|\mathrm{HF}\rangle$. Plotting the force parameter $q(Q)$ as function of the external parameter Q, stationary points, including saddle points, correspond to zeros of this function. The sign of the derivative in these points differentiates between minima and saddle points.

2.1.2 The Hartree-Fock-Bogolyubov theory

To include pairing correlations in the mean-field theory, the selfconsistent Hartree-Fock equations were generalized by Baranger [116]. He used the Bogolyubov transformation [117]

$$
\begin{aligned}
\hat{\alpha}_i &= \sum_{j=1}^{M} (u_{ij}^* \hat{a}_j + v_{ij}^* \hat{a}_j^+) \,, \\
\hat{\alpha}_i^+ &= \sum_{j=1}^{M} (u_{ij} \hat{a}_j^+ + v_{ij} \hat{a}_j) \,,
\end{aligned}
\tag{2.16}
$$

to introduce quasiparticle creation and annihilation operators $\hat{\alpha}_i^+$ and $\hat{\alpha}_i$, where M is the dimension of the single-particle space and u and v are transformation matrices. Introducing the $2M \times 2M$ matrix

$$
\hat{B} = \begin{pmatrix} \hat{u} & \hat{v}^* \\ \hat{v} & \hat{u}^* \end{pmatrix} \,,
\tag{2.17}
$$

the ortho-normalization of quasi-particle states requires that $\hat{B}$ is unitary

$$
\hat{B}\hat{B}^+ = \hat{B}^+\hat{B} = \hat{I} \,,
\tag{2.18}
$$

where $\hat{B}^+$ is the hermitian conjugate of the $\hat{B}$ matrix.

The transformation (2.16) and its inverse can now be written as

$$
\begin{pmatrix} \hat{\alpha} \\ \hat{\alpha}^+ \end{pmatrix} = \hat{B}^+ \begin{pmatrix} \hat{a} \\ \hat{a}^+ \end{pmatrix} \,; \qquad
\begin{pmatrix} \hat{a} \\ \hat{a}^+ \end{pmatrix} = \hat{B} \begin{pmatrix} \hat{\alpha} \\ \hat{\alpha}^+ \end{pmatrix} \,.
\tag{2.19}
$$

One assumes that the ground-state wave-function $|\text{HFB}\rangle$ of a nucleus with pairing correlations is a state of independent quasiparticles

$$
|\text{HFB}\rangle = \prod_{i=1}^{M} \hat{\alpha}_i |0\rangle \,.
\tag{2.20}
$$

This shows immediately that the state $|\text{HFB}\rangle$ is the quasiparticle vacuum

$$
\hat{\alpha}_i |\text{HFB}\rangle = 0 \,.
$$

From the hermitian density matrix

$$
\rho_{ij} = \langle \text{HFB}|\hat{a}_j^+ \hat{a}_i|\text{HFB}\rangle = \left[v^* v^T \right]_{ij}
\tag{2.21}
$$

and the skew-symmetric pairing tensor

$$
\kappa_{ij} = \langle \text{HFB}|\hat{a}_j \hat{a}_i|\text{HFB}\rangle = [v^* u^T]_{ij} = -\left[u v^+ \right]_{ij}
\tag{2.22}
$$

a generalised hermitian density matrix is formed

$$\hat{R} = \begin{pmatrix} \hat{\rho} & \hat{\kappa} \\ -\hat{\kappa}^* & 1 - \hat{\rho}^* \end{pmatrix} . \tag{2.23}$$

The expectation value of the Hamiltonian $\langle \text{HFB}|\hat{H}|\text{HFB}\rangle$ is given in terms of the matrices ρ_{ij} and κ_{ij} by

$$\langle \text{HFB}|\hat{H}|\text{HFB}\rangle = \sum_{ij} t_{ij}\rho_{ji} + \frac{1}{2}\sum_{ijkl}\langle ij|\overline{V}|kl\rangle \rho_{lj}\,\rho_{ki}$$

$$+ \frac{1}{4}\sum_{ijkl}\langle ij|\overline{V}|kl\rangle \kappa_{lk}\kappa_{ji}^* \tag{2.24}$$

and the expectation value of the particle number by

$$\langle \text{HFB}|\hat{N}|\text{HFB}\rangle = \sum_i \rho_{ii} . \tag{2.25}$$

Minimizing the total energy with the constraint that the expectation value of the particle number has the required value, leads to the stationarity condition for the HFB ground state,

$$\delta\left\{\langle \text{HFB}|\hat{H} - \lambda\hat{N}|\text{HFB}\rangle - \text{tr}\left[\Lambda(\hat{R}^2 - \hat{R})\right]\right\} = 0 , \tag{2.26}$$

where $\text{tr}[\Lambda(\hat{R}^2 - \hat{R})]$ is subtracted to take into account the Bogolyubov condition (2.18) expressed here in terms of the generalized density (2.23). The variation is to be done with respect to $\delta\rho$, $\delta\rho^*$, $\delta\kappa$ and $\delta\kappa^*$ and Λ is a hermitian matrix whose elements are Lagrange multipliers.

After performing the variation and eliminating the matrix Λ by using its hermiticity, the condition (2.26) can be expressed by the commutator:

$$[\hat{\mathcal{H}}, \hat{R}] = 0 , \tag{2.27}$$

where $\hat{\mathcal{H}}$ is given by

$$\hat{\mathcal{H}} = \begin{pmatrix} \hat{h} & \hat{\Delta} \\ -\hat{\Delta}^* & -\hat{h}^* \end{pmatrix} , \tag{2.28}$$

with

$$h_{ij} = \frac{\partial}{\partial \rho_{ji}}\langle \text{HFB}|\hat{H} - \lambda\hat{N}|\text{HFB}\rangle \tag{2.29}$$

and

$$\Delta_{ij} = 2\frac{\partial}{\partial \kappa_{ji}^*}\langle \text{HFB}|\hat{H}|\text{HFB}\rangle . \tag{2.30}$$

Eq. (2.27) is the analogue of Eq. (2.15) in the HF theory. To solve equation (2.27) it is sufficient to choose as Bogolyubov transformation $\hat{B}$ the unitary matrix which diagonalizes $\hat{\mathcal{H}}$. This leads to the HFB equations in their standard form

$$\hat{\mathcal{H}}\hat{B} = \hat{B}\,\hat{\mathscr{E}} \quad \text{with} \quad \mathscr{E}_{ij} = \mathcal{E}_i\,\delta_{ij}\ , \tag{2.31}$$

where $\mathcal{E}_i$ denotes the quasi-particle energy.

With the definition (2.30) the explicit form of the pairing matrix becomes

$$\Delta_{ij} = \frac{1}{2}\sum_{kl}\langle ij|\overline{V}|kl\rangle\kappa_{lk}\ . \tag{2.32}$$

The matrix elements of h (2.29) can be obtained in a similar way. For density-dependent interactions one gets from Eqs. (2.29) and (2.8)

$$h_{ij} = t_{ij} - \lambda\,\delta_{ij} + \Gamma_{ij}\ , \tag{2.33}$$

where

$$\Gamma_{ij} = \sum_{kl}\langle ik|\overline{V}(\rho)|jl\rangle\rho_{lk}$$
$$+ \frac{1}{2}\sum_{klmn}\langle mn|\frac{\partial\overline{V}(\rho)}{\partial\rho}\varphi_i^*\,\varphi_j|kl\rangle\left[\rho_{ln}\,\rho_{km} + \frac{1}{2}\,\kappa_{lk}\,\kappa_{nm}^*\right] \tag{2.34}$$

is the quasiparticle mean-field.

The total binding energy is given by

$$E = \operatorname{tr}\left\{(\hat{t} + \frac{1}{2}\hat{\Gamma})\hat{\rho} + \frac{1}{2}\hat{\Delta}\hat{\kappa}^*\right\} + E_{\mathrm{R}}\ , \tag{2.35}$$

where

$$E_{\mathrm{R}} = -\frac{1}{4}\sum_{ijklmn}\langle mn|\frac{\partial\overline{V}(\rho)}{\partial\rho}\varphi_i^*\,\varphi_j|kl\rangle\left[\rho_{ln}\,\rho_{km} + \frac{1}{2}\,\kappa_{lk}\,\kappa_{nm}^*\right]\rho_{ij}$$

is the rearrangement energy which originates from the dependence of the effective interaction potential on the density [118].

It is seen from Eq. (2.34) that in the case of density-dependent interactions the mean-field depends not only on the density matrix (2.21), but also on the pairing tensor (2.22). The Hartree-Fock equations follow from the HFB equations when the pairing-dependent terms are omitted and in the whole calculation one assumes that the matrices $\kappa = 0$ and $\Delta = 0$. In particular, the selfconsistent HF field depends on the density matrix only. In contrast to the HF theory the results of the HFB theory depend on the particle-hole as well as on the particle-particle channel of the effective interaction potential.

2.1.3 The Bardeen-Cooper-Schrieffer (BCS) theory

In view of the need to solve constrained HFB equations thousands of times to construct potential-energy surfaces, it is desirable to have a numerically less cumbersome alternative to account for pairing effects. Often one therefore uses the numerically simpler HF+BCS approximation in which one first generates single-particle energies ϵ_i and eigenstates $|i\rangle$ selfconsistently in the Hartree-Fock frame without accounting for the pairing contribution to the mean-field and then one includes the pairing correlations in a BCS step. This approach is in particular necessary when the effective interaction potential, from which the mean field is derived, does not properly describe the particle-particle channel.

In the BCS theory one assumes that the pairing potential acts only between states $|i\rangle$ and their time-reflected counterparts $|\bar{\imath}\rangle = \mathcal{T}|i\rangle$. For spherically symmetric systems $|i\rangle \equiv |nljm\rangle$ and the time-reflected state is given by [83]

$$|\bar{\imath}\rangle = \mathcal{T}|nljm\rangle = (-1)^{j+m}|nlj - m\rangle. \tag{2.36}$$

These pairs of degenerate states are therefore singlets in their relative spin and orbital angular momenta, and must then be isotriplets to satisfy the Pauli principle. In nonaxially symmetric, but time-reflection symmetric systems the single-particle states $|i\rangle$ may be expanded in a spherical basis and the rule Eq. (2.36) extended to such states. We will restrict the discussion in the following to time-reversal invariant systems with pairs of degenerate states (Kramers degeneracy). In particular rotating systems are not time-reversal invariant. For their discussion in the framework of the BCS theory we refer to Section 7.7 of Ref. [108].

The BCS Hamiltonian is

$$\hat{H} = \sum_{i \lessgtr 0} \epsilon_i \hat{a}_i^+ \hat{a}_i + \frac{1}{2} \sum_{i,j>0} \overline{V}_{i\bar{\imath},j\bar{\jmath}} \hat{a}_i^+ \hat{a}_{\bar{\imath}}^+ \hat{a}_{\bar{\jmath}} \hat{a}_j. \tag{2.37}$$

(The summation limits $i \lessgtr 0$ indicate a summation over all states i and their conjugates $\bar{\imath}$, $i > 0$ means summation over the pairs $(i, \bar{\imath})$ in this order only.) The transformation to quasiparticle operators simplifies in the BCS theory compared to the HFB definition, Eqs. (2.16). One obtains

$$\hat{\alpha}_i = u_i \hat{a}_i - v_i \hat{a}_{\bar{\imath}}^+ \ , \qquad\qquad \hat{\alpha}_i^+ = u_i \hat{a}_i^+ - v_i \hat{a}_{\bar{\imath}} \ ,$$
$$\hat{\alpha}_{\bar{\imath}} = u_i \hat{a}_{\bar{\imath}} + v_i \hat{a}_i^+ \ , \qquad\qquad \hat{\alpha}_{\bar{\imath}}^+ = u_i \hat{a}_{\bar{\imath}}^+ + v_i \hat{a}_i \tag{2.38}$$

with real u_i and v_i. With this definition we follow the convention of Ref. [108]. The normalization condition, which guarantees that the quasiparticle operators obey Fermi commutation rules, becomes

$$u_i^2 + v_i^2 = 1 \tag{2.39}$$

instead of Eq. (2.18) in the HFB theory. The ansatz for the BCS ground-state of an even proton-number and even neutron-number state is

$$|\text{BCS}\rangle = \prod_{i>0}(u_i + v_i \hat{a}_i^+ \hat{a}_{\bar{\imath}}^+)|0\rangle \ , \tag{2.40}$$

where $i > 0$ on the product sign indicates that all single-particle states $|i\rangle$, except their time-reversed counterparts should be taken. It is easy to show that the state $|\text{BCS}\rangle$ is the quasiparticle vacuum, $\hat{\alpha}_k|\text{BCS}\rangle = 0$. The BCS state is not an eigenstate of the particle-number operator. In terms of the u_i and v_i the expectation value of the particle-number operator $\hat{N} = \sum_i \hat{a}_i^+ \hat{a}_i$ is

$$N = \langle\text{BCS}|\hat{N}|\text{BCS}\rangle = 2\sum_{i>0} v_i^2 \tag{2.41}$$

and the expectation value of the Hamiltonian is

$$E_{\text{BCS}} = \langle\text{BCS}|\hat{H}|\text{BCS}\rangle = 2\sum_{i>0} \epsilon_i v_i^2 + \sum_{i>0} \overline{V}_{i\bar{i}i\bar{i}} v_i^4$$

$$+ \frac{1}{2}\sum_{i,j>0} \overline{V}_{i\bar{i}j\bar{j}} u_i v_i u_j v_j \ . \tag{2.42}$$

Minimization of the expectation value of the Routhian $\hat{H}' = \hat{H} - \lambda\hat{N}$ with respect to variations of the v_i, observing the constraint (2.39), yields the equation

$$\left(\frac{\partial}{\partial v_i} + \frac{\partial u_i}{\partial v_i}\frac{\partial}{\partial u_i}\right)\langle\text{BCS}|\hat{H} - \lambda\hat{N}|\text{BCS}\rangle = 0 \ ,$$

from which the BCS equations

$$2\tilde{\epsilon}_i u_i v_i + \Delta_i(v_i^2 - u_i^2) = 0 \ , \qquad i > 0 \ , \tag{2.43}$$

follow, where the abbreviations

$$\tilde{\epsilon}_i = \epsilon_i - \lambda + \overline{V}_{i\bar{i}i\bar{i}} v_i^2 \tag{2.44}$$

and

$$\Delta_i = -\sum_{j>0} \overline{V}_{i\bar{i}j\bar{j}} u_j v_j \ . \tag{2.45}$$

have been used. From Eqs. (2.39) and (2.43) one obtains

$$v_i^2 = \frac{1}{2}\left(1 - \frac{\tilde{\epsilon}_i}{\sqrt{\tilde{\epsilon}_i^2 + \Delta_i^2}}\right)$$

$$\tag{2.46}$$

$$u_i^2 = \frac{1}{2}\left(1 + \frac{\tilde{\epsilon}_i}{\sqrt{\tilde{\epsilon}_i^2 + \Delta_i^2}}\right) \ .$$

Inserting Eqs. (2.46) into Eq. (2.45) the gap equation

$$\Delta_i = -\frac{1}{2}\sum_{j>0} \overline{V}_{i\bar{i},j\bar{j}} \frac{\Delta_j}{\sqrt{\tilde{\epsilon}_j^2 + \Delta_j^2}} \tag{2.47}$$

is derived. Together with the constraint (2.41)

$$N = 2 \sum_{i>0} v_i^2 \qquad (2.48)$$

and Eq. (2.44) the gap equation (2.47) allows to determine the Lagrange multiplier λ and the v_i in terms of the ϵ_i and the matrix elements $\overline{V}_{i\bar{i},j\bar{j}}$ by an iterative procedure.

For the pairing potential in Eq. (2.37) sometimes a zero-range interaction is used [119, 120]

$$V^q(\boldsymbol{r}_1, \sigma_1; \boldsymbol{r}_2, \sigma_2) = V_0^q \frac{1 - \boldsymbol{\sigma}_1 \cdot \boldsymbol{\sigma}_2}{4} \delta(\boldsymbol{r}_1 - \boldsymbol{r}_2), \qquad q = n, p. \qquad (2.49)$$

With this interaction one obtains:

$$V_{i\bar{i}j\bar{j}}^q = V_0^q \int d^3r \rho_i^q(\boldsymbol{r}) \rho_j^q(\boldsymbol{r}),$$

where

$$\rho_i^q(\boldsymbol{r}) = |\varphi_i^q(\boldsymbol{r})|^2$$

and V_0^q is the pairing strength, adjusted to experimental odd-even mass differences e.g. in Ref. [120].

The term proportional to v_i^2 in Eq. (2.44) leads only to a shift of the energies ϵ_i, i.e. it renormalizes the single-particle potential. Since one assumes that the ϵ_i correspond already to the total mean field, it is reasonable to drop these terms to avoid double-counting. For the calculation of Strutinsky's pairing correction to be discussed in Sec. 2.3, this term does not make any difference.

The states of an odd nucleus with particle number $N + 1$ are given in the BCS theory by

$$\hat{\alpha}_k^+ |\text{BCS}\rangle = \hat{a}_k^+ \prod_{i>0; i \neq k} (u_i + v_i \hat{a}_i^+ \hat{a}_{\bar{i}}^+)|0\rangle,$$

where $|\text{BCS}\rangle$ is the BCS ground-state of the even nucleus with N particles. The expectation value of the Routhian $\hat{H}' = \hat{H} - \lambda \hat{N}$ of an odd nucleus is

$$E_{\text{BCS}}^{(k)} = \langle \text{BCS}|\hat{\alpha}_k \hat{H}' \hat{\alpha}_k^+ |\text{BCS}\rangle = \langle \text{BCS}|\hat{H}'|\text{BCS}\rangle + \mathcal{E}_k \qquad (2.50)$$

with the quasi-particle energy

$$\mathcal{E}_k = \sqrt{\tilde{\epsilon}_k^2 + \Delta_k^2}. \qquad (2.51)$$

It is assumed that the correlation structure of the paired N particles is not affected by adding an extra odd particle. This is only approximately true if many particles participate in the formation of the correlated state, cf. Sec. 6.3.4 of Ref. [108]. A more accurate treatment of an odd system is given below for the model with a state-independent pairing matrix-element.

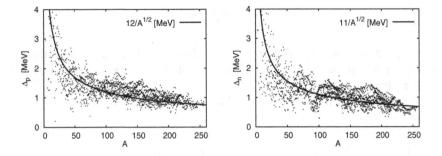

Fig. 2.1. Dependence of the proton and neutron gaps on the mass number, data from the mass table of Ref. [121].

The ground state k_0 corresponds to $\epsilon_{k_0} \approx \lambda$. From Eq. (2.44) follows, neglecting again the v_i^2 terms, $\mathcal{E}_{k_0} = 0$. To obtain the odd-even mass difference, defined usually as second finite difference

$$E_{\rm oe} = -(1/2)\,[\,E_{N+2}^{\rm GS} - E_{N+1}^{\rm GS} - (E_{N+1}^{\rm GS} - E_N^{\rm GS})]\,, \quad N \text{ even}, \qquad (2.52)$$

one uses the following relations between ground-state energies

$$E_{N+2}^{\rm GS} \approx E_N^{\rm GS} + 2\lambda\,, \qquad E_{N+1}^{\rm GS} \approx E_N^{\rm GS} + \lambda + \mathcal{E}_{k_0}\,,$$

and obtains

$$E_{\rm eo} = \mathcal{E}_{k_0} \approx \Delta_{k_0}. \qquad (2.53)$$

Note that there is an odd-even staggering of nuclear binding energies in HF approximation, even without pairing, connected with the breaking of time-reversal symmetry in odd systems. For a discussion of extracting both, the pairing and the mean-field odd-even effects independently from the data we refer to the investigation of Satuła, Dobaczewski, and Nazarewicz [122].

Fig. 2.1 shows that in the average $\Delta_p = 12 \cdot A^{-1/2}$ MeV and $\Delta_n = 11 \cdot A^{-1/2}$ MeV describes roughly the data for protons and neutrons, respectively. In Eq. (2.52) the assumption is made that the ground-state binding-energies of the three nuclei on the right-hand-side depend smoothly on N, except for pairing effects. This is not always the case, for example when the ground-state deformation changes rapidly as function of N; see Möller and Nix [123] for a detailed discussion of this problem. Naturally this difficulty occurs even more frequently when forth differences are used instead of the second differences of Eq. (2.52).

To calculate pairing corrections to ground-state binding-energies a rather schematic BCS calculation is mostly used, where the interaction potential in Eq. (2.37) is assumed to have state-independent matrix elements

$$\frac{1}{2} \sum_{i,j>0} \overline{V}_{i\bar{i}j\bar{j}}\hat{a}_i^+ \hat{a}_{\bar{i}}^+ \hat{a}_{\bar{j}}\hat{a}_j = -G \sum_{i,j>0}^{(\omega)} \hat{a}_i^+ \hat{a}_{\bar{i}}^+ \hat{a}_{\bar{j}}\hat{a}_j\,. \qquad (2.54)$$

The superscript ω on the sum shall indicate that the sum is restricted to states i, j, which lie in a band of width 2ω around the HF Fermi-energy. In this case the gap parameter Δ from Eq. (2.45) becomes also state-independent

$$\Delta = G \sum_{i>0}^{(\omega)} u_i v_i$$

and Eq. (2.44) simplifies to

$$\tilde{\epsilon}_i = \epsilon_i - \lambda - G v_i^2 . \tag{2.55}$$

The expectation value of the Routhian $\hat{H}' = \hat{H} - \lambda \hat{N}$ is

$$\langle \mathrm{BCS} | \hat{H}' | \mathrm{BCS} \rangle = 2 \sum_{i>0}^{(\omega)} \tilde{\epsilon}_i v_i^2 + \sum_{i>0}^{(\omega)} G v_i^4 - \Delta^2 / G \tag{2.56}$$

and the gap equation

$$\Delta = \frac{G}{2} \sum_{i>0}^{(\omega)} \frac{\Delta}{\sqrt{\tilde{\epsilon}_i^2 + \Delta^2}} . \tag{2.57}$$

Again, the term $G v_i^2$ in Eq. (2.55) and $G v_i^4$ in Eq. (2.56) may be dropped. Inserting

$$v_i^2 = \frac{1}{2} \left(1 - \frac{\tilde{\epsilon}_i}{\sqrt{\tilde{\epsilon}_i^2 + \Delta^2}} \right) \tag{2.58}$$

into Eqs. (2.48) and (2.55), one obtains together with Eq. 2.57 two nonlinear coupled equations for Δ and λ.

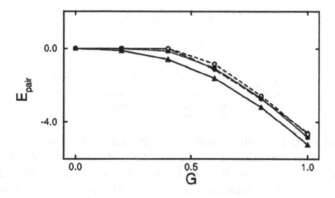

Fig. 2.2. The pairing energy for six particles in a deformed $j = 11/2$ orbital in a pairing-plus-quadrupole model. Results are shown for the HFB approach (full circles •), projection after variation (dashed line ◦), the Lipkin-Nogami procedure (full squares, indistinguishable in this figure from the HFB result), and the method of projection before variation (full triangles). Energies in units of G, (from Ref. [124]).

There is always the "trivial", HF solution of these equations

$$u_i = 0, \quad v_i = 1 \quad \text{for} \quad i \leq i_{\text{Fermi}}$$
$$u_i = 1, \quad v_i = 0 \quad \text{for} \quad i > i_{\text{Fermi}}$$

for which $\Delta = 0$. Nontrivial solutions, describing pairing-correlated states, require a large level density around the Fermi energy and also a sufficiently large pairing strength G. It turns out that for realistic parameters no pairing solution exists for magic nuclei, whereas for mid-subshell nuclei there is generally such a solution. Between these extreme situations a sharp transition occurs for fixed level density as a function of G, cf. Fig. 2.2, or for fixed G as function of N (or Z), which in turn determines the level density at the Fermi energy. Particle-number-projected HFB calculations have shown that the sharp phase transition is spurious and connected with the finite particle-number fluctuation in the BCS state (2.40). Various proposals have been made to cope with this problem, in particular, projection of the BCS wave function on to eigenstates of the particle-number operator; see Ref. [108] for a detailed discussion.

In connection with Strutinsky's shell correction, however, the Lipkin-Nogami approach [125–127] is most frequently used. In this approach a constraint on the variance of the particle number is introduced in the Routhian $\hat{H}'$ besides the constraint on the particle number. Calling the Lagrange multiplier for the additional constraint λ_2, Eq. (2.55) is modified to

$$\tilde{\epsilon}_i^{\text{LN}} = \epsilon_i - \lambda + (4\lambda_2 - G)v_i^2 \tag{2.59}$$

and the quasi-particle energy becomes

$$\mathcal{E}_i^{\text{LN}} = \sqrt{\epsilon_i^2 + \Delta^2} + \lambda_2 \tag{2.60}$$

with [127]

$$\lambda_2 = \frac{G}{4} \frac{\left(\sum_i^{(\omega)} u_i^3 v_i\right)\left(\sum_i^{(\omega)} u_i v_i^3\right) - \sum_i^{(\omega)} u_i^4 v_i^4}{\left(\sum_i^{(\omega)} u_i^2 v_i^2\right)^2 - \sum_i^{(\omega)} u_i^4 v_i^4}. \tag{2.61}$$

The gap and particle-number equations (2.38) and (2.48) have to be solved with the modified definition of $\tilde{\epsilon}_i$. In the light of results obtained for the weak pairing case (i.e. for small G) by Dobaczewski and Nazarewicz [128] and by Sheikh et al. [124], summarized in Fig. 2.2, the improvement gained by the Lipkin-Nogami method compared to plain HFB is only modest. On the other hand there are no systematic investigations on the quality of the Lipkin-Nogami correction in the context of Strutinsky's shell plus pairing correction.

Rather sophisticated procedures have been devised to determine the pairing-strength constant G for a given range ω from empirical mass differences within the BCS and the Lipkin-Nogami schemes [123]. However, all these efforts cannot hide the fact that the basic assumption of a state-independent pairing matrix element G does not describe the physical situation correctly all along the fission path: Around scission the single particle states separate into those localized in either one of the nascent fragments and those, above the Fermi energy, which are still distributed over the entire nuclear volume. One therefore expects four different characteristic values for G. After scission the states above the Fermi energy will also become localized in one of the fragments, leading, for asymmetric fission, to different pairing strengths in the two fragments and consequently to different chemical potentials and pairing gaps. Therefore there is a phase transition in the pairing degree of freedom around scission [129]: after scission the particle numbers of the two fragments are separately constants of motion whereas for a compact nucleus only the total number of particles is a constant of motion. One should therefore not interpolate G between the ground state of the fissioning nucleus and separated fragments by means of the shape function B_{surf} as proposed in Ref. [130] nor assume a constant G.

For an odd system Eqs.

$$N' = 1 + \sum_{i \neq k_0}^{(\omega)} \left(1 - \frac{\tilde{\epsilon}_i}{\mathcal{E}_i} \right) \quad \text{and} \quad \frac{2}{G} = \sum_{i \neq k_0}^{(\omega)} \frac{1}{\mathcal{E}_i} \qquad (2.62)$$

have to be solved for Δ and λ, instead of Eqs. (2.48) and (2.57). Here again the notation $\mathcal{E}_i = \sqrt{\tilde{\epsilon}_i^2 + \Delta^2}$ is used and N' is the number of particles occupying states in the 2ω band. The gap parameter Δ is smaller than for the neighboring even systems because the state k_0 is "blocked" and is not available for the set-up of the pairing correlations. Neglecting the v_i^4 term in Eq. 2.56 one finds for the ground-state expectation-value of the Routhian of an odd system

$$\langle \mathrm{BCS} | \hat{H}' | \mathrm{BCS} \rangle = 2 \sum_{i \neq k_0}^{(\omega)} \tilde{\epsilon}_i v_i^2 + (\epsilon_k - \lambda) - \Delta^2 / G \,.$$

If the pairing gap Δ is large compared to the level spacing, one obtains simple, analytic expressions for the pairing energy and the pairing gap by introducing a continuous, smooth single-particle level-density $g(e)$ (which shall include the time-reflected states) [97]. The gap equation (2.57) then becomes

$$\frac{2}{G} = \int_{\lambda-\omega}^{\lambda+\omega} \frac{\frac{1}{2} g(e) de}{\sqrt{(e-\lambda)^2 + \Delta^2}} \approx \frac{g(\lambda)}{2} \ln \frac{\sqrt{\omega^2 + \Delta^2} + \omega}{\sqrt{\omega^2 + \Delta^2} - \omega} \,. \qquad (2.63)$$

Since the pairing-window half-width ω should be large compared to the pairing gap, $\omega \gg \Delta$, Eq. (2.63) can be simplified to

$$\frac{2}{G} \approx g(\lambda) \ln \left(\frac{2\omega}{\Delta} \right) \,. \qquad (2.64)$$

This approximation is sometimes called the "uniform model". The relation (2.64) allows to estimate the pairing strength G when the average pairing gap is known from experimental data. One can also see that G depends on the width of the window 2ω, i.e. on the number of states included in the sum in Eq. (2.54): G decreases with increasing number of states, included in the pairing Hamiltonian.

The BCS energy-gain with respect to the energy of a system without pairing correlations is

$$\Delta E = \langle \mathrm{BCS}|\hat{H}'+\lambda\hat{N}|\mathrm{BCS}\rangle - 2\sum_{i\leq i_{\mathrm{Fermi}}} \epsilon_i = 2\sum_{i>0}^{(\omega)} \epsilon_i v_i^2 - \Delta^2/G - 2\sum_{i\leq i_{\mathrm{Fermi}}} \epsilon_i \quad (2.65)$$

for an even particle number, where the Gv^4 term in Eq. (2.56) is neglected and $N = 2\sum_i v_i^2$ is used. The quantity ΔE can be calculated analytically in the uniform model. Using Eqs. (2.55) and (2.58) we obtain

$$\widetilde{\Delta E} = \frac{1}{2}\int\limits_{\lambda-\omega}^{\lambda+\omega} e\left(1 - \frac{e-\lambda}{\sqrt{(e-\lambda)^2+\Delta^2}}\right) g(e)de - \frac{\Delta^2}{G} - \int\limits_{\lambda-\omega}^{\lambda} eg(e)de \ .$$

After evaluation of the integrals and eliminating $1/G$ with Eq. (2.63), the pairing energy becomes

$$\widetilde{\Delta E} = \frac{1}{2}g(\lambda)\omega^2\left(1 - \sqrt{1+\left(\frac{\Delta}{\omega}\right)^2}\right) \ . \quad (2.66)$$

With $\omega \gg \Delta$ the last equation can be approximated by

$$\widetilde{\Delta E} \approx -\frac{1}{4}g(\lambda)\Delta^2 \ . \quad (2.67)$$

The Lipkin-Nogami approach was also reformulated in the uniform model by Möller and Nix [123]. The expression for λ_2, Eq. (2.61), becomes rather lengthy, we refer to Ref. [123] for details.

2.1.4 Thomas-Fermi and extended Thomas-Fermi theories

Of particular importance for the theory of fission is the Thomas-Fermi (TF) approximation because its expression for the binding energy admits a lepto-dermous expansion, leading to a liquid-drop formula. In TF approximation one assumes that the kinetic energy term in the HF energy, Eq. (2.7), can be approximated by the Fermi-gas expression for neutrons ($q = n$) and protons ($q = p$) [114]

$$T_q = \frac{2}{h^3}\int_{\mathrm{vol}} d^3r \ 4\pi\int dp\,p^2 \frac{p^2}{2M_{\mathrm{nucl}}} = \frac{\hbar^2}{2M_{\mathrm{nucl}}}\alpha\int_{\mathrm{vol}} \rho_q^{5/3}(\mathbf{r})\,d^3r \quad (2.68)$$

with $\alpha = (3/5)(3\pi^2)^{2/3}$. The Fermi-gas relation $\rho = (8\pi/3)(p_f/h)^3$ was used in the last equation.

The resulting equation is of little use as long as the exchange term of the potential contains the full density matrix ρ_{ij}. Therefore the Thomas-Fermi scheme has only been used either in connection with zero-range interactions or the Skyrme interactions [32], where the exchange and the direct term contribute with the same radial matrix elements to the potential energy, or momentum-dependent effective interactions, like the Seyler-Blanchard interaction [131], which are supposed to account in the direct term already for exchange effects.

A better approximation of the kinetic energy would take the variation of the potential in the surface into account. The resulting expansion in increasingly higher order derivatives of the potential leads to the extended Thomas-Fermi (ETF) scheme [114,132]. Defining the kinetic energy-density expression by a sum over occupied states

$$\tau_q(\mathbf{r}) = \sum_{i\sigma} |\nabla\varphi_i(\mathbf{r}, \sigma, q)|^2 \tag{2.69}$$

in analogy to the density

$$\rho_q(\mathbf{r}) = \sum_{i\sigma} |\varphi_i(\mathbf{r}, \sigma, q)|^2 \, , \tag{2.70}$$

one derives an expansion

$$\tau_q(\rho, \nabla\rho) = \tau_q^{(\text{TF})}(\rho) + \tau_q^{(2)}(\rho, \nabla\rho) + \tau_q^{(4)}(\rho, \nabla\rho) \, , \tag{2.71}$$

where $\tau_q^{(\text{TF})}(\rho)$ is the Thomas-Fermi term

$$\tau_q^{(\text{TF})}(\rho_q) = \alpha\rho_q^{5/3}(\mathbf{r}) \, . \tag{2.72}$$

It yields the approximation (2.68) for the kinetic energy. The next term contains second order derivatives of the density (or squares of first order derivatives). If the effective potential does not depend on spin and momentum, the result is

$$\tau_q^{(2)}(\rho, \nabla\rho) = \frac{1}{36}\frac{(\nabla\rho_q)^2}{\rho_q} + \frac{1}{3}\Delta\rho_q \, . \tag{2.73}$$

The frequently used Skyrme and Gogny effective potentials (to be presented in Section 2.1.7) are momentum and spin dependent, which yields additional contributions to $\tau_q^{(2)}$

$$\tau_q^{(2)}[\rho_q] = \frac{1}{36}\frac{(\nabla\rho_q)^2}{\rho_q} + \frac{1}{3}\Delta\rho_q + \frac{1}{6}\frac{\nabla\rho_q \cdot \nabla f_q}{f_q} + \frac{1}{6}\rho_q\frac{\Delta f_q}{f_q}$$
$$- \frac{1}{12}\rho_q\left(\frac{\nabla f_q}{f_q}\right)^2 + \frac{1}{2}\left(\frac{2m}{\hbar^2}\right)^2\rho_q\left(\frac{\mathbf{W}_q}{f_q}\right)^2 \, , \tag{2.74}$$

where $f_q(\mathbf{r}) = M_{\text{nucl}}/M^*_{\text{nucl}}(\mathbf{r})$ is the effective mass form-factor, defined in Eq. (2.112), and $\mathbf{W}_q(\mathbf{r})$ is the spin-orbit form-factor given in Eq. (2.113) in terms of the parameters of the effective interaction potential. The next term in the ETF expansion of the kinetic energy $\tau_q^{(4)}(\rho, \boldsymbol{\nabla}\rho)$ is a rather lengthy expression containing fourth order derivatives of ρ, cf. Ref. [133, 134].

Using a zero-range interaction like the Skyrme potential, the result (2.74) allows to express the total energy as a functional of the densities $\rho_n(r)$, $\rho_p(r)$, and their derivatives. Minimization of the energy involves then a variation of densities, rather than wave functions as in the Hartree-Fock approach. The variation is subject to the constraint that the volume integrals over the densities should be equal to the neutron and proton numbers.

It is instructive to calculate some basic liquid-drop parameters with a Skyrme energy-functional of the general form (cf. Eq. (2.105))

$$\mathcal{E}(\rho, \tau) = \epsilon\tau + C\rho^2 + F\rho^\nu + B(\nabla\rho)^2 + G\rho\tau \qquad (2.75)$$

with $\epsilon = \hbar^2/(2M_{\text{nucl}})$ and using the second order ETF approximation of the general structure [135]

$$\tau \approx \alpha\rho^{5/3} + \beta(\nabla\rho)^2/\rho + \gamma\nabla^2\rho \ .$$

In terms of the parameters C, F, and G one obtains for the volume energy

$$c_1 = \frac{\mathcal{E}(\rho_0)}{\rho_0} = \epsilon\alpha\rho_0^{2/3} + C\rho_0 + F\rho_0^{\nu-1} + G\alpha\rho_0^{5/3} \ ,$$

where the saturation density ρ_0 is the solution of the equation

$$d\mathcal{E}/d\rho|_{\rho_0} = (5/3)\epsilon\alpha\rho_0^{2/3} + 2C\rho_0 + \nu F\rho_0^{\nu-1} + (8/3)G\alpha\rho_0^{5/3} = 0.$$

The nuclear incompressibility constant

$$K = 9\rho_0^2 \left.\frac{\partial^2(\mathcal{E}(\rho)/\rho)}{\partial\rho^2}\right|_{\rho=\rho_0}$$

is given in terms of ρ_0 by

$$K = 10\epsilon\alpha\rho_0^{5/3} + 18C\rho_0^2 + 9\nu(\nu-1)F\rho_0^\nu + 40G\alpha\rho_0^{8/3}$$

and the effective mass M_{nucl}^* by

$$\frac{M_{\text{nucl}}}{M_{\text{nucl}}^*} = \frac{1}{\epsilon}\frac{\partial\mathcal{E}}{\partial\tau} = 1 + \rho_0\frac{G}{\epsilon}$$

and therefore the volume part of the level-density parameter (cf. Chapter 3, Eq. (3.93)).

$$\frac{a_{\text{vol}}}{A} = \frac{2\pi}{9}(9\pi)^{1/3}\frac{M_{\text{nucl}}\, r_0^2}{\hbar^2}\frac{\epsilon}{\epsilon + G\rho_0} \ .$$

The four parameters C, F, ν, and G, relevant for homogeneous nuclear matter, correspond to the four quantities c_1, ρ_0, K, and a_{vol}/A. To the extend that

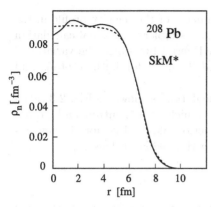
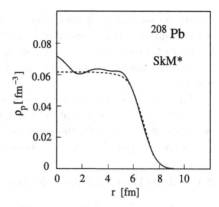

Fig. 2.3. Selfconsistent ETF neutron and proton densities (dashed lines) obtained with the Skyrme interaction SkM* for the nucleus ^{208}Pb compared with the corresponding Hartree-Fock densities (solid lines) (after Ref. [134]).

the latter are empirically known, the parameters in the Skyrme functional, except B, can be fitted.

For a plane infinite surface one obtains the surface-energy coefficient [136]

$$c_2 = 2(36\pi)^{1/3} c_1 \int_0^1 dx\, x\{[k_E + (1 - k_E)x^2][b + c(1 - x)]\}^{1/2} \qquad (2.76)$$

with

$$k_E = K/(18c_1), \qquad b = \epsilon\beta\rho_0^{2/3}/c_1, \qquad c = [B + G(\beta - \gamma)]\rho_0^{5/3}/c_1 .$$

The equation can be used to determine also the parameter B for given ETF parameters β and γ. To derive Eq. (2.76) the simplified equation of state for homogeneous matter

$$\mathcal{E}_{\text{hom}} = c_1 \frac{\rho}{\rho_0} \left[-1 + k_E \left(\frac{\rho - \rho_0}{\rho_0} \right)^2 + (1 - k_E) \left(\frac{\rho - \rho_0}{\rho_0} \right)^4 \right]$$

was used, which yields the same equilibrium values for ρ_0, c_1, and K as Eq. (2.75).

In general the density profile and the surface energy has to be determined numerically from minimizing the ETF energy with respect to the density $\rho(\mathbf{r})$. One expects densities of leptodermous type, i.e. without shell oscillations. In fact, the Euler-Lagrange equations of the variational procedure were solved numerically and it was shown [132, 137] that the resulting densities are very close to modified Fermi functions of the form

$$\rho(r) = \frac{\rho_c}{\left(1 + \exp\left([r - R^{(\rho)}]/d^{(\rho)}\right)\right)^\gamma}, \qquad (2.77)$$

where the parameter γ allows for an asymmetry of the density profile in the nuclear surface. When one performs the much simpler variational calculation in the restricted subspace of these modified Fermi functions, the variational parameters are the coefficients ρ_c, $R^{(\rho)}$, $d^{(\rho)}$ and γ for both, protons and neutrons.

An example of the ETF variational calculation is shown in Fig. 2.3 taken from Ref. [134], where the selfconsistent semiclassical neutron and proton densities obtained with the Skyrme interaction SkM* [138] for the nucleus ^{208}Pb are compared with the corresponding Hartree-Fock densities.

2.1.5 Relativistic mean field theory

Calculations of the lifetime of superheavy nuclei are very sensitive to the shell corrections to the liquid-drop background. These corrections depend crucially on the strength of the spin-orbit coupling in these nuclei. In nonrelativistic mean-field theories the spin-orbit potential depends on certain terms in the effective two-particle interaction which are fitted to properties of known nuclei. Their extrapolation to exotic nuclei is not quite unambiguous. It is therefore interesting to treat fission, in particular of superheavy nuclei, in a relativistic field theory were the spin-orbit coupling comes out naturally, without being added ad hoc from outside. In the following we discuss the relativistic, classical field theory of Walecka [139].

In this field-theoretical approach to the nuclear many-body problem the nucleus is treated as a set of nucleons described by Dirac spinors, which interact by the exchange of mesons. The attraction of the nucleons is caused by an effective scalar meson field σ $(I^\pi = 0^+, T = 0)$, which is to simulate the effect of the real scalar mesons. The short-ranged, repulsive interaction is connected with the exchange of a vector meson ω $(I^\pi = 1^-, T = 0)$ and an isovector meson denoted here by ρ $(I^\pi = 0^-, T = 1)$. The electromagnetic interaction is carried by photons with the vector field A. The Walecka model is derived from the phenomenological, Lorentz-invariant, renormalizable Lagrangian [113,140]

$$
\begin{aligned}
\mathcal{L} = &\, \bar{\psi}_i \{i\,\gamma^\mu\,\partial_\mu - M_{\text{nucl}}\}\psi_i \\
&+ \frac{1}{2}\partial^\mu\sigma\,\partial_\mu\sigma - \frac{1}{2}m_\sigma\sigma^2 - g_\sigma\,\bar{\psi}_i\psi_i\,\sigma \\
&- \frac{1}{4}\Omega^{\mu\nu}\Omega_{\mu\nu} + \frac{1}{2}m_\omega^2\,\omega^\mu\omega_\mu - g_\omega\,\bar{\psi}_i\,\gamma^\mu\,\psi_i\,\omega_\mu \\
&- \frac{1}{4}\boldsymbol{R}^{\mu\nu}\boldsymbol{R}_{\mu\nu} + \frac{1}{2}m_\rho^2\,\boldsymbol{\rho}^{\,\mu}\boldsymbol{\rho}_\mu - g_\rho\bar{\psi}_i\,\gamma^\mu\boldsymbol{\tau}\,\psi_i\,\boldsymbol{\rho}_\mu \\
&- \frac{1}{4}F^{\mu\nu}F_{\mu\nu} - e\bar{\psi}_i\,\gamma^\mu\,\frac{(1-\tau_3)}{2}\,\psi_i\,A_\mu\,,
\end{aligned}
\tag{2.78}
$$

where M_{nucl} is the nucleon mass, $m_\sigma, m_\omega, m_\rho$ are meson masses, and $g_\sigma, g_\omega, g_\rho$ are meson coupling constants. The isovector quantities are indicated by bold symbols. Following the usual practice we use in this subsection units such that $\hbar = c = 1$ and the four 4×4 matrices γ^μ, $\mu = 0 \ldots 3$, are defined by

$$\gamma^0 = \beta = \begin{pmatrix} I & 0 \\ 0 & -I \end{pmatrix}, \qquad \gamma^i = \begin{pmatrix} 0 & \boldsymbol{\sigma}^i \\ -\boldsymbol{\sigma}^i & 0 \end{pmatrix}, \qquad i = 1,2,3 \,,$$

where I is the 2×2 unit matrix and $\boldsymbol{\sigma}^i$, $i = x, y, z$, are the standard 2×2 Pauli matrices. The notation $\bar{\psi} = \psi^+ \gamma^0$ is used. The field tensors of the vector bosons are given by

$$\begin{aligned} \Omega^{\mu\nu} &= \partial^\mu \omega^\nu - \partial^\nu \omega^\mu, \\ \mathbf{R}^{\mu\nu} &= \partial^\mu \boldsymbol{\rho}^\nu - \partial^\nu \boldsymbol{\rho}^\mu - g_\rho(\boldsymbol{\rho}^\mu \times \boldsymbol{\rho}^\nu) \\ F^{\mu\nu} &= \partial^\mu A^\nu - \partial^\nu A^\mu \,. \end{aligned} \tag{2.79}$$

In the non-linear σ-ω-ρ model of Boguta and Bodmer [141] realistic compression and surface properties of nuclei could be obtained by replacing the mass term of the σ field by a nonlinear, self-interaction potential of the σ-field

$$\frac{1}{2}m_\sigma^2 \sigma^2 \to U(\sigma) = \frac{1}{2}m_\sigma^2 \sigma^2 + \frac{1}{3}g_2 \sigma^3 + \frac{1}{4}g_3 \sigma^4 \tag{2.80}$$

with additional phenomenological parameters g_2 and g_3.

So far the field variables in Eq. (2.78) are creation and destruction operators in Fock space. In the Walecka model one introduces a mean-field approximation by treating all fields as classical fields

$$\psi \to \langle \psi \rangle, \quad \sigma \to \langle \sigma \rangle, \quad \omega_\mu \to \langle \omega_\mu \rangle, \quad \boldsymbol{\rho}_\mu \to \langle \boldsymbol{\rho}_\mu \rangle, \quad A_\mu \to \langle A_\mu \rangle$$

and assuming that the meson and the electric fields are time-independent in the rest frame of the nucleus and that the time-dependence of the spinor can be split off as a phase factor $\exp(i\epsilon t)$. This allows us to omit all boson currents in the Euler-Lagrange equations of $\mathcal{L}$. Note that the term "mean field" is used here not with the same meaning as in the previously considered nonrelativistic theories.

With these approximations the Euler-Lagrange equations of the Lagrangian (2.78) yield a stationary Dirac equation for the spinor field ψ

$$\{-i\alpha\nabla + \beta M^*_{\mathrm{nucl}}(\boldsymbol{r}) + V(\boldsymbol{r})\}\psi_i(\boldsymbol{r}) = \varepsilon_i \psi_i(\boldsymbol{r}) \,, \tag{2.81}$$

stationary Klein-Gordon equations for the meson fields,

$$\begin{aligned} (-\Delta + m_\sigma^2)\, \sigma(\boldsymbol{r}) &= -g_\sigma \rho_s(\boldsymbol{r}) - g_2\sigma^2 - g_3\sigma^3 \,, \\ (-\Delta + m_\omega^2)\, \omega^0(\boldsymbol{r}) &= g_\omega j^0(\boldsymbol{r}) \,, \\ (-\Delta + m_\rho^2)\, \rho^0(\boldsymbol{r}) &= g_\rho \boldsymbol{j}^0(\boldsymbol{r}) \,, \end{aligned} \tag{2.82}$$

and the Poisson equation for the photon field

$$-\Delta\, A^0 = e j_p^0(\boldsymbol{r}) \,, \tag{2.83}$$

where the source terms of the last four equations are the scalar density

$$\rho_s = \sum_{i=1}^{A} \bar{\psi}_i \psi_i \tag{2.84}$$

and the time-like components of the current, the isovector current, and the charge current of the nuclear field. They are defined by

$$j^0 = \sum_{i=1}^{A} \psi_i^+ \psi_i \, , \quad \boldsymbol{j}^0 = \sum_{i=1}^{A} \psi_i^+ \boldsymbol{\tau} \, \psi_i \, , \quad j_p^0 = \sum_{i=1}^{A} \psi_i^+ \frac{(1-\tau_3)}{2} \psi_i \, . \tag{2.85}$$

The space-like components of these currents were assumed to vanish in the Walecka model. In the Dirac equation (2.81) β and $\alpha^i = \beta\gamma^i$ ($i = 1, 2, 3$) are the Dirac matrices. The mass in Eq. (2.81) is renormalized by the σ field and becomes an effective mass

$$M_{\text{nucl}}^*(\boldsymbol{r}) = M_{\text{nucl}} + g_\sigma \, \sigma(\boldsymbol{r}) \, , \tag{2.86}$$

and the so-called vector potential

$$V(\boldsymbol{r}) = g_\omega \, \omega^0(\boldsymbol{r}) + g_\rho \tau_3 \, \rho^0(\boldsymbol{r}) + \frac{e(1-\tau_3)}{2} A^0(\boldsymbol{r}) \tag{2.87}$$

is the time-like component of a Lorentz vector. The eigenvalues ϵ_i of the Dirac equation (2.81) are labeled by quantum numbers i. For a spherical system $i = \{njm_j\kappa t_z\}$, where n is the radial quantum number, j and m_j are the total angular momentum and its z component, $\kappa = \pm(j + 1/2)$, and t_z is an isospin quantum-number. It is assumed that in the ground state the nucleons occupy the lowest A eigenstates of Eq. (2.81). Eqs. (2.81)-(2.85) have to be solved selfconsistently by an iterative procedure [142, 143]

Treating the spinor field as a classical field implies the neglect of exchange contributions to the total energy

$$E = \sum_{i=1}^{A} \epsilon_i - \frac{1}{2} \int d^3r \left(g_\sigma \rho_s(\mathbf{r})\sigma(\mathbf{r}) + \frac{1}{3}g_2\sigma^3(\mathbf{r}) \right.$$
$$\left. + \frac{1}{2}g_3\sigma^4(\mathbf{r}) + g_\omega j^0(\mathbf{r})\omega^0(\mathbf{r}) + g_\rho \boldsymbol{j}^0(\mathbf{r})\rho^0(\mathbf{r}) + ej_p^0(\mathbf{r})A^0(\mathbf{r}) \right). \tag{2.88}$$

This approximation is therefore called Hartree approximation in field theory. For a treatment of the quantized nucleon field and inclusion of exchange terms in the energy we refer to the work of Brockmann [144] and numerical calculations also for spherical nuclei by Bouyssy et al. [145].

The cheapest way to account for pairing correlations in this context is the use of the BCS scheme of subsection 2.1.3 with a constant pairing matrix-element G, fitted to odd-even mass differences, using the ϵ_i as single-particle energies. From Eq. (2.46) occupation numbers v_i^2 are obtained and all sums over i in Eqs. (2.84), (2.85), and (2.88) are replaced by sums weighted with

v_i^2: $\sum_{i=1}^{A} \rightarrow \sum_{i}^{(\omega)} v_i^2$. The system consisting of the BCS equations and Eqs. (2.81)-(2.85) is solved selfconsistently [146]. Considerably more involved is an attempt to obtain pairing correlations from the coupling of the meson fields to the nucleons since it requires to treat all fields in their quantized form [140].

There are several parameter sets of the RMF theory and frequently used ones are the NL1, NL-Sh, and NL-3 sets [147] and the NL-Z2 parameters [148]. The latter set is

$$
\begin{array}{ll}
M_{\mathrm{nucl}} = 939.0 \text{ MeV} & g_\sigma = 10.217 \\
m_\sigma = 508.194 \text{ MeV} & g_2 = \text{-}10.431 \\
m_\omega = 782.501 \text{ MeV} & g_3 = \text{-}28.885 \\
m_\rho = 763.000 \text{ MeV} & g_\omega = 12.868 \\
& g_\rho = 4.474
\end{array}
$$

and was fitted to ground-state masses and electric formfactors of a selection of spherical nuclei between ^{16}O and ^{208}Pb, where a correction for the spurious center-of-mass motion was included.

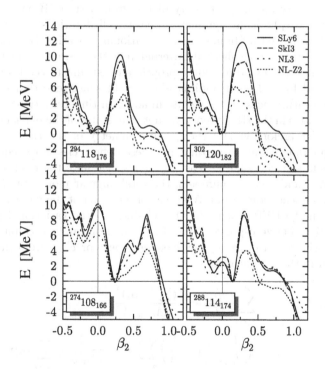

Fig. 2.4. The energetically favored fission barriers of four superheavy nuclei normalized to the ground-state energy. They were evaluated in Ref. [149] for two Skyrme interactions (SLy6, SkI3) and for the RMF theory with parameter sets NL-3 and NL-Z2 (from Ref [149]).

The RMF theory provides a reasonably good description of many features of nuclei, but it yields rather poor fission barrier heights and widths. In Fig. 2.4, taken from Ref. [149], the lowest selfconsistent fission barriers of four different superheavy nuclei are shown. Results obtained with two sets of RMF parameters (NL-3 and NL-Z2) are compared with estimates obtained with the Skyrme forces SLy6 and SkI3. It is seen that the barrier heights evaluated in these four models differ significantly from each other. Generally, the present parametrizations of the RMF theory yield too low barriers.

2.1.6 Spontaneous symmetry breaking

In all mean-field approaches the ground-state wave function violates fundamental symmetries of the Hamiltonian. A Slater determinant of bound states violates translational and Galilean invariances and for deformed ground states it also violates rotational symmetry. It is worth noting that in the iterative solution procedure of the HF equations the initial state should already contain the symmetry-breaking property of the ground state. If one starts, for example, with a spherically symmetric density, all iterations remain spherically symmetric since the Hamiltonian is rotationally invariant, even when the actual ground state is nonspherical. Alternatively, the desired symmetry violation may be induced by adding an appropriate constraint to the Hamiltonian H. Octupole-deformed local minima also violate space-reflection symmetry and non-spin-saturated systems violate in addition time-reflection symmetry.

Ground states in the HFB theory are not eigenstates of the particle-number operator and therefore violate the gauge symmetry of the Hamiltonian. A simple, approximate way to account for these deficiencies is to require that the mean-field solution shall at least in the average have the required values for the proton and neutron numbers, the total angular momentum, and the square of the linear momentum. Average conservation of the particle number is achieved in the HFB scheme by the $\lambda \hat{N}$-term in the variational principle Eq. (2.26). Spurious zero-point center-of-mass motion can be accounted for by subtracting the center-of-mass kinetic energy operator $\hat{E}_{\text{c.m.}}$ from the kinetic energy term in the Hamiltonian

$$
\hat{E}_{\text{kin}} - \hat{E}_{\text{c.m.}} = \sum_{i=1}^{A} \frac{\hat{\mathbf{p}}_i^2}{2M_{\text{nucl}}} - \frac{(\sum_{i=1}^{A} \hat{\mathbf{p}}_i)^2}{2M_{\text{nucl}}A}
$$
$$
= \frac{1}{2M_{\text{nucl}}} \left(1 - \frac{1}{A}\right) \sum_{i=1}^{A} \hat{\mathbf{p}}_i^2 - \frac{1}{2M_{\text{nucl}}A} \sum_{i>j} \hat{\mathbf{p}}_i \hat{\mathbf{p}}_j \ . \quad (2.89)
$$

The second term in the above equation is numerically involved as it includes double integrals and is neglected in most applications. A more detailed discussion of this problem is given e.g. in Ref. [150].

Similarly, to approximate the effect of the angular momentum projection on the energy of a deformed nucleus, one subtracts the zero-point rotational energy E_{rot}^0 from the total HFB energy (2.35)

$$E_{rot}^0 = \frac{1}{2\mathcal{J}} \langle 0 | \sum_i \hat{\mathbf{j}}_i^2 | 0 \rangle , \qquad (2.90)$$

where $\mathcal{J}$ is the moment of inertia of the nucleus. For its choice see for example Refs. [151, 152].

Within the framework of the mean-field theory this is the best one can do. However, the wave function of the ground state is still not an eigenstate of the symmetry operators, which is very undesirable when discussing spectroscopic observables. Let us consider the special case of rotational bands, either over a deformed ground state, or over secondary minima or fission saddle-points. The HF or HFB states in space representation $\Phi(\mathbf{x}_1, \mathbf{x}_2, \ldots \mathbf{x}_A)$ depend on the coordinates $\mathbf{x}_i$ in space, spin and isospin spaces of all A nucleons. They refer to an arbitrary space-fixed coordinate system. A wave function $\mathcal{D}(\theta_1, \theta_2, \theta_3)\Phi$ referring to a coordinate system, rotated by Euler angles $\theta_1, \theta_2, \theta_3$, would also be a solution of the HF or HFB equations to the same energy. Peierls and Yoccoz [153] therefore extended the mean-field scheme by considering test wave functions of the form

$$\Psi = \frac{1}{8\pi^2} \int_0^{2\pi} d\theta_1 \int_0^\pi \sin\theta_2 d\theta_2 \int_0^{2\pi} d\theta_3 \, g(\theta_1, \theta_2, \theta_3) \mathcal{D}(\theta_1, \theta_2, \theta_3)\Phi \qquad (2.91)$$

in the variational principle (2.26), where $g(\theta_1, \theta_2, \theta_3)$ is a weight function to be determined from the variational principle

$$\delta \frac{\langle \Psi | \hat{H} | \Psi \rangle}{\langle \Psi | \Psi \rangle} = 0. \qquad (2.92)$$

It turns out that the function g becomes a representation of the rotational group (Zeh's theorem [154]) thus making Ψ an eigenfunction of the angular momentum operator.

One may either vary the weight function g and the single-particle states in the HF state or the quasiparticle states in the HFB state at the same time, which is called "variation after projection" (VAP), or the single-particle or quasiparticle states are first determined by the variational principle (2.26) and only the function g is determined from Eq. (2.92). This is the "variation before projection". The latter approach is numerically less demanding, but leads to higher, i.e. less satisfactory, ground-state energies [155]. It can be shown that the difference between the binding energies in the two approaches becomes smaller when the symmetry breaking gets larger, that is, when the deviation from sphericity of the HF or HFB state increases. For a more detailed discussion of symmetry restoration by projection techniques we refer to Chap. 11 of Ref. [108]. For approximation schemes to VAP see the review by Flocard and Onishi [156]. Their relation to the generator-coordinate method in connection with the Gaussian overlap approximation (see Ref. [108], Chap. 10) has been studied in the work of Góźdź, Pomorski, Brack, and Werner [157–159], in particular in cases of simultaneous symmetry violation in several degrees of freedom.

To obtain states with well-defined particle number from the BCS wave function, the definition (2.40) is generalized by introducing a phase angle ϕ

$$|\text{BCS}(\phi)\rangle = \prod_{i>0}(u_i + e^{i\phi}v_i\hat{a}_i^+\hat{a}_{\bar{i}}^+|0\rangle .$$

One can convince oneself that expectation values like $\langle\text{BCS}(\phi)|\hat{N}|\text{BCS}(\phi)\rangle$ and $\langle\text{BCS}(\phi)|\hat{H}|\text{BCS}(\phi)\rangle$ or Δ are independent of ϕ and are still given by Eqs. (2.41), (2.42), and (2.47), respectively. Putting $e^{i\phi} = \zeta$ the state

$$|\Psi(p)\rangle = \frac{1}{2\pi i}\oint \frac{d\zeta}{\zeta^{p+1}}\prod_{i>0}(u_i + \zeta v_i\hat{a}_i^+\hat{a}_{\bar{i}}^+|0\rangle \tag{2.93}$$

consists of exactly p pairs, i.e. is an eigenstate of the particle number $N = 2p$, because, due to the residue theorem, the integral picks out just the term with p factors $\zeta v_i\hat{a}_i^+\hat{a}_{\bar{i}}^+$ in the numerator of the integrand [160, 161]. Using Wick's theorem [108] one can show that in terms of the operator of the number of pairs $\hat{N}/2$

$$|\text{BCS}(\phi)\rangle = e^{i\phi\hat{N}/2}|\text{BCS}\rangle .$$

Therefore Eq. (2.93) can also be written

$$|\Psi(p)\rangle = \int_0^{2\pi} d\phi\, e^{i\phi(\hat{N}/2-p)}|\text{BCS}\rangle . \tag{2.94}$$

One should note the analogy with Eq. (2.91) when the latter is restricted to rotations by an angle θ_3 around a symmetry axis: The expression $\mathcal{D}(\theta_3)\Phi = \Phi(\theta_3)$ corresponds to the transition from $|\text{BCS}\rangle$ to $|\text{BCS}(\phi)\rangle$. Since the degeneracy with respect to ϕ is due to a one-dimensional Abelian group, its representations are the phase functions $e^{-ip\phi}$ with integer p. Identifying the weight function $g(\theta_3)$ in Eq. (2.91) according to Zeh's theorem with this phase function, one obtains the integral (2.94).

2.1.7 Effective interactions

It is assumed, somewhat arbitrarily, that the central part of the nuclear effective two-body interaction is hermitian and – in space representation – short-ranged, compared to a typical inverse wave number $1/k_f$ of bound-state nuclear wave functions. It is further assumed to be rotational invariant in space, spin space, and isospin space and translational invariant. A prototype of such potential, local in the space variables $\mathbf{r}_1$ and $\mathbf{r}_2$ and in the spin and isospin variables, has the general form

$$V(\mathbf{r}_1,\mathbf{r}_2,\boldsymbol{\sigma}_1,\boldsymbol{\sigma}_2,\boldsymbol{\tau}_1,\boldsymbol{\tau}_2) = f(|\mathbf{r}_1 - \mathbf{r}_2|/\beta)\hat{A}(\hat{P}_\sigma,\hat{P}_\tau) \tag{2.95}$$

with a short-range function $f(x)$, for example $f = \exp(-x^2)$, $x = |\mathbf{r}_1 - \mathbf{r}_2|/\beta$ and a range parameter $\beta \ll 1/k_f$. The dependence on spin and isospin variables is given by the operator $\hat{A}$ of Eq. (2.2).

Skyrme interactions

The fast variation of the effective two-body interaction potential on the scale of the nuclear radius suggests a moment expansion of the matrix element $\langle l'm'|f(\mathbf{r}_1 - \mathbf{r}_2)|lm\rangle$ [162]. Introducing coordinates $\mathbf{r} = \mathbf{r}_1 - \mathbf{r}_2$ and $\mathbf{R} = \frac{1}{2}(\mathbf{r}_1 + \mathbf{r}_2)$ one expands the integrand in the spatial integral up to second order in $\mathbf{r}$

$$
\langle l'm'|f(r)|lm\rangle = \int \varphi_{l'}^*(\mathbf{r}_1)\varphi_{m'}^*(\mathbf{r}_2)f(r)\varphi_l(\mathbf{r}_1)\varphi_m(\mathbf{r}_2)d^3r\,d^3R
$$

$$
\approx \int f(r)\left[1 + (\mathbf{r}\boldsymbol{\nabla}_\mathbf{r}) + \frac{1}{2}(\mathbf{r}\boldsymbol{\nabla}_\mathbf{r})^2\right][\varphi_{l'}^*(\mathbf{r}_1)\varphi_{m'}^*(\mathbf{r}_2)]_{\mathbf{r}=0}
$$

$$
\times \left[1 + (\mathbf{r}\boldsymbol{\nabla}_\mathbf{r}) + \frac{1}{2}(\mathbf{r}\boldsymbol{\nabla}_\mathbf{r})^2\right][\varphi_l(\mathbf{r}_1)\varphi_m(\mathbf{r}_2)]_{\mathbf{r}=0}\,d^3r\,d^3R. \tag{2.96}
$$

Since for fixed $\mathbf{R}$

$$
\boldsymbol{\nabla}_\mathbf{r}[\varphi_l(\mathbf{r}_1)\varphi_m(\mathbf{r}_2)] = \frac{\partial\varphi_l(\mathbf{r}_1)\varphi_m(\mathbf{r}_2)}{\partial\mathbf{r}_1}\frac{\partial\mathbf{r}_1}{\partial\mathbf{r}} + \frac{\partial\varphi_l(\mathbf{r}_1)\varphi_m(\mathbf{r}_2)}{\partial\mathbf{r}_2}\frac{\partial\mathbf{r}_2}{\partial\mathbf{r}}
$$

$$
= \frac{1}{2}[\partial_{\mathbf{r}_1} - \partial_{\mathbf{r}_2}][\varphi_l(\mathbf{r}_1)\varphi_m(\mathbf{r}_2)]\,,
$$

one defines operators $\overleftarrow{\boldsymbol{\nabla}}_{12} = \boldsymbol{\nabla}_1 - \boldsymbol{\nabla}_2$ and $\overrightarrow{\boldsymbol{\nabla}}_{12} = \boldsymbol{\nabla}_1 - \boldsymbol{\nabla}_2$ acting on the bra-states $\varphi_{l'}^*(\mathbf{r}_1)\varphi_{m'}^*(\mathbf{r}_2)$ and on the ket-states $\varphi_l(\mathbf{r}_1)\varphi_m(\mathbf{r}_2)$, respectively. Performing the integration with respect to $\mathbf{r}$ first and calling the first nonvanishing moments of the function $f(r)$

$$
V_0 = \int f(r)d^3r, \quad \text{and} \quad 2V_2 = \int r^2 f(r)d^3r
$$

the integral of Eq. (2.96) can be rewritten as

$$
\int d^3R\,[\varphi_{l'}^*(\mathbf{r}_1)\varphi_{m'}^*(\mathbf{r}_2)]_{\mathbf{r}=0}\left[V_0 + \frac{V_2}{4}(\overleftarrow{\boldsymbol{\nabla}}^2 + \overrightarrow{\boldsymbol{\nabla}}^2 + 2\overleftarrow{\boldsymbol{\nabla}}\overrightarrow{\boldsymbol{\nabla}})\right][\varphi_l(\mathbf{r}_1)\varphi_m(\mathbf{r}_2)]_{\mathbf{r}=0}
$$

$$
= \int d^3r_1 d^3r_2 \delta(\mathbf{r}_1 - \mathbf{r}_2)\,[\varphi_{l'}^*(\mathbf{r}_1)\varphi_{m'}^*(\mathbf{r}_2)]\left[V_0 + \frac{V_2}{4}(\overleftarrow{\boldsymbol{\nabla}}^2 + \overrightarrow{\boldsymbol{\nabla}}^2 + 2\overleftarrow{\boldsymbol{\nabla}}\overrightarrow{\boldsymbol{\nabla}})\right]
$$

$$
\times [\varphi_l(\mathbf{r}_1)\varphi_m(\mathbf{r}_2)]. \tag{2.97}
$$

Note that all odd moments of $f(r)$ vanish. The exchange operator $\hat{A}$ simplifies considerably for zero-range potential-functions f. Introducing the operator of spatial exchange $\hat{P}_r$, we have

$$\hat{P}_r^2 = \hat{P}_\sigma^2 = \hat{P}_\tau^2 = 1 \, .$$

When applied to an antisymmetrized two-nucleon state,

$$\hat{P}_r \hat{P}_\sigma \hat{P}_\tau = -1 \, .$$

From these relations follows

$$\hat{P}_r \delta(\mathbf{r}_1 - \mathbf{r}_2) = \delta(\mathbf{r}_1 - \mathbf{r}_2), \qquad \hat{P}_r \overleftarrow{\nabla}^2 \delta(\mathbf{r}_1 - \mathbf{r}_2) = \overleftarrow{\nabla}^2 \delta(\mathbf{r}_1 - \mathbf{r}_2),$$

$$\hat{P}_r \overleftarrow{\nabla} \delta(\mathbf{r}_1 - \mathbf{r}_2) = - \overleftarrow{\nabla} \delta(\mathbf{r}_1 - \mathbf{r}_2), \qquad \hat{P}_\tau \overleftarrow{\nabla} \delta(\mathbf{r}_1 - \mathbf{r}_2) = \hat{P}_\sigma \overleftarrow{\nabla} \delta(\mathbf{r}_1 - \mathbf{r}_2),$$

$$\hat{P}_\tau \delta(\mathbf{r}_1 - \mathbf{r}_2) = -\hat{P}_\sigma \delta(\mathbf{r}_1 - \mathbf{r}_2), \qquad \hat{P}_\tau \overleftarrow{\nabla}^2 \delta(\mathbf{r}_1 - \mathbf{r}_2) = -\hat{P}_\sigma \overleftarrow{\nabla}^2 \delta(\mathbf{r}_1 - \mathbf{r}_2)$$

and the same equations with $\overrightarrow{\nabla}$. From these results, together with Eq. (2.97), the general form of the two-particle effective potential becomes, in the approximation of very short range,

$$V_{12} = t_0(1 + x_0 \hat{P}_\sigma) \, \delta(\mathbf{r}_1 - \mathbf{r}_2)$$
$$- \frac{1}{8} t_1 (1 + x_1 \hat{P}_\sigma) \left[\overleftarrow{\nabla}_{12}^2 \, \delta(\mathbf{r}_1 - \mathbf{r}_2) + \delta(\mathbf{r}_1 - \mathbf{r}_2) \, \overrightarrow{\nabla}_{12}^2 \right]$$
$$+ \frac{1}{4} t_2 (1 + x_2 \hat{P}_\sigma) \, \overleftarrow{\nabla}_{12} \, \delta(\mathbf{r}_1 - \mathbf{r}_2) \, \overrightarrow{\nabla}_{12} \qquad (2.98)$$

with

$$\begin{aligned}
t_0 &= (W + M)V_0, & t_0 x_0 &= (B + H)V_0 \\
-t_1 &= 2(W + M)V_2, & -t_1 x_1 &= 2(B + H)V_2 \\
t_2 &= 2(W - M)V_2, & t_2 x_2 &= 2(B - H)V_2 \, .
\end{aligned}$$

To control the size of the spin-orbit splitting in the HF single-particle potential, derived from the two-body interaction (2.98), Bell and Skyrme [163] added a spin-orbit term

$$V_{LS} = i(1/4)W_{LS}\left[\overleftarrow{\nabla}_{12} \times \delta(\mathbf{r}_1 - \mathbf{r}_2) \, \overrightarrow{\nabla}_{12} \right] \cdot (\boldsymbol{\sigma}_1 + \boldsymbol{\sigma}_2) \qquad (2.99)$$

to the central potential (2.98), where $\boldsymbol{\sigma}_i$ is twice the spin operator of nucleon i. To enforce proper saturation density and energy density in bulk nuclear matter, Skyrme added also a repulsive three-body force of zero range $V_{123} = t_3 \delta(\mathbf{r}_1 - \mathbf{r}_2)\delta(\mathbf{r}_2 - \mathbf{r}_3)$ to the two-body interaction [164]. For spin-saturated systems this term was shown by Bäckmann, Jackson, and Speth [165] to be equivalent to a density-dependent two-particle potential

$$V_{123} = \frac{t_3}{6}(1 + x_3 \hat{P}_\sigma)\rho(\mathbf{R})\delta(\mathbf{r}_1 - \mathbf{r}_2) \, ,$$

where $\rho(\mathbf{R})$ is the sum of proton and neutron densities. To obtain more flexibility to reproduce the position of giant monopole resonance-energies, one uses generally the slightly more general form

$$V_{123} = \frac{t_3}{6}(1 + x_3 \hat{P}_\sigma)[\rho(\mathbf{R})]^\gamma \delta(\mathbf{r}_1 - \mathbf{r}_2) \tag{2.100}$$

with additional parameters x_3 and γ. Finally, the Coulomb two-body potential between protons is given by

$$V_{\text{Coul}} = \frac{e^2}{|\mathbf{r}_1 - \mathbf{r}_2|} . \tag{2.101}$$

The total effective interaction is the sum of the potentials (2.98)-(2.101)

$$V_{\text{Skyrme}} = V_{12} + V_{LS} + V_{123} + V_{\text{Coul}} . \tag{2.102}$$

The parameters of the Skyrme interaction are: $t_0, t_1, t_2, t_3, x_0, x_1, x_2, x_3, W_{LS}$ and γ. There are several sets of these parameters adjusted to binding energies of various mass regions and additional observables, in particular RPA-type excitations of monopole, dipole or quadrupole character. Frequently used sets are, e.g., SkIII [166] (fairly good binding energies, less perfect radii), SkM* [138] (good binding energies, including fission barriers, good radii), SkP [167] (gives realistic pairing with Skyrme interactions), SLy6 [168] (fitted to describe neutron-rich nuclei and neutron stars), SkI3 [169] (gives isovector spin-orbit splitting and electromagnetic form factors), or BSk1 [170] (based on an astrophysically biased, large-scale mass-fit); see Fig. 2.4 for fission-barrier heights of superheavy nuclei, calculated with Skyrme interactions SLy6 and SkI3.

For fission studies the SkM* potential has often been used. Its set of parameters

$$
\begin{array}{ll}
t_0 = \text{-2645 MeV fm}^3 & x_0 = 0.09 \\
t_1 = \text{410 MeV fm}^5 & x_1 = 0 \\
t_2 = \text{-135 MeV fm}^5 & x_2 = 0 \\
t_3 = \text{15595 MeV fm}^{(3+3\gamma)} & x_3 = 0 \\
W_{LS} = \text{130 MeV fm}^5 & \gamma = 1/6
\end{array}
$$

was chosen to reproduce in particular experimental fission barrier heights.

Using Slater determinants to calculate the expectation value of the Hamiltonian with Skyrme interactions, leads to the energy functional

$$E = \int \mathcal{E}(\rho(\mathbf{r}), \boldsymbol{\nabla}\rho(\mathbf{r}), \tau(\mathbf{r}), \mathbf{J}) \, d^3r , \tag{2.103}$$

where the energy density $\mathcal{E}$ is a function of the nucleon densities ρ_q, Eq. (2.70), the kinetic energy densities τ_q, Eq. (2.69), and spin-orbit densities

$$\mathbf{J}_q = -i \sum_{i,\sigma,\sigma'} \varphi_i(\mathbf{r}, \sigma, q)[\boldsymbol{\nabla}\varphi_i(\mathbf{r}, \sigma', q) \times \langle \sigma | \boldsymbol{\sigma} | \sigma' \rangle] . \tag{2.104}$$

Assuming that there is no isospin mixing in the HF basis and that the ground state is time-reversal invariant, the explicit form of $\mathcal{E}$ was derived by Vautherin and Brink [171]

$$\mathcal{E}[\rho_q(\boldsymbol{r}), \tau_q(\boldsymbol{r}), \boldsymbol{J}_q(\boldsymbol{r})] =$$

$$\frac{\hbar^2}{2M_{\text{nucl}}}\tau + \frac{1}{2}t_0\left[\left(1+\frac{x_0}{2}\right)\rho^2 - \left(x_0+\frac{1}{2}\right)(\rho_n^2 + \rho_p^2)\right]$$

$$+\frac{1}{12}t_3\rho^\gamma\left[\left(1+\frac{x_3}{2}\right)\rho^2 - \left(x_3+\frac{1}{2}\right)(\rho_n^2 + \rho_p^2)\right]$$

$$+\frac{1}{4}\left[t_1\left(1+\frac{x_1}{2}\right) + t_2\left(1+\frac{x_2}{2}\right)\right]\tau\rho$$

$$-\frac{1}{4}\left[t_1\left(x_1+\frac{1}{2}\right) - t_2\left(x_2+\frac{1}{2}\right)\right](\tau_n\rho_n + \tau_p\rho_p) \qquad (2.105)$$

$$+\frac{1}{16}\left[3t_1\left(1+\frac{x_1}{2}\right) - t_2\left(1+\frac{x_2}{2}\right)\right](\boldsymbol{\nabla}\rho)^2$$

$$-\frac{1}{16}\left[3t_1\left(x_1+\frac{1}{2}\right) + t_2\left(x_2+\frac{1}{2}\right)\right][(\boldsymbol{\nabla}\rho_n)^2 + (\boldsymbol{\nabla}\rho_p)^2]$$

$$+\frac{W_{LS}}{2}[\boldsymbol{J}\cdot\boldsymbol{\nabla}\rho + \boldsymbol{J}_n\cdot\boldsymbol{\nabla}\rho_n + \boldsymbol{J}_p\cdot\boldsymbol{\nabla}\rho_p] + \mathcal{E}_{\text{Coul}} ,$$

where $\rho = \rho_p + \rho_n$, $\tau = \tau_p + \tau_n$, and $\mathcal{E}_{\text{Coul}}$ is the Coulomb energy-density. The latter can be written as sum of a direct and an exchange contribution. The exchange contribution is usually approximated by the Slater form [162, 172] (cf. Eq. (2.142) below)

$$\mathcal{E}_{\text{Coul}}(\boldsymbol{r}) = \frac{e^2}{2}\rho_p(\boldsymbol{r})\int d^3r'\,\frac{\rho_p(\boldsymbol{r}')}{|\boldsymbol{r}-\boldsymbol{r}'|} - \frac{3}{4}e^2\left(\frac{3}{\pi}\right)^{1/3}\rho_p^{4/3}(\boldsymbol{r}) . \qquad (2.106)$$

For systems whose ground-state wave-function is not time-reversal invariant, for example, for odd systems, the energy density has additional terms [115, 173]. The terms in the first two lines on the left-hand-side of Eq. (2.105) determine essentially properties of bulk nuclear matter, the saturation density ρ_0, the volume-energy coefficient of the liquid-drop formula c_1, and the incompressibility parameter K. The terms in the third and fourth line control the effective mass M^*_{nucl} and those in the fifth and sixth line contribute to the surface energy.

Minimizing the total energy of Eq. (2.103) with respect to variations of the single-particle wavefunctions $\varphi_j^{(q)}$, one obtains the Skyrme HF-equation

$$\hat{h}_q\varphi_j^{(q)} = \varepsilon_j^{(q)}\varphi_j^{(q)} . \qquad (2.107)$$

for the mean-field single-particle Hamiltonian

$$\hat{h}_q = -\boldsymbol{\nabla}\frac{\hbar^2}{2M_{\text{nucl}}^{(q)*}(\boldsymbol{r})}\boldsymbol{\nabla} + V_q(\boldsymbol{r}) - i\boldsymbol{W}_q(\boldsymbol{r})\cdot(\boldsymbol{\nabla}\times\boldsymbol{\sigma}) . \qquad (2.108)$$

The central nuclear potential $V_q(\boldsymbol{r})$ is obtained as functional derivative of the energy (2.103) with respect to the density

$$V_q(\boldsymbol{r}) = \frac{\delta E}{\delta \rho_q(\boldsymbol{r})} = \frac{\partial \mathcal{E}}{\partial \rho_q} - \boldsymbol{\nabla} \left(\frac{\partial \mathcal{E}}{\partial (\boldsymbol{\nabla} \rho_q)} \right) . \tag{2.109}$$

For Skyrme potentials this yields

$$\begin{aligned}
V_q(\boldsymbol{r}) = {} & t_0 \left(1 + \frac{x_0}{2}\right) \rho - t_0 \left(x_0 + \frac{1}{2}\right) \rho_q \\
& + \frac{t_3}{12} \gamma \rho^{\gamma-1} \left[\left(1 + \frac{x_3}{2}\right) \rho^2 - \left(x_3 + \frac{1}{2}\right) (\rho_n^2 + \rho_p^2)\right] \\
& + \frac{t_3}{6} \rho^\gamma \left[\left(1 + \frac{x_3}{2}\right) \rho - \left(x_3 + \frac{1}{2}\right) \rho_q\right] \\
& + \frac{1}{4} \left[t_1 \left(1 + \frac{x_1}{2}\right) + t_2 \left(1 + \frac{x_2}{2}\right)\right] \tau \\
& - \frac{1}{4} \left[t_1 \left(x_1 - \frac{1}{2}\right) + t_2 \left(x_2 + \frac{1}{2}\right)\right] \tau_q \\
& - \frac{1}{8} \left[3t_1 \left(1 + \frac{x_1}{2}\right) - t_2 \left(1 + \frac{x_2}{2}\right)\right] (\boldsymbol{\nabla}^2 \rho) \\
& + \frac{1}{8} \left[3t_1 \left(x_1 + \frac{1}{2}\right) + t_2 \left(x_2 + \frac{1}{2}\right)\right] (\boldsymbol{\nabla}^2 \rho_q) \\
& - \frac{W_{LS}}{2} [\boldsymbol{\nabla} \cdot \boldsymbol{J} + \boldsymbol{\nabla} \cdot \boldsymbol{J}_q] + V_{\text{Coul}} ,
\end{aligned} \tag{2.110}$$

where the Coulomb potential is given by

$$V_{\text{Coul}}(\boldsymbol{r}) = e^2 \int \frac{\rho_p(\boldsymbol{r}')}{|\boldsymbol{r} - \boldsymbol{r}'|} d^3 r' - e^2 \left(\frac{3}{\pi}\right)^{1/3} \rho_p^{1/3}(\boldsymbol{r}) . \tag{2.111}$$

The effective-mass form-factor is defined as

$$\begin{aligned}
f_q(\boldsymbol{r}) = {} & \frac{M_{\text{nucl}}^{(q)}}{M_{\text{nucl}}^{(q)*}(\boldsymbol{r})} = \frac{2 M_{\text{nucl}}^{(q)}}{\hbar^2} \frac{\delta E}{\delta \tau_q(\boldsymbol{r})} \\
= {} & 1 + \frac{2 M_{\text{nucl}}^{(q)}}{\hbar^2} \left\{ \frac{1}{4} \left[t_1(1 + \frac{x_1}{2}) + t_2(1 + \frac{x_2}{2})\right] \rho(\boldsymbol{r}) \right. \\
& \left. - \frac{1}{4} \left[t_1(x_1 + \frac{1}{2}) - t_2(x_2 + \frac{1}{2})\right] \rho_q(\boldsymbol{r}) \right\}
\end{aligned} \tag{2.112}$$

and the spin-orbit potential as

$$\boldsymbol{W}_q(\boldsymbol{r}) = \frac{\delta E}{\delta \boldsymbol{J}_q(\boldsymbol{r})} = \frac{1}{2} W_{LS} \boldsymbol{\nabla} (\rho + \rho_q) . \tag{2.113}$$

With the exception of the potential SkP, Skyrme potentials are unrealistic in the particle-particle channel and therefore cannot be used alone in the HFB theory. Pairing correlations have to be accounted for e.g. in the framework of the HF+BCS scheme with specially fitted pairing matrix elements. An example for a global fit of the ground-state binding energies of the 1995 mass table [174] is the work of Goriely, Tondeur, and Pearson [152]. They use the Skyrme potential MSk7 for the HF part (the particle-hole channel) and the δ potential of Eq. (2.49) for the BCS part (the particle-particle channel) with slightly different parameters V_0^q for even and odd systems. Later these authors used the full HFB equations with specially fitted Skyrme potentials plus a delta potential (2.49) and a phenomenological Wigner term (cf. Eq. (2.156)) for a comprehensive mass fit [170, 175].

Gogny interaction

Gogny proposed an effective interaction, suitable for HFB calculations, which could be fitted to ground-state binding-energies, single-particle levels in light nuclei and RPA-type modes [118, 176]. He combined the pair of Gauss functions from the Brink-Bocker interaction potential [177] with the three-body and the spin-orbit terms of the Skyrme interaction. The resulting potential is

$$V_{12} = \sum_{i=1}^{2} \exp\left[-\frac{|\boldsymbol{r}_1 - \boldsymbol{r}_2|^2}{\mu_i^2}\right] \cdot (W_i + B_i \hat{P}_\sigma - H_i \hat{P}_\tau - M_i \hat{P}_\sigma \hat{P}_\tau)$$
$$+ t_3(1 + x_0 \hat{P}_\sigma)\, \delta(\boldsymbol{r}_1 - \boldsymbol{r}_2) \left[\rho\left(\frac{\boldsymbol{r}_1 + \boldsymbol{r}_2}{2}\right)\right]^\gamma$$
$$+ i W_{\text{LS}}(\boldsymbol{\sigma}_1 + \boldsymbol{\sigma}_2)\cdot \overleftarrow{\boldsymbol{\nabla}}_{12} \times \delta(\boldsymbol{r}_1 - \boldsymbol{r}_2)\, \overrightarrow{\boldsymbol{\nabla}}_{12} + V_{\text{Coul}}\ .$$

In the first line are two finite range potentials $(i = 1, 2)$, one attractive and the other repulsive, with the usual superposition of Wigner, Bartlett, Heisenberg and Majorana spin-isospin contributions. $\hat{P}_\sigma$ and $\hat{P}_\tau$ are the exchange operators of spin and isospin variables, respectively. The remaining terms are taken over from the standard Skyrme interaction.

Dechargé and Gogny adopted the following set of parameters in 1980 [178] and called it D1S

$$\mu_1 = 0.7 \text{ fm} \qquad\qquad \mu_2 = 1.2 \text{ fm}$$
$$W_1 = -1720.3 \text{ MeV} \qquad\qquad W_2 = 103.639 \text{ MeV}$$
$$B_1 = 1300 \text{ MeV} \qquad\qquad B_2 = -163.483 \text{ MeV}$$
$$H_1 = -1813.53 \text{ MeV} \qquad\qquad H_2 = 162.812 \text{ MeV}$$
$$M_1 = 1397.60 \text{ MeV} \qquad\qquad M_2 = -223.934 \text{ MeV}$$
$$t_3 = 1390.60 \text{ MeV fm}^{3(1+\gamma)} \qquad x_0 = 1$$
$$\gamma = 1/3 \qquad\qquad W_{LS} = 130 \text{ MeV fm}^5$$

This parameter set describes binding energies and pairing properties of nuclei across the periodic table. Its predictive power is remarkable: Over the past 25 years, since the force was fitted, the number of accurately measured nuclear binding energies has increased considerably. These new data are still well described by the parameter set D1S.

Seyler-Blanchard interactions

Seyler and Blanchard [131] proposed a momentum-dependent, finite-range interaction

$$V_\nu(r, p) = -C_\nu \frac{e^{-r/\beta)}}{r/\beta} \left[1 - \left(\frac{p}{p_D}\right)^2 \right], \tag{2.114}$$

to be used in the Thomas-Fermi approximation. It is supposed to include already the effect of the exchange term in the potential energy. In this definition $r = |\mathbf{r}_1 - \mathbf{r}_2|$, $p = |\mathbf{p}_1 - \mathbf{p}_2|$, which means that the interaction is translational and rotational invariant in ordinary and, independently, in momentum space. The strength parameter C_ν has one value for proton-proton and for neutron-neutron interaction, $\nu = $ like, and another value for proton-neutron interaction, $\nu = $ unlike. The interaction becomes repulsive for large relative momenta, which corresponds in TF-theory to large density. The parameter p_D therefore controls the saturation density of nuclear matter. To obtain the potential U_q, seen by a proton ($q = 1$) or a neutron ($q = 2$), the interaction (2.114) has to be folded into the proton and neutron densities in phase space, which are $2/h^3$ in TF approximation

$$U_q(\mathbf{r}, \mathbf{p}) = \frac{2}{h^3} \sum_{\nu=1,2} \int_{\text{vol}} d^3r' \left[\int_{(p_f^\mu)} d^3p' V_\nu(|\mathbf{r} - \mathbf{r}'|, |\mathbf{p} - \mathbf{p}'|) \right] + \delta_{q1} V_{\text{Coul}} ,$$

where $\mu = (\nu q \bmod 3)$ and the p-integration is over the Fermi sphere of the protons for $\mu = 1$ or the neutrons for $\mu = 2$. The total nuclear energy is given again as a phase-space integral in terms of U_q by

$$E = \frac{2}{h^3} \sum_{q=1,2} \int_{\text{vol}} d^3r \int_{(p_f^{(q)})} d^3p \left\{ \frac{p^2}{2M_{\text{nucl}}^{(q)}} + \frac{1}{2} U_q(\mathbf{r}, \mathbf{p}) \right\}$$

$$= \frac{2}{h^3} \sum_{q=1,2} \int d^3r \left\{ \frac{4\pi}{10} \frac{(p_f^{(q)}(\mathbf{r}))^5}{M_{\text{nucl}}^{(q)}} + \frac{C_q}{h^3} \int d^3r' \frac{e^{|\mathbf{r}-\mathbf{r}'|/\beta}}{|\mathbf{r} - \mathbf{r}'|/\beta} \sum_{\mu=1,2} I_{q\mu} \right\}$$

$$+ \frac{2e^2}{h^6} \left(\frac{4\pi}{3}\right)^2 \int d^3r \int d^3r' \frac{[p_f^{(1)}(\mathbf{r}) p_f^{(1)}(\mathbf{r}')]^3}{|\mathbf{r} - \mathbf{r}'|} + E_{\text{Coulex}} , \tag{2.115}$$

where the Coulomb exchange-energy E_{Coulex} is treated in the Slater approximation (2.141). The two integrations in momentum space

$$I_{q\mu}(\mathbf{r}, \mathbf{r}') = \int_{(p_f^{(q)})} d^3p \int_{(p_f^{(\mu)})} d^3p' \left[1 - \frac{(\mathbf{p} - \mathbf{p}')^2}{p_D^2} \right]$$

can be done in an elementary way:

$$I_{q\mu} = \left(\frac{4\pi}{3}\right)^2 \left(p_f^{(q)} p_f^{(\mu)}\right)^3 \left\{ 1 - \frac{3}{5} \left[\left(\frac{p_f^{(q)}(\mathbf{r})}{p_D}\right)^2 + \left(\frac{p_f^{(\mu)}(\mathbf{r})}{p_D}\right)^2 \right] \right\} .$$

In the last equations $p_f^{(q)}$ is the Fermi momentum of nucleons of type q.

The proton and neutron numbers Z and N can be written as functionals of $p_f^{(q)}(\mathbf{r})$

$$n_q = \frac{2}{h^3} \int_{\text{vol}} d^3r \int_{p_f^{(q)}} d^3p = \frac{8\pi}{3h^3} \int_{\text{vol}} d^3r (p_f^{(q)})^3 , \qquad q = 1, 2 .$$

The energy has to be minimized with respect to $p_f^{(1)}$ and $p_f^{(2)}$ with the constraint that $n_1 = Z$ and $n_2 = N$

$$\frac{\delta(E\{p_f(\mathbf{r})\} - \lambda_f^{(1)} n_1\{p_f(\mathbf{r})\} - \lambda_f^{(2)} n_2\{p_f(\mathbf{r})\})}{\delta p_f^{(q)}(\mathbf{r})} = 0, \qquad q = 1, 2 .$$

The resulting two coupled, nonlinear integral equations for the functions $p_f^{(1)}(\mathbf{r})$ and $p_f^{(2)}(\mathbf{r})$ have to be solved numerically. In general the integrals are three-dimensional with integral kernels

$$\frac{\exp(|\mathbf{r} - \mathbf{r}'|/\beta)}{(|\mathbf{r} - \mathbf{r}'|/\beta)}$$

from the Seyler-Blanchard interaction and $|\mathbf{r} - \mathbf{r}'|^{-1}$ from the Coulomb potential. For spherical shapes these integrals can be reduced to one-dimensional integrals [179]. The Lagrange multipliers $\lambda_f^{(1)}$ and $\lambda_f^{(2)}$ are the separation energies for protons and neutrons, respectively.

The Seyler-Blanchard interaction (2.114) was used in Myers and Swiatecki's droplet model [180], where the parameter set

$$\begin{aligned}
C_{\text{like}} &= 367.56 \text{ MeV} & C_{\text{unlike}} &= 289.66 \text{ MeV} \\
p_D^2/(2M_{\text{nucl}}) &= 82.030 \text{ MeV} & \beta &= 0.62567 \text{ fm}
\end{aligned}$$

was fitted to the liquid-drop mass-formula of Ref. [37].

Myers and Swiatecki later extended the Seyler-Blanchard interaction (2.114) to have seven, instead of the four original adjustable parameters

$$V(r,p) = -C \frac{e^{-r/\beta}}{r/\beta}$$
$$\times \left\{ (1 \mp \xi) - (1 \mp \zeta) \left[\left(\frac{p}{p_D} \right)^2 - \frac{c_\gamma}{p} + c_\sigma (\rho(\mathbf{r}_1)^{2/3} + \rho(\mathbf{r}_2)^{2/3}) \right] \right\} ,$$

where the upper sign refers to like nucleons and the lower sign to unlike nucleons. A set of fit parameters given in Ref. [181] is

$$\begin{aligned}
C &= 86.98 \text{ MeV} & p_D/\hbar &= 4.8067 \text{ fm}^{-1} & c_\gamma/\hbar &= 68.459 \text{ MeV fm}^{-1} \\
c_\sigma &= 155.7 \text{ MeV fm}^2 & \xi &= 0.27976 & \zeta &= 0.55665 \\
\beta &= 0.5929 \text{ fm}.
\end{aligned}$$

2.1.8 Mean-field potential and other leptodermous distributions

One of the most remarkable properties of nuclei is their very large incompress-ibility. As a consequence, the density in the nuclear interior is in Thomas-Fermi approximation constant and nearly independent of the size and shape of the nucleus and it falls off to zero in a surface layer with a thickness of the order of one femtometer. Distributions in space which are constant in the interior and decrease monotonically to zero in a thin surface layer have been called leptodermous by Myers and Swiatecki [180, 182]. For a spherical nucleus the equivalent sharp radius is defined as the radius of a density distribution with a sharp surface, the same saturation density in the interior and the same volume integral as the actual nucleus. Because of the large nuclear incompressibility it is this radius rather than the half-density radius which is proportional to $A^{1/3}$ [183] and in fact, not only in Thomas-Fermi, but also in Hartree-Fock ap-proximation. Experimentally this is well confirmed for the charge distribution by elastic electron scattering data [184]. There is an obvious generalization of the equivalent sharp-surface sphere of spherical nuclei to the equivalent sharp surface of deformed nuclei as a general reference surface defining the nuclear shape.

The representation of a leptodermous distribution ρ most frequently used is the Fermi function. For spherical nuclei

$$\rho(r) = \frac{\rho_0}{1 + \exp\left([r - R_0]/d\right)} \tag{2.116}$$

with the saturation density ρ_0, half-density radius R_0 and diffuseness pa-rameter d. The equivalent sharp radius is $R_{\text{eq}} = R_0\{[1 + (\pi d/R_0)^2]\}^{1/3} + \mathcal{O}(\exp[-R_0/d])$ [183, 185]. For non-spherical, but axially symmetric nuclei the equivalent sharp surface may be defined in cylindrical coordinates r, z by a function $\pi(r, z; \beta_i) = 0$, where the β_i are deformation parameters. The dif-ference $r - R_0$ in the spherical Fermi function (2.116) is then replaced by $l = \pi(r, z)/|\nabla\pi(r, z)|$ [186]. This ansatz guaranties that the surface thickness, measured in the direction of the surface normal, remains constant over the surface. In addition, an overall deformation-dependent scaling of r and z in $\pi(r, z)$ is needed to account for the fact that $l = 0$ refers to the half-density surface in the Fermi function rather than the equivalent sharp surface [132].

In connection with a leptodermous ansatz for a variational density in ETF calculations the generalized Fermi function

$$\rho(r) = \frac{\rho_0}{\{1 + \exp\left([r - R_0]/d\right)\}^\gamma} \tag{2.117}$$

with variational parameters ρ_0, d, and γ was found to be useful [132, 137]. To account also for a central density depression because of Coulomb repulsion effects, Chu, Jennings, and Brack [187] employed the five-parameter ansatz

$$\rho(r) = \frac{\rho_0}{1 + \rho_1} \frac{1 + \rho_1 e^{-r^2/\beta^2}}{\{1 + \exp\left([r - R_0]/d\right)\}^\gamma}, \tag{2.118}$$

where ρ_0 is the central density.

An alternative method to construct a leptodermous function is to fold a step function Θ, which is constant inside the equivalent sharp surface and zero outside, into a short-range function S, normalized so that its volume integral is one and to multiply the folding product by the central depression factor $(1 + wr^2)$ [188]

$$\rho(r) = \rho_0(1 + wr^2) \int \Theta(\mathbf{r'})S(|\mathbf{r} - \mathbf{r'}|)d^3r' . \tag{2.119}$$

Friedrich, Voegler, and Reinhard [189] obtained a very convenient representation of the scattering form factor $F(q)$ by taking the Fourier transform of this expression

$$F(q) = \mathcal{F}_{r \to q}\{\rho(r)\} = \rho_0 \mathcal{F}_{r \to q}\{S\}(1 - w\Delta_q)\mathcal{F}_{r \to q}\{\Theta\} ,$$

directly comparable with the elastic cross section of electron-scattering experiments.

Another example for a leptodermous distribution is the Woods-Saxon shell-model potential [190], approximating the selfconsistent mean field. It consists of a central part V_{cent}, the spin-orbit term V_{so}, and the Coulomb potential V_{Coul} for protons:

$$V^{WS}(\mathbf{r}, \mathbf{p}, \mathbf{s}; \beta_i) = V_{\text{cent}}(\mathbf{r}; \beta_i) + V_{\text{so}}(\mathbf{r}, \mathbf{p}, \mathbf{s}; \beta_i) + V_{\text{Coul}}(\mathbf{r}; \beta_i) . \tag{2.120}$$

The central part is defined by

$$V_{\text{cent}}(\mathbf{r}; \beta_i) = \frac{V_0[1 \pm \kappa I]}{[1 + \exp(l(\mathbf{r}; \beta_i)/d)]} , \tag{2.121}$$

where $I = (N - Z)/A$ is the relative neutron excess. The plus sign in this formula holds for protons, the minus sign for neutrons. There are three empirical parameters. V_0, κ, and d. The function $l(\mathbf{r}, \beta_i)$ has to be determined numerically.

The spin-orbit potential is taken in the form

$$V_{\text{so}}(\mathbf{r}, \mathbf{p}, \mathbf{s}; \beta_i) = i\lambda^q \left(\frac{\hbar}{2M_{\text{nucl}}c}\right)^2 (\nabla V_{\text{cent}} \times \nabla) \cdot \boldsymbol{\sigma}, \quad q = p, n , \tag{2.122}$$

where the linear momentum is $\mathbf{p} = -i\hbar\nabla$ and the spin vector is $\mathbf{s} = (1/2)\hbar\boldsymbol{\sigma}$. For spherical mean fields Eq. (2.122) can be written

$$V_{\text{so}}(\mathbf{r}, \mathbf{l}, \boldsymbol{\sigma}) = -\lambda^q \left(\frac{\hbar}{2M_{\text{nucl}}c}\right)^2 \frac{\partial V_{\text{cent}}}{r\partial r} (\mathbf{l} \cdot \boldsymbol{\sigma}) ,$$

where $\mathbf{l} = \mathbf{r} \times \mathbf{p}/\hbar$. The Coulomb potential for protons is assumed to be that of a uniform charge distribution with a sharp surface and radius $R_{\text{Coul}} = r_p A^{1/3}$ in the spherical case.

Table 2.1. Frequently used sets of Woods-Saxon potential parameters. The radii R_0 in the definition of the function $l(r, z; \beta_i)$ in Eq. (2.121) are related to the r_q parameters in the table by $R_0^{(q)} = r_q A^{1/3}$, q =p,n. The diffuseness parameters of the Woods-Saxon potential d_V and d^{so} refer to the central and spin-orbit potentials, respectively.

Parameter	units	Universal [191]	Blomqvist [192]	Rost [193]	Chepurnov [194]	Myers/Pauli [195, 196]
V_0	MeV	49.6	51.0	49.65	53.3	51.4
κ	-	0.86	0.649	0.86	0.63	$0.829\,\bar{\delta}/I$
d_V	fm	0.70	0.67	0.70	0.63	0.66
d^{so}	fm	0.70	0.67	0.70	0.63	0.55
r_n	fm	1.347	1.27	1.347	1.24	$1.16\,\alpha\beta_n\gamma_n$
λ^n	-	35.0	32.0	31.5	$23.8\cdot(1+2I)$	$1087/(51.4 - 42.6\bar{\delta})$
r_n^{so}	fm	1.31	1.27	1.28	1.24	$1.16\,\alpha(1 - 0.98/R_0^2)$
r_p	fm	1.275	1.27	1.275	1.24	$1.16\,\alpha\beta_p\gamma_p$
λ^p	-	36.0	32.0	17.8	$23.8\cdot(1+2I)$	$1087/(51.4 + 42.6\bar{\delta})$
r_p^{so}	fm	1.20	1.27	0.932	1.24	$1.16\,\alpha(1 - 0.98/R_0^2)$

A few commonly used parameter sets for the Woods-Saxon potential are listed in Table 2.1. The parameter sets reported by Blomqvist and Wahlborn [192] and by Rost [193] were fitted to single-particle and single-hole energies of odd nuclei in the vicinity of ^{208}Pb, corrected for some couplings of these states to collective excitations, whereas the set "universal" of Ref. [191] was adjusted to the single-particle levels of all odd-A nuclei with $A \geq 40$. The latter was quite successful in generating single-particle spectra of nuclei across the periodic table. The potential depth and the spin-orbit strength of the parameter set used by Pauli [196] were fitted to single-particle levels in the lead region. The radius parameters were taken from the droplet-model potential of Myers [195]. For a detailed discussion of the various effects incorporated in the droplet model we refer to Sec. 2.2.3 of this chapter. The small quantities

$$\bar{\delta} = \frac{(N-Z)/A + 0.0112Z^2/A^{5/3}}{1 + 3.15/A^{1/3}}$$

$$\bar{\epsilon} = -0.147A^{-1/3} + 0.33\bar{\delta}^2 + 0.00248Z^2A^{-4/3} \ .$$

have the same form as in Eqs. (2.173) and (2.174) (except for the shape dependent functions which are dropped here). The parameter $\bar{\delta}$ plays the role of the relative neutron excess $I = (N-Z)/A$ in the droplet model and $\bar{\epsilon}$ accounts for an average density compression, due to the opposing effects of surface tension and Coulomb repulsion. The basic quantity of the droplet model is the equivalent sharp radius $R_0 = 1.16\alpha A^{1/3}$ of the density with $\alpha = (1-\bar{\epsilon})$. The equivalent sharp radius of the shell-model potential is $R_V^q = R_0\beta_q$ with $\beta_q = 1 + \Delta R/R_0 \pm 0.22\bar{\delta}/R_0$. Here $\Delta R = 0.82 - 0.56R_0^{-1}$ accounts for the fact that the radius of the potential is by ΔR larger than the density radius because of nonlinear saturation effects of the effective two-body interaction. The term $\pm 0.22\bar{\delta}$ accounts for the difference in the radius between neutron (upper sign) and proton (lower sign) nuclear potentials. The factor $\gamma_q = 1 - (\pi^2/3)(d_V/R_0^q)^2$ converts the equivalent sharp radius into the slightly smaller half-value radius of the Woods-Saxon potential.

Alternatively Bolsterli, Fiset, Nix, and Norton [96] used a folding product of the step function Θ, introduced in Eq. (2.119) and a Yukawa function as short-range function S

$$Y(r) = \frac{e^{-|r|/a}}{4\pi a^2 |r|} \tag{2.123}$$

to generate the central part of a leptodermous shell-model potential by a folding product

$$V_{\text{cent}}(\mathbf{r}) = V_0 \int \Theta(\mathbf{r}')Y(|\mathbf{r} - \mathbf{r}'|)d^3r' , \tag{2.124}$$

for which we shall use in the following the short-hand notation

$$V(\mathbf{r}) = V_0 \int \Theta(\mathbf{r}')S(|\mathbf{r} - \mathbf{r}'|)d^3r' \stackrel{\text{def}}{=} V_0 \, \Theta(\mathbf{r}) * S(r) \ .$$

The central depth is denoted by V_0. The volume enclosed in the equivalent sharp surface is not changed by folding

$$\int \Theta(\mathbf{r})d^3r = \int \Theta(\mathbf{r}) * S(r)d^3r, \tag{2.125}$$

independent of the shape of the equivalent sharp surface for any folding function S, normalized to unit volume. To prove this theorem we define three-dimensional Fourier transforms

$$\mathcal{F}_{\mathbf{r}\to\mathbf{k}}\{\Theta\} = \int \Theta(\mathbf{r})e^{i\mathbf{r}\mathbf{k}}d^3r$$

and notice that for $\mathbf{k} \to 0$ one obtains the volume integral of Θ. Using the folding theorem for the Fourier transformation

$$\mathcal{F}_{\mathbf{r}\to\mathbf{k}}\{\Theta * S\} = \mathcal{F}_{\mathbf{r}\to\mathbf{k}}\{\Theta\}\mathcal{F}_{\mathbf{r}\to\mathbf{k}}\{S\},$$

in the limit $\mathbf{k} \to 0$, Eq. (2.125) follows immediately.

As in Eq. (2.120) the complete shell-model potential consists of central V_{cent}, spin-orbit V_{so} and Coulomb V_{Coul} parts. The central part is written by Möller et al. [197]

$$V_{\text{cent}}(\mathbf{r}) = (V_s \pm V_a \bar{\delta}) \int_{\text{vol}} Y(|\mathbf{r} - \mathbf{r}'|)d^3 r',$$

where the Yukawa function Y has the range $a_{\text{pot}} = 0.8$ fm and the integration is extended over the volume enclosed by the equivalent sharp potential surface. The radius of the latter is $R_{\text{eq}} = (1 - \bar{\epsilon})(R_0 + 0.82 - 0.56/R_0 \pm 0.22\bar{\delta})$, where the upper and lower sign refers to protons and neutrons, respectively and the parameters $\bar{\epsilon}$ and $\bar{\delta}$ are taken from the droplet model [195]. They are the same as in the Pauli-Myers potential [196], discussed above. The strength parameters of the central potential are

$$V_s = 52.5 \text{ MeV}, \quad V_a = 48.7 \text{ MeV}.$$

The spin-orbit term is obtained in the same way from the central potential as in Eq. (2.122). The strength parameters, given in Ref. [197], are (in MeV)

$$\lambda^p = 6.0 \left(\frac{A}{240}\right) + 28.0 \,,$$

$$\lambda^n = 4.5 \left(\frac{A}{240}\right) + 31.5 \,.$$

The Coulomb potential is evaluated assuming a uniform density distribution inside the equivalent sharp surface

$$V_{\text{Coul}}(\mathbf{r}_1) = \frac{Ze^2}{(4\pi/3)Ar_0^3} \int_{\text{vol}} \frac{d^3 r_2}{|\mathbf{r}_1 - \mathbf{r}_2|} \,. \tag{2.126}$$

It is a disadvantage of folding products that one has to calculate the folding integral. For some important shapes closed-form expressions exist [188]. For example, for the sphere with the choice $S = Y$ [96] the integration in $\mathbf{r}$-space in Eq. (2.124) is elementary and yields

$$V_{\text{cent}}(r) = V_0 \begin{cases} 1 - \left(1 + \dfrac{R_{\text{eq}}}{a}\right) e^{-R_{\text{eq}}/a} \sinh\left(\dfrac{r}{a}\right) \Big/ (r/a) & r \leq R_{\text{eq}}, \\[2ex] \left\{\dfrac{R_{\text{eq}}}{a} \cosh\left(\dfrac{R_{\text{eq}}}{a}\right) - \sinh\left(\dfrac{R_{\text{eq}}}{a}\right)\right\} \dfrac{e^{(-r/a)}}{(r/a)} & r > R_{\text{eq}}. \end{cases} \tag{2.127}$$

In general, using the divergence theorem, the three-dimensional folding integral can be reduced to a surface integral [188, 198] which has to be evaluated numerically. Use of the Yukawa function (2.123) leads to leptodermous distributions with a discontinuous second normal derivative at the equivalent sharp surface as seen explicitly in formula (2.127). For some applications this is undesirable, e.g. when one wants to use analytic continuation techniques. Homomorphic folding distributions are obtained with the choice $S = (\pi^{1/2}a)^{-1}\exp(-r^2/a^2)$. Unfortunately, most applications require distributions with an exponential, rather than a Gaussian tail [184]. See Ref. [199] how to overcome this deficiency approximately. For an extensive presentation of other forms of leptodermous functions we refer to the book by Hasse and Myers [200].

2.2 Macroscopic mass formulae

2.2.1 Leptodermous expansions

In Thomas-Fermi approximation a local nuclear energy density $\mathcal{E}(\mathbf{r})$ is defined. Because of the saturation property of nuclear matter the quantity $g(\mathbf{r}) = \mathcal{E}(\mathbf{r}) - c_1\rho(\mathbf{r})$ vanishes in the bulk of a nucleus when c_1 is the volume-energy parameter of the liquid-drop formula. Outside the surface $\mathcal{E}(\mathbf{r})$ and the particle density $\rho(\mathbf{r})$ vanish individually so that $g(\mathbf{r})$ is different from zero only in a surface layer. It turns out that the shape of the profile of g in the direction of the surface normal is rather independent of the size and shape of the nucleus for which it is calculated. Writing now the nuclear binding energy in Thomas-Fermi approximation as $E = c_1 A + \int g(\mathbf{r})d^3r$, the volume integral over g gives rise to the surface energy in the liquid-drop formula.

To show this more explicitly we introduce surface coordinates u and v on the equivalent sharp surface Σ, leading to a surface-integration element $d\sigma$. In addition, we define a coordinate z in the direction of the surface normal. Starting from Σ we may construct a sequence of surfaces $\Sigma', \Sigma'', \ldots$ by shifting each point P on Σ by the amount $dz, 2dz, \ldots$ in the direction of the surface normal in P. The surface element $d\sigma$ may be mapped by this construction on to surface elements $d\sigma(dz), d\sigma(2dz), \ldots$ on the surfaces $\Sigma', \Sigma'', \ldots$. The same construction can be made to both sides of the reference surface Σ. The volume element d^3r may now be expanded in the form

$$d^3r = d\sigma(z)dz = (1 + \kappa z + \mathcal{K}z^2 \ldots)d\sigma(0)dz, \qquad (2.128)$$

with coefficients

$$\kappa = \frac{\partial[d\sigma(z)/d\sigma(0)]}{\partial z}\bigg|_{z=0} \quad \text{and} \quad \mathcal{K} = \frac{\partial^2[d\sigma(z)/d\sigma(0)]}{\partial z^2}\bigg|_{z=0}.$$

Keeping in mind that $g(\mathbf{r})$ depends only on the z coordinate in the direction of the surface normal, the volume integral of the surface energy may be expanded in moments of g

$$E_{\text{surf}} = \int g(\mathbf{r})d^3\mathbf{r} = \int g(z)dz \int d\sigma$$
$$+ \int g(z)zdz \int \kappa d\sigma + \int g(z)z^2dz \int \mathcal{K}d\sigma + \dots . \qquad (2.129)$$

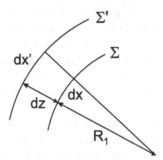

Fig. 2.5. Cut through the surfaces Σ and Σ' along the main curvature line in the x direction.

To exhibit the geometrical meaning of the coefficients κ and $\mathcal{K}$ we use the fact that in an infinitesimal neighborhood of a point P on Σ one can introduce orthogonal coordinates x, y instead of u, v such that the metric becomes Euclidean in x, y in this neighborhood. These coordinates point in the direction of the main curvature lines on Σ [201]. When R_1 is the curvature radius corresponding to the main curvature direction x, the line element dx on Σ is mapped onto $dx' = (1 + dz/R_1)dx$ on the surface Σ' as shown in Fig.2.5. Similarly dy is mapped onto $dy' = (1 + dz/R_2)dy$ and therefore

$$dx'dy' = dxdy \left[1 + (R_1^{-1} + R_2^{-1})dz + (R_1R_2)^{-1}(dz)^2\right.$$

$$\left. + \text{ additional terms of order } (dz)^2\right].$$

From this equation we see by comparison with Eq. (2.128) that $\kappa = R_1^{-1} + R_2^{-1}$ is twice the mean curvature and $\mathcal{K} = (R_1R_2)^{-1}$ is the Gaussian curvature, where we have dropped other terms of quadratic order in the main curvatures proportional to κ^2 [41]. We can now interpret the first term on the right-hand-side of Eq. (2.129) as the surface-energy constant $c_2 = \int g(z)dz$ times the surface area $\int d\sigma$, the second term is the curvature-energy constant $c_3 = \int g(z)zdz$ times twice the mean curvature, integrated over the surface Σ, and the last term kept is proportional to the Gaussian curvature, integrated over Σ. According to the Gauß-Bonnet theorem [201] this integral is 4π for a smooth, simply connected surface Σ, independent of its shape, and it jumps to 8π beyond scission.

For spherical nuclei the first term of the expansion (2.129) is proportional to R^2, the second to R, and the third to R^0. Formula (2.129) is therefore often seen as the beginning of an asymptotic expansion in decreasing powers of R_{eq} or $A^{1/3}$. Therefore, the coefficient c_2 is obtained from a Thomas-Fermi treatment of an infinite, plane surface [132], c_3 follows by subtracting the volume and the surface term from the Thomas-Fermi binding energy of a spherical nucleus in the limit $R_{eq} \to \infty$, and so on, recursively (see e.g. Erdelyi [202] for this expansion technique). For any finite R_{eq} (or finite $A^{1/3}$) the asymptotic series has to be truncated to achieve an optimal representation of the generalized surface energy. It has to be truncated the earlier, the smaller R_{eq} is. In practice one does however not really consider the asymptotic series, but truncates the expansion (2.129) with the first or the second term and determines the coefficients c_i by a least-squares fit to a finite set of nuclei. Note that the c_i from this fit must not be identical with the coefficients of the asymptotic series.

There is a clear indication that the asymptotic series should be terminated before the term with the Gaussian curvature: The jump of this term at scission does certainly not reflect the behavior of the Thomas-Fermi model which the liquid-drop expansion is supposed to represent.

It is convenient to introduce the ratio of the surface energy to the surface energy of the sphere with the same volume $B_{surf} = (\text{surface area})/(4\pi R^2)$ and similarly for the curvature energy $B_{curv} = \int \kappa \, dS/(8\pi R)$. Both quantities are independent of the nuclear size and depend only on the shape of the equivalent sharp surface.

The liquid-drop expansion (2.129) does not account for the contribution of proximity forces to the total binding energy in strongly necked-in configurations ("crevices") and between fragments beyond scission ("gaps"). Therefore an extra term must be added to the liquid-drop expansion for the proximity energy. In Ref. [41] such term is derived as a generalization of Eq. (1.8) for arbitrarily shaped, but smooth juxtaposed surfaces.

There is an alternative approach to a parameterization of the surface-energy integral on the left-hand-side of Eq. (2.129): One can try a simple, two-parameter ansatz for the function $g(\mathbf{r})$ [188]: Since the gradient of the folding product $\Theta * Y$ is surface-peaked, one can construct a surface-peaked scalar by taking the square of it. One is therefore lead to the ansatz $g(\mathbf{r}) = 4a\sigma[\nabla(\Theta * Y)]^2$ with the range parameter a in the Yukawa function and a strength parameter σ. (In the limit of a sharp surface, $a \to 0$, the expression $4a \int [\nabla(\Theta * Y)]^2 d^3r \to$ surface area [188]. Therefore the parameter σ has the meaning of the surface tension in this limit.) The generalized surface energy $E_{surf} = E - E_{vol} = \int g(\mathbf{r}) d^3r$ may therefore be written in short-hand notation as

$$E_{surf} = \int g(\mathbf{r}) d^3r = 4a\sigma \nabla(\Theta * Y) \cdot \nabla(Y * \Theta), \qquad (2.130)$$

where the dot shall indicate integration over the spatial coordinates as well as vector multiplication of the gradients. The multiple integrations in this expression can be simplified somewhat. Partial integration yields $E_{surf} = -4a\sigma\Theta * Y * \Delta Y \cdot \Theta$, where use has been made of the fact that differentiation of a folding product can be done by differentiating either factor and that the multiple product in Eq. (2.130) is associative as well as commutative, which is easily verified by writing the implied integrations explicitly. We now use the identity[1] $Y * \Delta Y = [1/(2a)]\partial_a Y$, which can be proved by taking the Fourier transforms of both sides of the identity and using $\mathcal{F}_{\mathbf{r}\to\mathbf{k}}\{Y\} = (1 + a^2 k^2)^{-1}$. With this identity one obtains

$$E_{\text{surf}} = -2\sigma\partial_a\Theta * Y \cdot \Theta, \qquad (2.131)$$

which is the representation of the nuclear surface energy proposed by Krappe, Nix, and Sierk [203] with $\sigma = c_2/(4\pi r_0^2)$ in terms of the constant c_2, used in Eq. (1.1). Performing the differentiation with respect to a, one gets a linear combination of a Yukawa and an exponential function, folded into $\Theta(\mathbf{r})$

$$E_{\text{surf}} = -\frac{4\sigma}{a}\Theta * (E_{\text{nor}} - Y) \cdot \Theta$$

$$\equiv -\frac{4\sigma}{a}\int d^3r\, d^3r'[E_{\text{nor}}(\mathbf{r} - \mathbf{r}') - Y(\mathbf{r} - \mathbf{r}')], \qquad (2.132)$$

where the integration is in $\mathbf{r}$ and $\mathbf{r}'$ over the volume enclosed by the equivalent sharp surface and E_{nor} denotes the normalized exponential function $\exp(-r/a)/(8\pi a^3)$. The representation (2.131) for the surface energy has therefore been referred to as "Yukawa-plus-exponential". The derivation of this expression is based on the assumption that the local curvature radius is always larger than the surface-thickness parameter a. For high-order multipole vibrations this assumption is not satisfied and the ansatz (2.131) cannot be used (if such modes make sense for a drop with a diffuse surface in the first place). If the condition is satisfied, all shape information is contained in the function $\Theta(\mathbf{r})$ and the profile of g in the direction of the surface normal is given by Y.

An advantage of this representation over the liquid-drop expansion is that it automatically includes proximity effects. For instance, for two nuclei with a center-of-mass distance $\mathbf{D}$, whose shapes are given by the shape functions $\Theta_1(\mathbf{r})$ and $\Theta_2(\mathbf{r} - \mathbf{D})$ and which do not overlap, insertion of $\Theta = \Theta_1 + \Theta_2$ into Eq. (2.131) yields the proximity energy as a double folding product

$$E_{\text{prox}}(\mathbf{D}) = -\frac{c_2}{2\pi r_0^2}\partial_a\, \Theta_1 * Y * \Theta_2. \qquad (2.133)$$

Multiple folding products as in Eq. (2.133) are often conveniently calculated by taking the Fourier transforms of the folding factors and transforming their product back into ordinary space [204]. For two spheres with radii R_1 and R_2 and $R_i \gg a$ it is shown in Ref. [203] that their interaction potential becomes

---

[1] We use the short-hand notation $\partial_a \equiv \partial/\partial a$

$$E_{\text{prox}}(D) = -\frac{a(R_1 - a)(R_2 - a)\,c_2}{r_0^2\,D(2 + s/a)}\left(4 - \frac{R_1}{R_1 - a} - \frac{R_2}{R_2 - a} + \frac{s}{a}\right)e^{-s/a},$$

where the exponentially small terms $\exp(-R_1/a)$ and $\exp(-R_2/a)$ are neglected compared to unity. Neglecting also a/R_1, a/R_2, and $s/(R_1 + R_2)$ compared to 1, one obtains

$$E_{\text{prox}}(D) \approx -\frac{a\overline{R}c_2}{r_0^2}e^{-s/a} \qquad (2.134)$$

with $s = D - R_1 - R_2$. This expression is compatible with Eq. (1.8) for the proximity potential, except that the reduced radius $\overline{R} = R_1 R_2/(R_1 + R_2)$ refers there to the half-density radii and in this case to the equivalent sharp radii.

Note that the interaction energy (2.133) is based on a "frozen-density" assumption, which is often called "sudden approximation" in scattering theory. Another advantage of the folding ansatz is a lesser sensitivity of the deformation energy to high multipole wiggles on the surface compared to sharp-surface formulae, which is somewhat unphysical. In a radial moment expansion of the folding expression for g all odd moments vanish. One may compensate this deficiency by adding at least a curvature energy term to the expression (2.131). However, liquid-drop model fits show that this term is rather small [41].

A disadvantage of the Yukawa-plus-exponential ansatz is the need to calculate the six-dimensional integral in Eq. (2.132). Although closed expressions for the integral exist for some important shape classes, like a sphere or two spheres at a given distance [203] and expansions of the interaction of deformed nuclei in terms of their multipole moments [188], in general the double volume integral can be reduced only to a double surface integral which must then be evaluated numerically [203]. To see how the reduction is achieved, use is made of the possibility to represent the folding kernel $K(|\mathbf{r} - \mathbf{r}'|)$ in Eq. (2.132) by a double divergence [188]

$$K(|\mathbf{r} - \mathbf{r}'|) = \sum_{i,j}^{1\ldots3} \frac{\partial}{\partial x_i}\frac{\partial}{\partial x_j'}(x_i - x_i')(x_j - x_j')F(|\mathbf{r} - \mathbf{r}'|),$$

so that the application of Gauss' theorem to the $\mathbf{r}$-integration as well as to the $\mathbf{r}'$-integration yields

$$\Theta * K \cdot \Theta = \int\int \{d\mathbf{S}(\mathbf{r} - \mathbf{r}')\}\{d\mathbf{S}'(\mathbf{r} - \mathbf{r}')\}F(|\mathbf{r} - \mathbf{r}'|). \qquad (2.135)$$

The function $F(r)$ satisfies the differential equation

$$r^2 F''(r) + 8r F'(r) + 12F(r) = -K(r)$$

with boundary conditions $\lim_{r\to 0} r^4 F(r) = 0$ and $\lim_{r\to\infty} F(r) = 0$. Simple closed-form solutions of the differential equation exist e.g. for the Yukawa function $K(r) \equiv Y(r) = (4\pi a^3)^{-1}e^{-x}/x$ with $x = r/a$:

$$F(r) = -(4\pi a^3 x^4)^{-1}[x - 2 + (x + 2)e^{-x}] \tag{2.136}$$

and for $K(r) = 1/r$:

$$F(r) = -1/(6r).$$

One defines the shape function B_{surf} for the Yukawa-plus-exponential model as the ratio of E_{surf} to the surface energy of a sphere with the same volume and a sharp surface. Using Eqs. (2.131), (2.135), and (2.136) we write B_{surf} as a double surface integral to be extended in both variables over the equivalent sharp surface [205]

$$B_{\text{surf}} = \frac{1}{8\pi^2 r_0^2 A^{2/3}} \int \int \left\{ 2 - \left[\frac{(\mathbf{r} - \mathbf{r}')^2}{a^2} + 2\frac{|\mathbf{r} - \mathbf{r}'|}{a} + 2 \right] e^{-|\mathbf{r} - \mathbf{r}'|/a} \right\}$$

$$\times \frac{\{(\mathbf{r} - \mathbf{r}') \cdot d\mathbf{S}\} \{(\mathbf{r} - \mathbf{r}') \cdot d\mathbf{S}'\}}{|\mathbf{r} - \mathbf{r}'|^4}. \tag{2.137}$$

The quantity B_{surf} does now depend on the size of the system and it is not unity for a sphere as in the case of a sharp surface.

We have so far discussed only the nuclear part of the liquid-drop formula. The Coulomb part consists of a direct and an exchange contribution. In the direct part $E_{\text{Coul}} = (1/2)\rho_p * (e^2/r) \cdot \rho_p$ the Thomas-Fermi proton-number density ρ_p is conveniently represented by a folding product $\rho_p^{(0)} \Theta(\mathbf{r}) * Y$ with the charge density constant $\rho_p^{(0)}$ and a charge-diffuseness parameter a_{den} in the Yukawa function. The Fourier transform of the resulting double folding product $Y*(1/r)*Y$ is, according to the folding theorem, given by the ordinary product of the Fourier transforms of the three folding factors Y, $1/r$, and Y

$$\frac{4\pi}{k^2} \frac{1}{(1 + a_{\text{den}}^2 k^2)^2} = 4\pi \left[\frac{1}{k^2} - \frac{a_{\text{den}}^2}{1 + a_{\text{den}}^2 k^2} - \frac{a_{\text{den}}^2}{(1 + a_{\text{den}}^2 k^2)^2} \right],$$

where a partial fraction expansion with respect to k^2 has been made on the right-hand-side. The bracket may also be written $k^{-2} - a_{\text{den}}^2[(a_{\text{den}}/2)\partial_{a_{\text{den}}} + 2](1 + a_{\text{den}}^2 k^2)^{-1}$. Transforming back into ordinary space, one obtains

$$E_{\text{Coul}} = \frac{[\rho_p^{(0)} e]^2}{2} \left\{ \Theta * \frac{1}{r} \cdot \Theta - 4\pi a_{\text{den}}^2 \left[\frac{a_{\text{den}}}{2} \partial_{a_{\text{den}}} + 2 \right] \Theta * Y \cdot \Theta \right\}$$

$$= \frac{[\rho_p^{(0)} e]^2}{2} \Theta * \left\{ \frac{1}{r} - \left(\frac{1}{r} + \frac{1}{2a_{\text{den}}} \right) \exp(-r/a_{\text{den}}) \right\} \cdot \Theta. \tag{2.138}$$

The direct part of the Coulomb energy is seen to consist of a sum of the Coulomb energy of a sharp-surface charge distribution and a surface-diffuseness correction in the form of some Yukawa-plus-exponential expression. The double volume integration implied in these formulae can again be converted into a double surface integration by writing $1/r$ and Y as double divergence. Using this transformation the ratio of the direct Coulomb energy to the Coulomb energy of the sphere with equal volume and sharp surface is given in terms of the ratio $\xi = |\mathbf{r} - \mathbf{r}'|/a_{\text{den}}$ by

$$B_{\text{Coul}} = -\frac{15}{32\pi^2 R^5 a_{\text{den}}} \int \int \left[\frac{1}{6\xi} - \frac{2\xi - 5 + (5 + 3\xi + \xi^2/2)\exp(-\xi)}{\xi^4} \right]$$

$$\times \{(\mathbf{r} - \mathbf{r}') \cdot d\mathbf{S}\} \{(\mathbf{r} - \mathbf{r}') \cdot d\mathbf{S}'\} . \tag{2.139}$$

To calculate the volume contribution to the Coulomb exchange energy

$$E_{\text{Coulex}} = -\frac{1}{2} \int \frac{e^2}{|\mathbf{r} - \mathbf{r}'|} |\rho_p(\mathbf{r}, \mathbf{r}')|^2 d^3r d^3r'$$

one needs the mixed proton density for a homogeneous, infinite fermion system,

$$\rho_p(\mathbf{r}, \mathbf{r}') = \frac{2}{(2\pi)^3} \int_{|k|<k_f} \exp i[\mathbf{k}(\mathbf{r} - \mathbf{r}')] d^3k ,$$

where $\hbar k_f$ is the Fermi momentum of the protons [206]. Introducing the Fourier transform of the Coulomb potential

$$\frac{e^2}{|\mathbf{r} - \mathbf{r}'|} = 4\pi e^2 \int \frac{d^3q}{(2\pi)^3} \frac{e^{i\mathbf{q}(\mathbf{r}-\mathbf{r}')}}{q^2} ,$$

the Coulomb exchange energy becomes

$$E_{\text{Coulex}} = -\frac{8\pi e^2}{(2\pi)^9} \int d^3(\mathbf{r}-\mathbf{r}') \int d^3R \int \frac{d^3q}{q^2} \int_{|k'|<k_f} d^3k \int_{|k|<k_f} d^3k' e^{i(\mathbf{q}+\mathbf{k}-\mathbf{k}')(\mathbf{r}-\mathbf{r}')} ,$$

where the change of variables $\int d^3r \int d^3r' = \int d^3R \int d^3(\mathbf{r} - \mathbf{r}')$ was made. The spatial integrals can be easily performed. With the normalization volume $v = \int d^3R$ one obtains

$$E_{\text{Coulex}} = -\frac{8\pi e^2 v}{(2\pi)^6} \int_{|k|<k_f} d^3k \int_{|k'|<k_f} d^3k' \frac{1}{(\mathbf{k} - \mathbf{k}')^2}$$

$$= -\frac{16\pi e^2 v p_f^4}{h^4} \int_0^1 x dx \int_0^1 x' \ln \frac{x + x'}{x - x'} dx'$$

$$= -4\pi e^2 p_f^4 h^{-4} v, \tag{2.140}$$

where $x = k/k_f$. Using the relation $p_f^3 = 3\pi^2\hbar^3 Z/v$ for the Fermi momentum p_f of the protons in Thomas-Fermi approximation, the volume contribution of the Coulomb exchange energy becomes

$$E_{\text{Coulex}} = -3e^2 p_f Z/(4\pi\hbar) = -(3/4)(3/2\pi)^{2/3}(e^2/r_0)Z^{4/3}A^{-1/3} . \tag{2.141}$$

In the mass formula it is referred to as Slater term [172], though it was first derived by Felix Bloch [207]. In empirical fits of the liquid-drop formula this term is sometimes absorbed in the nuclear volume energy. The Coulomb exchange energy-density used in Eq. (2.106) follows immediately from Eq. (2.140)

$$E_{\text{Coulex}}/v = \mathcal{E}_{\text{Coulex}}(\mathbf{r}) = -(3/4)e^2(3/\pi)^{1/3}\rho_p^{4/3}(\mathbf{r}) \qquad (2.142)$$

with $(p_f(\mathbf{r})/\hbar)^3 = 3\pi^2\rho_p(\mathbf{r})$.

To obtain the surface contribution to the Coulomb exchange energy we consider the local Fermi momentum $p_f(\mathbf{r})$ in Thomas-Fermi approximation in terms of the single-particle potential $V(\mathbf{r})$ and the Fermi energy ϵ_f

$$p_f^2(\mathbf{r}) = 2M_{\text{nucl}}\left[\epsilon_f - V(\mathbf{r})\right]\Theta(\epsilon_f - V(\mathbf{r})),$$

the volume v in terms of the saturation density $\rho_p^{(0)}$ of the protons

$$v = \int \rho_p(\mathbf{r})/\rho_p^{(0)}\, d^3r = \int \left(\frac{\epsilon_f - V(\mathbf{r})}{\epsilon_f - V(0)}\right)^{3/2}\Theta(\epsilon_f - V(\mathbf{r}))\, d^3r,$$

and the quantity $q_2 = \int g_{\text{Coulex}}(\mathbf{r})d^3r$ with the surface-peaked integrand

$$g_{\text{Coulex}}(\mathbf{r}) = p_f^4(\mathbf{r})/p_f^4(0) - \rho(\mathbf{r})/\rho(0)$$
$$= \left[\left(\frac{\epsilon_f - V(\mathbf{r})}{\epsilon_f - V(0)}\right)^2 - \left(\frac{\epsilon_f - V(\mathbf{r})}{\epsilon_f - V(0)}\right)^{3/2}\right]\Theta(\epsilon_f - V(\mathbf{r})).$$

Inserting these definitions into Eq. (2.140) yields

$$E_{\text{Coulex}} = -4\pi e^2 h^{-4}\int p_f^4(\mathbf{r})\, d^3r = -4\pi e^2 h^{-4}p_f^4(\mathbf{r} = 0)\,(v + q_2). \qquad (2.143)$$

An expansion of q_2 in moments of $g_{\text{Coulex}}(r)$ analogous to Eq. (2.129) leads to the surface (and curvature) contributions of the Coulomb exchange energy.

For some representations of a spherical, leptodermous charge distribution other than in terms of a folding product, the Coulomb energy was evaluated as a power series in the ratio between the surface diffuseness and the radius, see Hasse and Myers, Ref. [200] Chap. 4.2.

2.2.2 Liquid-drop mass-formulae

From the leptodermous expansion of the nuclear binding energy in Thomas-Fermi approximation follows the classical three-term mass formula [14, 29] discussed in the second section of the introduction

$$E = E_{\text{vol}} + E_{\text{surf}} + E_{\text{Coul}} \qquad (2.144)$$

with the volume energy

$$E_{\text{vol}} = a_V(1 - \kappa_V I^2)A\,, \qquad (2.145)$$

the surface energy

$$E_{\text{surf}} = a_S(1 - \kappa_S I^2)A^{2/3}B_{\text{surf}}, \qquad (2.146)$$

and the direct Coulomb energy

$$E_{\text{Coul}} = \frac{1}{2}\left(\frac{3Ze}{4\pi R_0^3}\right)^2 \int_{\text{vol}} d^3r \int_{\text{vol}} d^3r' \frac{1}{|\mathbf{r}-\mathbf{r}'|} = \frac{3}{5}\frac{e^2}{r_0}Z^2 A^{-1/3}B_{\text{Coul}}, \quad (2.147)$$

where the dependence of the nuclear binding energy on the relative neutron excess $I = (N - Z)/A$ is to be seen as a power-series expansion, which should contain only even powers of I because of the charge symmetry of the nuclear forces. Empirically it appears to be sufficient to retain only the quadratic term in this expansion. The shape-dependence is contained in the shape functions B_{surf} and B_{Coul}. In the liquid-drop model B_{surf} is the ratio of the surface area of the nucleus and the surface of a sphere of equal volume. In cylindrical coordinates one has for axially symmetric shapes

$$B_{\text{surf}} = \frac{1}{2R_0^2}\int_{z_{\text{min}}}^{z_{\text{max}}} \rho\sqrt{1+\rho'^2}\,dz. \qquad (2.148)$$

In the finite-range liquid-drop model it is given by Eq. (2.137). The Coulomb shape-function is represented by the double-surface integral

$$B_{\text{Coul}} = -\frac{5}{64\pi^2 R^5}\int\int \frac{1}{|\mathbf{r}-\mathbf{r}'|}\{(\mathbf{r}-\mathbf{r}')\cdot d\mathbf{S}\}\,\{(\mathbf{r}-\mathbf{r}')\cdot d\mathbf{S}'\} \qquad (2.149)$$

in a model with a sharp surface and by Eq (2.139) for a diffuse charge distribution. The volume energy is shape-independent along the fission path, including the scission point. This holds also for the neutron-excess term if one assumes that the charge density Z/A stays constant at scission, $Z/A = Z_1/A_1$, which implies $Z_1/A_1 = Z_2/A_2$ and $I = I_1 = I_2$.

The diffuseness correction to the Coulomb energy and the Coulomb exchange-energy are often simplified by retaining only the volume terms of their leptodermous expansions [37]. From Eq. (2.138) one obtains

$$E_{\text{diff}} = -\frac{\rho_p^2 e^2}{2}\frac{4\pi}{3}R^3 \int d^3r \left(\frac{1}{r} + \frac{1}{2a_{\text{den}}}\right)e^{-r/a_{\text{den}}}$$

$$= -3\frac{e^2}{r_0}\left(\frac{a_{\text{den}}}{r_0}\right)^2 \frac{Z^2}{A} \qquad (2.150)$$

and for the Coulomb exchange-energy, E_{Coulex}, Eq. (2.141). These terms are therefore shape-independent. Note that the charge diffuseness parameter d, used in Ref. [37], refers to the diffuseness parameter of the Fermi function (2.116), which is related to the Yukawa parameter a by $a^2 = (\pi^2/6)d^2$ [200].

So far we have only considered the first order effects of the Coulomb force on the binding energy. But one has to keep in mind that even if the free nucleon-nucleon interaction V_{nn} is charge symmetric, the effective force may violate that symmetry since it is a nonlinear functional of $V_{nn} + V_{\text{Coul}}$. One has therefore to expect a contribution to the nuclear symmetry energy which violates charge symmetry. In leading order this effect yields a term linear in I in the nuclear volume energy

$$E_a = c_a I A = c_a(N - Z) . \tag{2.151}$$

The next term in the leptodermous expansion is the curvature energy

$$E_{\text{curv}} = a_{\text{curv}}(1 - \kappa_{\text{curv}} I^2) A^{1/3} B_{\text{curv}} \tag{2.152}$$

with the curvature form-factor

$$B_{\text{curv}} = \frac{1}{8\pi R_0} \int \kappa \, dS ,$$

where the integral is to be extended over the equivalent sharp surface of the nucleus and the integrand is twice the mean local curvature. For axially symmetric shapes, given in cylindrical coordinates by $\rho(z)$, one obtains

$$\begin{aligned}
B_{\text{curv}} &= \frac{1}{4R_0} \int_{z_{\min}}^{z_{\max}} \left(1 - \frac{\rho\rho''}{1 + \rho'^2}\right) dz \\
&= \frac{1}{4R_0} \int_{z_{\min}}^{z_{\max}} (1 + \rho' \arctan \rho') dz ,
\end{aligned} \tag{2.153}$$

where R_0 is the radius of the sphere with equal volume. Expressions for B_{curv} in other coordinate systems are given in the book by Hasse and Myers [200]. It is worth noting that the representation of the profile function $g(\mathbf{r})$ in terms of a folding product yields no odd surface moments. It is therefore compatible with the folding ansatz for the surface energy to add a curvature term.

Often a number of terms are included in mass formulae which do not follow from leptodermous expansions. Among these is the odd-even term [197]

$$E_{\text{oe}} = \begin{cases} \overline{\Delta}_p + \overline{\Delta}_n - \delta_{np} & Z \text{ and } N \text{ odd} \\ \overline{\Delta}_p & Z \text{ odd} \quad N \text{ even} \\ \overline{\Delta}_n & Z \text{ even} \quad N \text{ odd} \\ 0 & Z \text{ and } N \text{ even} \end{cases} \tag{2.154}$$

with

$$\overline{\Delta}_n = \frac{r_m}{N^{1/3}} , \qquad \overline{\Delta}_p = \frac{r_m}{Z^{1/3}} , \qquad \delta_{np} = \frac{r'_m}{A^{2/3}}$$

and empirical parameters r_m, r'_m (see Table 2.2). An alternative form is

$$E_{\text{oe}} = \begin{cases} \Delta - \delta/2 & Z \text{ and } N \text{ odd} \\ \delta/2 & Z \text{ or } N \text{ odd} \\ -\Delta + \delta/2 & Z \text{ and } N \text{ even} \end{cases} \tag{2.155}$$

with $\Delta = 12$ MeV$/A^{1/2}$ and $\delta = 20$ MeV$/A$ [208].

Plotting the difference between the experimental binding energies and the liquid-drop formula Eqs. (2.144), (2.127) plus some shell correction versus I, the relative neutron excess, one obtains a cusp at $N = Z$. This motivated Myers and Swiatecki [37] to add a term

$$E_{\text{Wigner}} = -t \exp(-\alpha_W |I|) \text{ MeV} \tag{2.156}$$

to the mass formula. It was named after Wigner who was the first to derive an expression for the symmetry energy which contained a term proportional to $|N - Z|$ [209]. Myers and Swiatecki gave parameters $t=7$ MeV, $\alpha_W=10$ for the Wigner energy in Ref. [37]. In a later paper [210] they used $t=10$ MeV and $\alpha_W=4.2$. Möller and Nix [211] reported $t=11.19$ MeV and $\alpha_W=13$. The form of the Wigner term was changed in Refs. [208] and [212] to

$$E_{\text{Wigner}} = W \left(|I| + \begin{array}{l} 1/A \\ 0 \end{array} \right) \begin{array}{l} N = Z, \text{ both odd} \\ \text{otherwise} \end{array} \tag{2.157}$$

and eventually renamed "congruence energy" [210].

Satula and Wyss [213] found that for $N = Z$ nuclei and their nearest neighbors the isosinglet pn-pairing might outweigh the usually dominant isotriplet pairing which leads to a cusp in the mass formula around $N = Z$. Since a priori little is known about the relative strength of the isosinglet to the isotriplet part of the effective interaction in the particle-particle channel, it is not quite decided yet whether the Wigner cusp is only due to isosinglet pairing [214]. Since the effect is restricted to $N = Z$ nuclei and their neighbors, it is at any rate not parametrized by either of the forms (2.156) or (2.157).

One may question whether it is reasonable to include terms (2.154) - (2.157) in the bulk part of the mass formula when microscopic shell and pairing corrections are later added anyway. In view of inherent uncertainties in the liquid-drop expansion one may also question whether one should include some other terms, which appear sometimes in mass formulae. There is for instance a correction term to account for the difference between charge and proton densities because of the finite size of the charge density of the proton, characterized by the range constant r_p of its electric form factor [215]

$$E_{\text{formfactor}} = f(k_f r_p)Z^2/A. \tag{2.158}$$

One can argue that this effect can be taken care of by the choice of the parameter a_{den} in the diffuseness correction to the Coulomb energy with sufficient accuracy.

Another term, sometimes included in the mass formula [215, 216], is a zero-point energy, supposedly connected with the fission degree of freedom

$$E_{\text{zp}} = (1/2)\hbar\omega_{\text{quadrupole}}. \tag{2.159}$$

However, the Thomas-Fermi binding energy, which the macroscopic mass formula is to represent, contains the zero-point energies of all $3A$ degrees of freedom, as does the Hartree-Fock approach. One may rather wonder that it contains too much zero-point energy because of the spurious center-of-mass motion. In fact, no allowance is made in mass formulae to correct for any of the symmetry-breaking features of the mean-field theory.

A further term of doubtful origin is a constant, i.e. a term of order A^0

$$E_0 = a_0 . \qquad (2.160)$$

Being of the same order as the shell corrections, it may in fact correct deficiencies of shell-correction recipes.

It is a common feature of the contributions (2.154)-(2.160) to the mass formula that they do not have a natural shape dependence. They are either considered as shape-independent with the undesirable consequence that they double at scission, or they are given rather ad hoc a shape dependence, for example for E_{Wigner} and E_0 in Ref. [217] or, in terms of the shape function B_{surf}, for the pairing energy E_{oe} by Madland and Nix [130].

Table 2.2. The liquid-drop parameters in the first line are fitted to ground-state masses from Ref. [218] and a selection of fission barriers, including some with neutron numbers around 40 and around 50. In the second line only ground-state masses (from Ref. [219]) were used in the fit. The numbers in the third line were taken from Ref. [197]. In the fourth line the parameters of the Wigner term (2.156) were taken from Ref. [210], the Coulomb radius and diffuseness parameters were obtained from the mass fit.

type	a_V[MeV]	κ_V	a_S[MeV]	κ_S	a_{curv}[MeV]	κ_{curv}	a_0[MeV]	c_a[MeV]
FRLDM	16.025	1.932	21.330	2.378	–	–	2.040	0.097
LDM	15.4920	1.8601	16.9707	2.2938	3.8602	-2.376	–	–

type	W[MeV]	r_m[MeV]	r'_m[MeV]	a_{den}[fm]	a[fm]	r_o[fm]	r_p[fm]	
FRLDM	30	4.8	6.6	0.7	0.68	1.16	0.8	

type	t[MeV]	α_W	r_{den}[fm]	a_{den}[fm]				
LDM	10	4.2	1.21725	0.62049				

A term of marginal importance in the mass formula is the total binding energy of the electrons, which is always smaller than 1 MeV [212]. It accounts for the fact that mass formulae refer to the masses of neutral atoms, rather than bound neutrons and hydrogen atoms. Its form is based on a result by Foldy [220], which he obtained in the framework of the Hartree approximation for the electrons,

$$E_{\mathrm{el}} = 1.43 \cdot 10^{-5} \, Z^{2.39} \text{ MeV}. \qquad (2.161)$$

Numerous fits of the liquid-drop parameters have been reported [197, 203, 215, 224, 226–228]. They depend not only on the continuously increas-

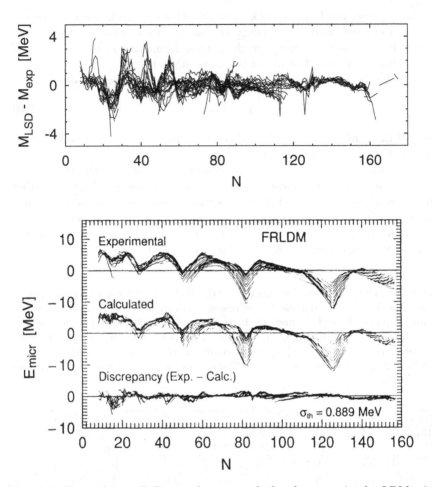

Fig. 2.6. Upper frame: Difference between calculated masses in the LDM with parameters from table 2.2 and measured masses for the 2766 nuclei from the tables of Antony [219]. Lines connect isotopes of the same element, from Ref. [221]. Lower frame: In the upper part the same for the FRLDM and experimental data for 1654 nuclides from Audi's midstream evaluation [218] (with four corrections). The calculated graphs refer to Strutinsky shell and pairing correction-energies, from Ref. [222].

ing empirical data base of very accurately measured ground-state binding-energies [121, 174, 219, 223, 229] and fission barriers, but also on the weight with which fission-barrier information is used in the fit, besides ground-state masses. It also depends on the selection of terms beyond the leptodermous expansion, Eqs. (2.154) - (2.161), which are retained in the fit. And it finally depends on the particular choice of the recipe to calculate shell and pairing corrections. In Table 2.2 we list the liquid-drop parameters obtained in re-

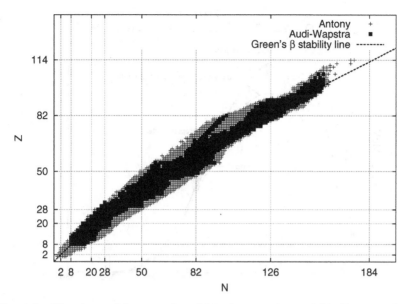

Fig. 2.7. The chart of isotopes for which the experimental binding energies are known. The crosses correspond to data from the compilation of Antony [219], while black squares to the data from Ref. [223] (from Ref. [228]).

cent fits. The finite-range liquid-drop model (FRLDM) was fitted by Möller Sierk, and Iwamoto [230] and the sharp-surface liquid-drop model (LDM) by Pomorski and Dudek [221, 228]. In both cases Strutinsky's shell and pairing corrections and ground-state deformation energies of Möller et al. [197] were taken into account. The Coulomb exchange-energy was expressed in the Slater approximation (2.141) in both fits. For the Coulomb surface-diffuseness correction in the LDM the expression (2.150) was used. The FRLDM fit includes a zero-point energy-term. Fig. 2.6 shows the deviation of experimental from calculated ground-state binding energies for the LDM fit of Ref. [221] and for the FRLDM fit of Ref. [222]. The quality of the fit is often characterized by the root-mean-square deviation σ of the fit. This would be an appropriate measure if the deviations had a random distribution. However, the figures show clearly systematic trends, presumably due to missing pieces of physics in the standard microscopic-macroscopic approach (to be discussed in the next section), which was used in both fits. The nuclides whose measured ground-state masses were used in some recent fits are shown in Fig. 2.7. Also indicated in the figure is Green's line of β stable nuclei [231]

$$Z = \frac{A}{2} \left(1 - \frac{0.4A}{A + 200} \right). \tag{2.162}$$

In Fig. 2.8 the fission barrier is plotted as function of the mass number A, calculated in various versions of the liquid-drop model (without shell and

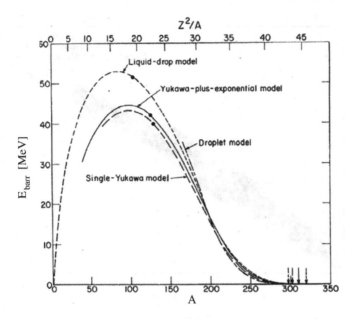

Fig. 2.8. Comparison of the fission barrier heights of β-stable nuclei obtained in four different macroscopic models. The parameter sets of the liquid-drop [224], droplet [208], single-Yukawa [225], and Yukawa-plus-exponential [203] are the same as in the original papers. The arrows indicate the vanishing of the macroscopic fission barrier in the four models. The thick black dots indicate the Businaro-Gallone point: to the left of these points the fission barrier is unstable with respect to left-right asymmetry. The four arrows indicate the A_{crit} at which the fission barrier disappears in the four models. The droplet model yields the smallest and the liquid-drop model the largest A_{crit}, the two finite-range models give the two values in between (after Ref. [203]).

pairing corrections). For medium mass and light nuclei sharp-surface models give substantially higher barriers than diffuse-surface models. These nuclei have rather strongly necked-in saddle-point shapes. Therefore the proximity interaction between the nascent fragments becomes important. This interaction is automatically included in diffuse-surface liquid-drop models, but was not considered in the sharp-surface models shown in Fig. 2.8.

Instead of adding the contribution from crevices to the liquid-drop energy as proposed in Ref. [41], more recently the A^0 terms of the liquid-drop formula, i.e. the Wigner and the Gaussian curvature terms, have been used to mimic proximity effects. In Fig. 2.9 experimental and theoretical barrier heights are compared. Results of two calculations are shown: with the liquid-drop model of Ref. [221] and with a selfconsistent Thomas-Fermi approach using the Seyler-Blanchard interaction [181]. In both calculations shell corrections were neglected at the saddle point and a shape-independent Wigner term (2.157) was used. Assuming some smoothly interpolating shape dependence of the Wigner term, it is easy to lower the calculated barriers for the four lightest systems in Fig. 2.9 [221]. These data points come from heavy-ion fusion-fission experiments [232, 233]. To extract from the measured data the symmetric, zero-temperature saddle-point of the non-rotating system, to which the diamonds in the Figure refer, quite a number of assumptions have to be made which are the source of systematic errors, not indicated in the figure. Besides, the macroscopic-microscopic approach used in Figs. 2.6 and 2.9 has its own sources of systematic errors: (a) too restricted choices of the deformation-parameter space for the ground and saddle-point states, and (b) ground state multiconfigurational mixing.

2.2.3 The droplet model

There are two rather small effects which were neglected in the leptodermous expansion, discussed above: First, nuclei are not completely incompressible with the consequence that large nuclei have a smaller central density than (neutral) bulk nuclear matter since they are dilated by the Coulomb repulsion of the protons. This effect is opposed by a compression effect due to the surface tension. The two effects roughly balance for medium-size nuclei, but for light nuclei a net increase of the central density results, compared to nuclear matter and a net decrease for heavy nuclei. A second deviation from the standard liquid-drop expansion is caused by the slightly smaller radius of the proton

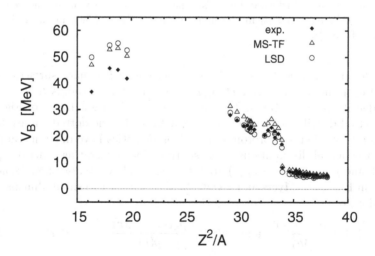

Fig. 2.9. The Myers and Swiatecki Thomas-Fermi (MS-TF, points) and the liquid-drop model [221] (LSD, circles) estimates of the fission barrier heights as functions of the parameter Z^2/A. The experimental data are denoted by diamonds. The Thomas-Fermi and the experimental values of the barrier heights are taken from Ref. [181].

density compared to the neutron density because of the excess of neutron numbers. The result is the development of a neutron skin.

Both effects lead to small changes in the liquid-drop formula. Myers and Swiatecki considered them in their droplet model [180] in a low-order power-series expansion. To simplify the discussion we restrict us in the following to spherical shapes. General shapes are treated in Ref. [182]. To define appropriate expansion parameters, Myers and Swiatecki define reference den-

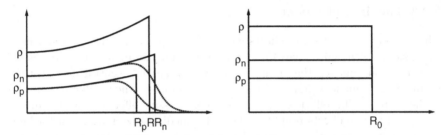

Fig. 2.10. Schematic plots of the droplet model densities as a functions of the nuclear radius are presented in the left figure. The solid lines show the bulk functions ρ_n, ρ_p, and ρ extrapolated out to the effective sharp neutron R_n, proton R_p, and liquid drop R radii. The dotted lines give the actual densities ρ_n^{actual} and ρ_p^{actual} with their realistic surface diffuseness. The corresponding plot for the generating densities $\rho_n^{(0)}$, $\rho_p^{(0)}$, and ρ_0 in the liquid drop model is shown in the right figure (after Ref. [180]).

sities for protons, neutrons, and for all nucleons with sharp surfaces. In the liquid drop model they are rectangular with a common radius $R^{(0)} = r_0 A^{1/3}$ and density $\rho_0 = 3/(4\pi r_0^3)$. They are shown in the right frame of Fig. 2.10. The actual distributions, given in the left frame by the dotted curves, have a central depression and proton and neutron densities have different radii. Besides the actual distributions, their sharp-surface generating functions $\rho_p(r)$, $\rho_n(r)$, and $\rho(r) = \rho_p(r) + \rho_n(r)$ are introduced. They are also shown schematically in Fig. 2.10. Myers and Swiatecki also defined two small, dimensionless functions of r

$$\epsilon(r) = \frac{\rho_0 - \rho(r)}{3\rho_0} \quad \text{and} \quad \delta(r) = \frac{\rho_n(r) - \rho_p(r)}{\rho(r)}, \qquad r \leq R. \qquad (2.163)$$

They describe the deviation of the density ρ from its nuclear matter value ρ_0 and the local density asymmetry. In this definition the proton density $\rho_p(r)$ is assumed to be extrapolated smoothly from its bulk value $\rho_p(R_p))$ to the reference radius R to be defined below.

Using Eqs. (2.163) and keeping only linear terms in ϵ, one finds the following expressions for the total, the proton, and the neutron densities

$$\rho(r) = \rho_0[1 - 3\epsilon(r)] \ ,$$

$$\rho_p(r) = \frac{1}{2}\rho_0[1 - 3\epsilon(r) - \delta(r)], \qquad (2.164)$$

$$\rho_n(r) = \frac{1}{2}\rho_0[1 - 3\epsilon(r) + \delta(r)], \qquad r \leq R.$$

The reference radius R and the proton and neutron radii R_p and R_n, respectively, are defined by

$$\frac{4}{3}\pi R^3 \bar{\rho} = A \ , \qquad \frac{4}{3}\pi R_p^3 \bar{\rho}_p = Z \ , \qquad \frac{4}{3}\pi R_n^3 \bar{\rho}_n = N \ . \qquad (2.165)$$

in terms of average densities $\bar{\rho}$, where the average is taken over a sphere with radius R

$$\bar{\rho} = \frac{3}{R^3}\int_0^R \rho(r)r^2 dr$$

and similarly for $\bar{\rho}_p$ and $\bar{\rho}_n$. Averaging in the same way Eqs. (2.164) yields, together with Eqs. (2.165) to linear order in $\bar{\epsilon}$ and $\bar{\delta}$,

$$R \ = r_0 A^{1/3}(1 + \bar{\epsilon}) \ ,$$

$$R_p = r_0(2Z)^{1/3}(1 + \bar{\epsilon} + \tfrac{1}{3}\bar{\delta}) \ , \qquad (2.166)$$

$$R_n = r_0(2N)^{1/3}(1 + \bar{\epsilon} - \tfrac{1}{3}\bar{\delta}) \ .$$

The thickness parameter of the neutron skin τ is defined by

$$\tau = \frac{R_n - R_p}{r_0}. \qquad (2.167)$$

To first order in small quantities one obtains

$$\tau \approx \frac{2}{3}(I - \bar{\delta})A^{1/3} \ , \qquad (2.168)$$

where $I = (N - Z)/A$ is the relative neutron excess.

The ansatz for the total energy of the nucleus

$$E = E_{\mathrm{vol}} + E_{\mathrm{surf}} + E_{\mathrm{curv}} + E_{\mathrm{Coul}} + E_{\mathrm{Coulex}}$$

$$= \bar{\rho}\int_R e(r)d^3r + \int_R \sigma(r)dS + \frac{a_{\mathrm{curv}}}{8\pi r_0}\int_R \kappa\, dS$$

$$+ \frac{e^2}{2}\int_{R_p} \frac{\rho_p(\mathbf{r})\rho_p(\mathbf{r}')}{|\mathbf{r} - \mathbf{r}'|}d^3r\, d^3r' - \frac{3e^2}{4}\left(\frac{3}{\pi}\right)^{1/3}\int_{R_p}\rho_p^{4/3}(r)d^3r \quad (2.169)$$

shall be expressed as functional of ϵ and δ. In this equation the first integral is to be taken over a sphere of radius R, the second and third integral is over the surface of the sphere with radius R, the fourth integral is in $\mathbf{r}$ and $\mathbf{r}'$ over spheres with radius $R_p(\bar{\epsilon}, \bar{\delta})$, and the last integral, the volume contribution to the Coulomb exchange term, extends over a sphere with radius R_p. To exhibit the dependence of the integrands on ϵ and δ the volume-energy density $e(r)$ is expanded in the droplet model in a Taylor series up to second powers of ϵ and δ^2

$$e = -a_V + J\delta^2 + \frac{1}{2}(K\epsilon^2 - 2L\epsilon\delta^2 + M\delta^4), \qquad (2.170)$$

where a_V is the volume-energy constant of the liquid-drop model, K is the nuclear incompressibility, and J, L, and M are model parameters. Similarly the surface energy density is expanded in powers of $\delta(R)$ and τ up to second order

$$4\pi r_0^2 \sigma = a_S + H\tau^2 + 2P\tau\delta(R) - G\delta^2(R), \qquad (2.171)$$

where a_S is the surface-energy constant of the liquid-drop model and H, P, and G are further model parameters of the droplet model. In this expansion of the binding energy E, terms up to the order $A^{1/3}$, $I^2 A^{2/3}$, and $I^4 A$ are kept in decreasing order of $A^{1/3}$ and increasing order of I^2.

Inserting the results (2.166), (2.168) and the expansions (2.170) and (2.171) into Eq. (2.169), yields E as functional of $\epsilon(r)$ and $\delta(r)$, besides its dependence on the liquid-drop parameters a_V, a_S, a_{curv}, and r_0. The functional is minimized with respect to ϵ and δ with the constraint that Eqs. (2.166) are satisfied for fixed A, Z, and N. After a lengthy calculation one obtains for the energy minimum [180, 182]

$$
\begin{aligned}
E = {} & \left(-a_V + J\bar{\delta}^2 - \frac{1}{2}K\bar{\epsilon}^2 + \frac{1}{2}M\bar{\delta}^4\right)A \\
& + \left(a_S + \frac{9}{4}\frac{J^2}{Q}\bar{\delta}^2\right)A^{2/3}B_{\mathrm{surf}} + a_{\mathrm{curv}}A^{1/3}B_{\mathrm{curv}} \\
& + c_1\frac{Z^2}{A^{1/3}}B_{\mathrm{Coul}} - c_2 Z^2 A^{1/3}B_r - c_5 Z^2 B_w - c_3\frac{Z^2}{A} - c_4\frac{Z^{4/3}}{A^{1/3}} .
\end{aligned}
\qquad (2.172)
$$

in terms of the minimum values of the parameters $\bar{\delta}$ and $\bar{\epsilon}$

$$\bar{\delta} = \left(I + \frac{3}{16}\frac{c_1}{Q}\frac{Z}{A^{2/3}}B_v\right)\Big/\left(1 + \frac{9}{4}\frac{J}{Q}A^{-1/3}B_{\mathrm{surf}}\right), \qquad (2.173)$$

and

$$\bar{\epsilon} = \frac{1}{K}\left(-2a_S A^{-1/3}B_{\mathrm{surf}} + L\bar{\delta}^2 + c_1\frac{Z^2}{A^{4/3}}B_{\mathrm{Coul}}\right), \qquad (2.174)$$

where we report the result for the more general case of not necessarily spherical nuclei [182]. It is shown in Ref. [180] that the parameter G can be expressed in terms of parameters H, P, and J and the parameters H and P enter in the final result only in the combination $Q = H/[1 - (2/3)(P/J)]$.

The model has nine independent, adjustable parameters, whose values, as given by Myers [212], are

$$\begin{array}{llll}
a_V = 15.96 \text{ MeV}, & a_S = 20.69 \text{ MeV}, & a_{\text{curv}} = 0 & \text{MeV}, \\
J = 36.8 \text{ MeV}, & Q = 17 \text{ MeV}, & K = 240 \text{ MeV}, & \text{(2.175)} \\
L = 100 \text{ MeV}, & M = 0 \text{ MeV}, & r_0 = 1.18 \text{ fm},
\end{array}$$

and five constants given in terms of them

$$c_1 = \frac{3}{5}\frac{e^2}{r_0}, \qquad\qquad c_2 = \frac{c_1^2}{336}\left(\frac{1}{J} + \frac{18}{K}\right),$$

$$c_3 = \frac{5}{2}c_1\left(\frac{b}{r_0}\right)^2, \quad \text{where } b = 0.99 \text{ fm},$$

$$c_4 = \frac{5}{4}c_1\left(\frac{3}{3\pi}\right)^{2/3}, \qquad c_5 = \frac{c_1^2}{64Q}.$$

The shape functionals B_{surf}, B_{Coul}, B_{curv}, B_r, B_v, and B_w describe the ratios of the surface, Coulomb, curvature, redistribution, and surface redistribution energies of the first and second kind to their corresponding values in the spherical case, respectively. They are given by the following integrals [182]

$$\begin{aligned}
B_{\text{surf}} &= \int_S d\sigma/(4\pi R^2), \\
B_{\text{Coul}} &= \int_V w(\boldsymbol{r})d\tau/(\tfrac{32}{15}\pi^2 R^5), \\
B_{\text{curv}} &= \int_S \kappa d\sigma/(8\pi R), \\
B_r &= \int_V [\tilde{w}(\boldsymbol{r})]^2 d\tau/(\tfrac{64}{1575}\pi^3 R^7), \qquad\qquad \text{(2.176)} \\
B_v &= -\int_S \tilde{w}(\boldsymbol{r})d\sigma/(\tfrac{16}{15}\pi^2 R^4), \\
B_w &= \int_S [\tilde{w}(\boldsymbol{r})]^2 d\sigma/(\tfrac{64}{225}\pi^3 R^6),
\end{aligned}$$

where R is the equivalent sharp radius of a spherical nucleus, $\kappa = R_1^{-1} + R_2^{-1}$ is twice the surface curvature, defined by the two local principal radii of curvature R_1 and R_2, $w(\boldsymbol{r}) = \int_V |\boldsymbol{r} - \boldsymbol{r}'|^{-1}d\tau$ is proportional to the Coulomb potential, and $\tilde{w} = w(\boldsymbol{r}) - \bar{w}$ is its deviation from the average value $\bar{w} = \int_V w(\boldsymbol{r})d\tau/(\tfrac{4}{3}\pi R^3)$.

The droplet model for the macroscopic part of the nuclear binding energy is quite successful in representing ground-state masses and geometrical parameters like charge radii and density diffusenesses, but it rather fails in the description of fission barrier heights [226]. In present applications the droplet expression is used in combination with the Yukawa-plus-exponential (YPE) model [203]. In this hybrid model [197, 234]), which was named "finite-range droplet-model" (FRDM), the binding energy is given by

$$E = E_{\text{droplet}} + E_a + E_{\text{formfactor}} + E_{\text{zp}} + E_0 + E_{\text{oe}} + E_{\text{Wigner}} + E_{\text{el}}, \quad \text{(2.177)}$$

where the droplet-energy term E_{droplet} is slightly modified compared to Eq. (2.172)

$$E_{\text{droplet}} = (-a_V + J\bar{\delta}^2 - (1/2)K\bar{\epsilon}^2)A$$
$$+ \left(a_S + \frac{9}{4}\frac{J^2}{Q}(\bar{\delta}\theta)^2\right)A^{2/3}B_{\text{surf}}^{\text{fr}} + a_{\text{curv}}A^{1/3}B_{\text{curv}}$$
$$+ c_1\frac{Z^2}{A^{1/3}}B_{\text{Coul}}^{\text{fr}} - c_2 Z^2 A^{1/3}B_r - c_5 Z^2 B_w\theta - c_4\frac{Z^{4/3}}{A^{1/3}}. \quad (2.178)$$

In this equation $B_{\text{surf}}^{\text{fr}}$ and $B_{\text{Coul}}^{\text{fr}}$ are the finite-range expressions (2.137) and (2.139). The latter includes also the diffuseness correction $c_3 Z^2/A$. Also the ratio $\theta = B_{\text{surf}}^{\text{fr}}/B_{\text{surf}}^{\text{sharp}}$ between the surface shape-functions in the finite-range and the sharp-surface models appears in Eq. (2.178). The definitions of $\bar{\delta}$ and $\bar{\epsilon}$ are sightly modified, compared to Eqs. (2.173) and (2.174)

$$\bar{\delta} = \left(I + \frac{3}{16}\frac{c_1}{Q}\frac{Z}{A^{2/3}}B_v\theta\right)\Big/\left(1 + \frac{9}{4}\frac{J}{Q}A^{-1/3}B_{\text{surf}}^{\text{fr}}\theta\right) \quad (2.179)$$

and

$$\bar{\epsilon} = \frac{1}{K}\left(Ce^{-\gamma A^{1/3}} - 2a_S A^{-1/3}B_2 + L\bar{\delta}^2 + c_1\frac{Z^2}{A^{4/3}}B_4\right), \quad (2.180)$$

with two additional shape functions

$$B_2 = \frac{1}{2x_0}\frac{d\,(x^2 B_{\text{surf}}^{\text{fr}})}{dx}\Big|_{x=x_0}, \qquad B_4 = -y_0^2\left[\frac{d}{dy}\left(\frac{B_{\text{Coul}}^{\text{fr}}}{y}\right)\right]_{y=y_0},$$

where $x = R/a$, $x_0 = (r_0/a)A^{1/3}$, $y = R/a_{\text{den}}$, and $y_0 = (r_0/a_{\text{den}})A^{1/3}$. In Ref. [235] it was felt to be useful to include a term $C\exp(-\gamma A^{1/3})$ in the expression (2.180) for $\bar{\epsilon}$. The term is nonanalytic in the expansion parameter $A^{-1/3}$ of the droplet model, where it can therefore not appear. But it does appear in the finite-range model for the sphere. No suggestion is made for its shape dependence. The remaining seven terms in Eq. (2.177) are given in Eqs. (2.151), (2.158) - (2.161), (2.154), and (2.157).

Of the many empirical parameters, which appear in the various terms of Eq. (2.177) the following constants were fitted to ground-state binding energies and binding-energy differences [197], all given in MeV

$a_V = 16.247$	volume-energy	$Q = 29.21$	surface stiffness
$a_S = 22.92$	surface energy	$C_a = 0.436$	charge asymmetry
$a_{\text{curv}} = 0$	curvature energy	$a_0 = 0$	A^0-term
$J = 32.73$	symmetry energy	$C = 60$	C parameter in $\bar{\epsilon}$
$L = 0$	density symmetry	$W = 30$	Wigner term
$r_m = 4.8$	average pairing gap	$r_m' = 6.6$	n-p interaction

and the dimensionless constant $\gamma = 0.831$ in the exponent of the compressibility term of Eq. (2.180). Eq. (2.177) contains also a number of parameters

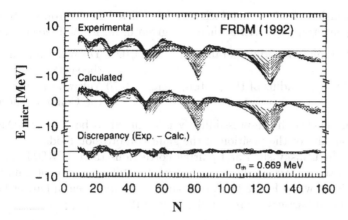

Fig. 2.11. Discrepancy between experimental and theoretical nuclear masses, calculated by adding shell and pairing corrections to the macroscopic energy. The latter was obtained from the FRDM of Ref. [222] (after Ref. [197]).

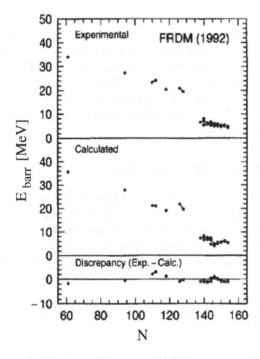

Fig. 2.12. Experimental and theoretical fission barriers, calculated by adding shell and pairing corrections to the FRDM of Ref. [222] (after Ref. [197]).

of the dimension of a length, which are either fitted to electron scattering or light-ion scattering data. All of them are given in fm in the following list

$r_0 = 1.16$ nuclear radius parameter, $a = 0.68$ Yukawa range in
                                                       the surface energy,
$r_p = 0.80$ rms radius of the proton, $a_{den} = 0.70$ Yukawa range in
                                                       the Coulomb energy.

Finally, the nuclear incompressibility was assumed to be $K = 240$ MeV to account roughly for the position of the giant compression mode.

Adding Strutinsky shell and pairing corrections to the FRDM energy of Eq. (2.177) the remaining discrepancy with the 1654 experimental masses of Ref. [218] is shown in Fig. 2.11. A comparison of experimental barrier heights with calculated barriers is presented in Fig. 2.12.

2.3 Shell and pairing corrections to the macroscopic mass formula

It is desirable to have a scheme for calculating ground-state binding energies and energy surfaces which is numerically less expensive than selfconsistent HF+BCS or even HFB calculations. We have seen that the ETF expression for the binding energy admits the numerically very convenient liquid-drop expansion. One may therefore ask whether the difference between ETF and HF can also be calculated in a sufficiently convenient way. Strutinsky showed that this is indeed possible [99].

2.3.1 The Strutinsky shell-correction

Strutinsky observed that the deviation of the binding energy from the liquid-drop prediction was large for those nuclides with a smaller than average single-particle level-density above the Fermi surface. He therefore introduced an averaged reference level-density $\tilde{g}_q(\epsilon)$ for protons ($q = p$) and neutrons ($q = n$) by averaging the actual HF level density $g_q(\epsilon) = \sum_i \delta(\epsilon - \epsilon_i^q)$ by a folding procedure

$$\tilde{g}_q(\epsilon) = \frac{1}{\gamma} \int g_q(\epsilon') f\left(\frac{\epsilon - \epsilon'}{\gamma}\right) d\epsilon' = \frac{1}{\gamma} \sum_i f\left(\frac{\epsilon - \epsilon_i^q}{\gamma}\right) . \qquad (2.181)$$

The folding function $f(\epsilon/\gamma)$ is required to have the range γ, of the order of the shell spacing $\hbar\omega \approx 41/A^{1/3}$ MeV and the folding operation (2.181) shall transform a smooth function, for example a polynomial Q_n of order $n \leq M$ into itself. Strutinsky proposed the ansatz $f(x) = w(x)P_M(x)$ with a weight function $w(x) > 0$ and a correction polynomial $P_M(x)$, which satisfies the equation

$$Q_n(x) = \int Q_n(x - x')P_M(x')w(x')dx' \tag{2.182}$$

for any polynomial Q_n with $n \leq M$.

An explicit representation of the polynomial P_M is obtained by expanding $Q_n(x - x')$ in Eq. (2.182) around x

$$Q_n(x) = \sum_{\mu=0}^{n} \frac{(-1)^\mu}{\mu!} \frac{d^\mu Q_n(x)}{dx^\mu} \int x'^\mu P_M(x')w(x')dx' \,,$$

which implies the condition [236]

$$\int x'^\mu P_M(x')w(x')dx' = \delta_{\mu 0} \quad \text{for all} \quad \mu \leq M. \tag{2.183}$$

One can show [237] that from this equation follows that all odd powers of the polynomial P_M vanish because the odd moments $\int_{-\infty}^{\infty} x^{2n+1}w(x)dx$ are zero. In this case Eq. (2.183) is equivalent to the requirement that for any polynomial $Q_n(x)$ of order $n \leq M$

$$\int Q_n(x')P_M(x')w(x')dx' = Q_n(0) \,. \tag{2.184}$$

For any weight function $w(x)$ there exists a set of orthonormal polynomials $p_n(x)$ with

$$\int p_n(x)p_m(x)w(x)dx = \delta_{nm}$$

(see e.g. Chap. 22 of Ref. [238]). In terms of these polynomials the ansatz

$$P_M(x) = \sum_{n=0}^{M} p_n(x)p_n(0) \tag{2.185}$$

is made. Expanding $Q_n(x)$ in terms of the first n polynomials $p_n(x)$, the ansatz (2.185) is seen to satisfy the condition (2.184).

For the weight function Strutinsky chose a Gaussian $w(x) = \pi^{-1/2}e^{-x^2}$ and the integration interval $-\infty < x < \infty$ in order to make the average independent of the integration boundaries. The corresponding orthogonal polynomials are the Hermite polynomials $H_n(x)$ with normalization factor $[n!2^n]^{-1}$ [238]. Because the Gaussian is symmetric, we can restrict ourselves to the subset of Hermite polynomials with even index. Since $H_{2n}(0) = (-1)^n(2n)!/n!$, the correction polynomial becomes

$$P_M(x) = \sum_{n=0}^{M/2}(-1)^n \frac{H_{2n}(x)}{2^{2n}n!} = L_{M/2}^{1/2}(x^2) \,, \quad M \text{ even}, \tag{2.186}$$

where $L_{M/2}^{1/2}$ is a generalized Laguerre polynomial. The last equality of (2.186) can for instance be proved by induction using Eqs. 22.4.7, 22.7.30, and 22.5.40

of Ref. [238]. The functions $f(x) = \pi^{-1/2}L_{M/2}^{1/2}e^{-x^2}$ with $M = 0, 2, 4, 6, 8$ are given below

$$f(x) = \begin{cases} \frac{1}{\sqrt{\pi}}e^{-x^2} \\ \frac{1}{\sqrt{\pi}}e^{-x^2}(\frac{3}{2} - x^2) \\ \frac{1}{\sqrt{\pi}}e^{-x^2}(\frac{15}{8} - \frac{5}{2}x^2 + \frac{1}{2}x^4) \\ \frac{1}{\sqrt{\pi}}e^{-x^2}(\frac{35}{16} - \frac{35}{8}x^2 + \frac{7}{4}x^4 - \frac{1}{6}x^6) \\ \frac{1}{\sqrt{\pi}}e^{-x^2}(\frac{315}{128} - \frac{105}{16}x^2 + \frac{63}{16}x^4 - \frac{3}{4}x^6 + \frac{1}{24}x^8) \,. \end{cases}$$

Other weight functions $w(x)$ have been proposed [236, 239, 240]. However, the Gaussian weight has almost exclusively been used in actual calculations.

In the following we will drop the index q and derive the shell correction for neutrons. For protons analogous equations hold. In analogy with the relation

$$N = 2\int_0^{\epsilon_f} \sum_i \delta(\epsilon_i - \epsilon)d\epsilon \,,$$

which defines the Fermi energy ϵ_f (the factor 2 accounts for the Kramers degeneracy of the single-particle levels), one introduces a chemical potential $\tilde{\lambda}$ by

$$N = 2\int_0^{\tilde{\lambda}} \tilde{g}(\epsilon)d\epsilon \tag{2.187}$$

for the reference nucleus. Similarly its density matrix is defined in the HF basis (cf. Eq. (2.12) by

$$\tilde{n}_i = \frac{1}{\gamma}\int_{-\infty}^{\tilde{\lambda}} f\left(\frac{\epsilon - \epsilon_i}{\gamma}\right)d\epsilon \tag{2.188}$$

and in an arbitrary basis $\chi_l(\mathbf{r}, \sigma, \tau)$, using the unitary matrix U of Eq. (2.4), the density matrix becomes $\tilde{\rho}_{kl} = \sum_i U_{ki}\tilde{n}_i U_{il}^{-1}$. For the reference nucleus the Hartree-Fock energy, Eq. (2.7), is

$$\tilde{E}^{HF} = \sum_{l_1 l_2} t_{l_1 l_2}\tilde{\rho}_{l_2 l_1} + 1/2 \sum_{l_1 l_1' l_2 l_2'} \tilde{\rho}_{l_1 l_1'}\overline{V}_{l_1' l_2' l_1 l_2}\tilde{\rho}_{l_2' l_2} \,. \tag{2.189}$$

In terms of the difference

$$\delta\rho_{l_2 l_1} = \rho_{l_2 l_1} - \tilde{\rho}_{l_2 l_1}$$

and using Eq. (2.10) the shell correction is, to linear order in $\delta\rho$

$$\delta E^{HF} = E^{HF} - \tilde{E}^{HF} = \sum_{l_1 l_2}(t_{l_1 l_2} + \Gamma_{l_1 l_2})\delta\rho_{l_2 l_1} + \mathcal{O}(\delta\rho^2) \,. \tag{2.190}$$

We now assume that a shell-model Hamiltonian $\hat{h}^{\text{shell}}$ can be found with a local potential $V^{\text{shell}}(\mathbf{r})$ having eigenvalues $\epsilon_i^{\text{shell}}$ and eigenstates φ_i^{shell} which approximate the eigenvalues ϵ_i and eigenstates φ_i of the HF Hamiltonian in the subspace $\mathcal{P}$ spanned by the states φ_i with eigenvalues ϵ_i in a vicinity of order γ on both sides of the Fermi energy ϵ_f

$$\sum_{l_1 l_2} (t_{l_1 l_2} + \Gamma_{l_1 l_2}) \hat{a}_{l_1}^+ \hat{a}_{l_2} \approx \hat{h}^{\text{shell}} \qquad \text{in } \mathcal{P} .$$

Then the shell correction energy can be written

$$\delta E^{\text{HF}} \approx \delta E^{\text{shell}} = \sum_{i \leq i_{\text{Fermi}}} \epsilon_i^{\text{shell}} - \int_{-\infty}^{\tilde{\lambda}} \epsilon \, \tilde{g}^{\text{shell}}(\epsilon) d\epsilon . \qquad (2.191)$$

To account for the effect of pairing correlations on the ground-state binding energy in the framework of the HF+BCS scheme, the pairing correction $\Delta P = \Delta E - \widetilde{\Delta E}$ from Eqs. (2.65) and (2.66) or (2.67) is added to the shell correction. If we assume that the smoothed HF energy $\tilde{E}^{\text{HF}}$ admits a leptodermous expansion, we arrive at the basic formula of the macroscopic-microscopic approach of calculating binding energies

$$E^{\text{HF+BCS}} = E_{\text{mac}} + \sum_{q=n,p} (\delta E_q^{\text{shell}} + \Delta P_q) . \qquad (2.192)$$

In this formula the macroscopic energy, usually identified with the liquid-drop energy, depends smoothly on the proton and neutron numbers and accounts for most of the binding energy. The shell correction shows a rather oscillatory behavior as function of $N^{1/3}$ and $Z^{1/3}$, as will be shown below and contributes less than 10 MeV to the binding energy as seen in the upper frames of Fig. 2.11.

The Hamiltonian $\hat{h}^{\text{shell}}(\mathbf{r})$, used to generate the shell correction, violates the translational invariance of the original N-body Hamiltonian. It may however be expected that the spurious energies connected with symmetry violations cancel in the average in the subtraction procedure (2.192).

Details of the potential V^{shell} far away from the Fermi energy do not influence the shell correction (2.191) at all. What matters is the size and shape of the surface, defined by $V^{\text{shell}}(\mathbf{r}_{\text{surf}}) = \epsilon_f$. The latter should approximately coincide with the shape of the equivalent sharp surface of the nucleus under consideration [241]. The size can be taken from the droplet model [196] or fitted to certain single-particle energies of odd nuclei, cf. Sec. 2.1.8 for results. Less sensitive is the shell correction to the slope of the potential at the Fermi surface $\nabla V_{\text{cent}} \cdot \hat{\mathbf{n}}|_{\mathbf{r}=\mathbf{r}_{\text{surf}}}$, where $\hat{\mathbf{n}}$ is the unit vector in the direction of the normal on the equivalent sharp surface. Shell corrections have indeed been calculated successfully with Woods-Saxon as well as Nilsson potentials if their size and shape are properly chosen. For a quantitative investigation of the appropriate choice of the shell-model potential we refer to Refs. [242], [243] and [244].

Plotting the smooth energy $\tilde{E}^{\text{shell}}(\gamma)$ for an harmonic oscillator potential versus the averaging width γ, one obtains a constant value when $\gamma \geq \hbar\omega$, provided the order of the correction polynomial is $M \geq 4$, since the smooth part of the oscillator spectrum is a polynomial of order 4 ("plateau behavior"). For other potentials this is not the case. The smooth part of their spectrum $\tilde{g}$ is therefore not exactly reproduced by the folding procedure of Eq. (2.184). To minimize the resulting error with respect to the choice of γ, one uses the plateau condition

$$\partial \tilde{E}^{\text{shell}} / \partial \gamma = 0 \ . \tag{2.193}$$

The resulting optimal γ depends on M. Brack and Pauli [239] showed that the plateau condition (2.193) implies the equality

$$\int_{-\infty}^{\tilde{\lambda}} \epsilon \tilde{g}(\epsilon) d\epsilon = \sum_i \epsilon_i \tilde{n}_i \ , \tag{2.194}$$

which can be used in Eq. (2.191).

The integration interval from $-\infty$ to $+\infty$, connected with the Gaussian weight function, causes problems for light nuclei because the Fermi energy may not be much more than by the width γ away from the lowest level, where the spectrum ends. Also in finite potentials the Fermi energy may not be much more than by γ away from the continuum limit, where the discrete spectrum also ends abruptly. In both cases the folding procedure leads to spurious edge effects because it reproduces polynomials, but not polynomials cut at two ends. Various recipes have been proposed to overcome this problem. Concerning the continuum limit, it is common practice to diagonalize the shell-model Hamiltonian in an oscillator basis, in which case one generates (spurious) discrete eigenstates $\epsilon_i > 0$, i.e. in the continuum. However, since they conveniently continue the sequence of the real (negative) eigenvalues smoothly, they dampen out the upper edge effect. The price to be paid is an uncontrollable effect on the smoothed properties of the reference nucleus, which becomes the more serious the smaller the distance of the Fermi energy (in units of γ) $|\epsilon_f|/\gamma$ is from the continuum limit. Moreover often the stationarity condition (2.193) cannot be satisfied. For neutron-rich nuclei this can become a serious problem. Another proposal is to mirror the spectrum at its lower and upper edges. A similar decrease of the reliability of the shell correction with decreasing $|\epsilon_f|/\gamma$ is again the drawback also of this procedure. Attempts to modify the averaging prescription by restricting it to a finite energy interval [240] were not satisfactory either since the resulting shell correction depends too strongly on the choice of the integration limits. Another recipe to cope with the edge effect was proposed by Tondeur [245]. He introduced energy-dependent coefficients in the polynomial $P_M(x)$ of Eq. (2.185) to improve the plateau condition (2.193) for sufficiently large γ.

In a series of papers Ivanyuk and Strutinsky [246–248] developed an alternative averaging procedure, which requires the knowledge of single-particle levels only up to the level $i = N_2$. In this approach the smooth background energy $\tilde{E}^{\text{shell}}(N)$ is obtained for a given potential by averaging the shell-model energy

$$E^{\text{shell}}(N) = \sum_{i \leq i_{\text{Fermi}}} \epsilon_i(N)$$

with respect to N between N_1 and N_2. $\tilde{E}^{\text{shell}}(N)$ is defined as a polynomial of order M in $N^{1/3}$ which minimizes the mean-square deviation $\sum_{N=N_1}^{N_2} [E^{\text{shell}}(N) - \tilde{E}^{\text{shell}}(N)]^2$ with respect to the coefficients of the polynomial (note: $M \ll N_2 - N_1$). To decrease the dependence of $\tilde{E}^{\text{shell}}(N)$ on N_1 and N_2 a Gaussian weight function was later introduced in Ref. [247] and the averaged energy can then be written

$$\tilde{E}^{\text{shell}}(N) = \sum_{N'=N_1}^{N_2} K\big(x(N), x(N')\big) e^{-x^2(N')} E^{\text{shell}}(N')$$

with $x(N) = (N^{1/3} - N_0^{1/3})/\Delta N$ and the "curvature correction" $K(x, x') = \sum_{k=1}^{M} p_k(x) p_k(x')$, where $p_k(x)$ are the normalized orthogonal polynomials for the Gaussian weight function in the discrete interval $N_1 \leq N \leq N_2$, see Ref. [238], Sec. 22.17, and $N_0 \approx N$, $\Delta N \approx 3/4$. The difference $\delta U(N) = E^{\text{shell}}(N) - \tilde{E}^{\text{shell}}(N)$ turns out to have systematically a larger absolute value than the shell correction $\delta E^{\text{shell}}(N)$ of Eq. (2.191). It was shown in Ref. [248] that

$$\delta E^{\text{shell}}(N) = \delta U(N) + \frac{1}{2}\, \tilde{g}(\mu_N) \langle (\delta \mu_N)^2 \rangle \,, \tag{2.195}$$

where $\tilde{g}(\mu_N)$ is the mean level density at the Fermi energy and $\langle (\delta \mu_N)^2 \rangle$ is the variance of the Fermi energy. Ivanyuk investigated the use of Eq. (2.195) to calculate δE^{shell} when the Fermi energy is too close to the continuum limit to use the standard Strutinsky averaging safely [249].

The macroscopic-microscopic approach, Eq. (2.192), can easily be extended to shape-constraint HF energies. The single-particle energy-spectra in the corresponding deformed shell-model potentials show a remarkable level bunching for axially and reflection symmetric shapes when the axis ratio is close to 1:2, and, somewhat lesser, when it is 1:3. This is analogous to the even stronger level bunching in spherical potentials, giving rise to the magic numbers. The situation is shown in Fig. 2.13 for the deformed oscillator potential (without spin-orbit coupling). The Fourier components of the oscillating part of the level density have been related to the classical closed orbits in the shell-model potential [114]. The orbits with the smallest action along the trajectory give rise to the largest Fourier amplitudes. It turns out that for axis ratios expressed by small integers especially short (in units of the action) closed orbits exist which lead to a level bunching and corresponding large shell corrections. For a detailed discussion of the relation between shell corrections and closed classical orbits we refer to Strutinsky's review article [251] and to Chaps. 4 and 6 of Ref. [114].

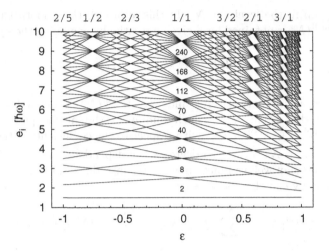

Fig. 2.13. Energy levels of an axially symmetric, harmonic oscillator potential as function of the spheroidal deformation parameter ϵ (after Ref. [250]).

Shell corrections are particularly large and negative when there is a large gap above the last occupied level. If the liquid-drop barrier has an extremum close to a maximum of the shell correction, the latter has a particularly strong effect on the topography of the total energy surface of such a nucleus: It can stabilize an unstable or almost unstable ground state of the liquid-drop model and convert a liquid-drop saddle-point into a local secondary minimum, corresponding to a fission-isomeric state.

2.3.2 Microscopic justification of the Strutinsky shell-correction

It remains to be shown that the smoothed HF energy $\tilde{E}^{\mathrm{HF}}$ can be identified with the ETF energy and therefore admits a leptodermous expansion. Migdal and Krainov [252] showed already that the energy of the atomic many-electron system in Hartree approximation becomes the Thomas-Fermi energy when sums over single-particle states are replaced by integrals over a smooth level density and the single-particle wave-functions are replaced by their WKB approximations. And in fact, Gross [253], Bhaduri and Ross [254], and Jennings [255] showed that the ETF energy approximates the Strutinsky-averaged HF energy.

The central tool in the prove of Refs. [254] and [255] is the single-particle partition-function

$$Z_{\mathrm{HF}}(\beta) = \sum_i e^{-\beta \epsilon_i} = \int_0^\infty g(\epsilon) e^{-\beta \epsilon} d\epsilon , \qquad (2.196)$$

where the ϵ_i are the Hartree-Fock eigenstates, and the ETF analogue of Z_{HF} is

$$Z_{\text{ETF}}(\beta) = Z_{\text{TF}}(1 + \hbar\chi_1(\beta)\ldots + \hbar^4\chi_4(\beta))\,. \tag{2.197}$$

This expression is the result of a Wigner-Kirkwood expansion of Z_{HF} [114] and is often misleadingly called an expansion in "powers of $\hbar$". It is in fact an asymptotic expansion in inverse powers of the single-particle action, measured in units of $\hbar$. The Thomas-Fermi partition function $Z_{\text{TF}}(\beta)$ in Eq. (2.197) is given by

$$Z_{\text{TF}} = \frac{2}{h^3} \int e^{-\beta h^{\text{shell}}(\mathbf{p},\mathbf{r})} d^3p\, d^3r \,,$$

where $h^{\text{shell}}(\mathbf{p},\mathbf{r})$ is the classical single-particle Hamiltonian and the factor 2 accounts for the spin degree of freedom. For the harmonic oscillator $Z_{\text{TF}} = 2/(\beta\hbar\omega)^3$. The expansion (2.197) is divergent for $\beta \to \infty$. It is to be treated as a semiconvergent series and should be truncated with the term $\hbar^4\chi_4$. One can show [114] that for small β one has for the harmonic oscillator

$$Z_{\text{ETF}}(\beta) \approx Z_{\text{HF}}(\beta) \qquad \text{if } \beta\hbar\omega \ll 1 \tag{2.198}$$

(high-temperature limit). For more general potentials $\hbar\omega$ has to be replaced in this formula by the average distance between main shells.

The single-particle level density $g_{\text{HF}}(\epsilon) = \sum_i \delta(\epsilon - \epsilon_i)$ is the inverse Laplace transform of the partition function

$$g_{\text{HF}}(\epsilon) = \mathcal{L}^{-1}_{\beta \to \epsilon}[Z_{\text{HF}}(\beta)] \equiv \frac{1}{2\pi i} \int_{c-i\infty}^{c+i\infty} Z_{\text{HF}}(\beta) e^{\epsilon\beta} d\epsilon \,, \tag{2.199}$$

which follows immediately by inserting the definition (2.196). Analogously the ETF level density is defined by

$$g_{\text{ETF}}(\epsilon) = \mathcal{L}^{-1}_{\beta \to \epsilon}[Z_{\text{ETF}}(\beta)]\,.$$

According to Eq. (2.181) the averaged HF level density is a folding product $g_{\text{HF}} * f(\epsilon/\gamma)/\gamma$. Using the folding theorem of Laplace transformations and the fact that the two-sided Laplace transform of $f(x)$ is

$$\int_{-\infty}^{\infty} e^{-\beta\epsilon} f(\epsilon\gamma) d\epsilon/\gamma = e^{(\gamma\beta/2)^2} \mathcal{P}_M(\beta\gamma)\,, \tag{2.200}$$

with some polynomial $\mathcal{P}_M$ of order M [255], the averaged HF level density can be written in terms of the partition function Z_{HF}

$$\tilde{g}_{\text{HF}}(\epsilon) = \mathcal{L}^{-1}_{\beta \to \epsilon}[Z_{\text{HF}}(\beta) e^{(\gamma\beta/2)^2} \mathcal{P}_M(\gamma\beta)]$$
$$= \frac{1}{2\pi i} \int_{c-i\infty}^{c+i\infty} Z_{\text{HF}}(\beta) e^{(\gamma\beta/2)^2} \mathcal{P}_M(\gamma\beta) e^{\epsilon\beta} d\beta\,.$$

In the limit $c \to +0$ and introducing the new integration variable $x = -i\beta$, one obtains

$$\tilde{g}_{\mathrm{HF}}(\epsilon) = \frac{1}{2\pi} \int_{-\infty}^{\infty} Z_{\mathrm{HF}}(ix) e^{-(\gamma x/2)^2} \mathcal{P}_M(i\gamma x) e^{i\epsilon x} dx \,.$$

Because of the Gaussian in the integrand the most appreciable contributions to the integral come from $x < 1/\gamma$. The averaging width γ is of the order $\hbar\omega$. The partition function Z_{HF} is therefore only needed for $\beta\hbar\omega < 1$, where it can be replaced by its semiclassical approximation Z_{ETF}. We therefore conclude that $\tilde{g}_{\mathrm{HF}} \approx \tilde{g}_{\mathrm{ETF}}(\epsilon)$.

To complete the proof that the difference between the HF and ETF energies is well approximated by the Strutinsky shell-correction, we have to show that

$$N = \int_0^\mu g_{\mathrm{ETF}}(\epsilon) d\epsilon = \int_{-\infty}^\mu \tilde{g}_{\mathrm{ETF}}(\epsilon) d\epsilon$$

and   (2.201)

$$E = \int_0^\mu g_{\mathrm{ETF}}(\epsilon) \epsilon d\epsilon = \int_{-\infty}^\mu \tilde{g}_{\mathrm{ETF}}(\epsilon) \epsilon d\epsilon$$

holds for the same value of μ. It is plausible that $g_{\mathrm{ETF}}(\beta)$ is a smooth function of β and may therefore be well represented by a polynomial. Since polynomials (of low order) are reproduced by the folding procedure, Eq. (2.182), one expects $\tilde{g}_{\mathrm{ETF}}(\epsilon) = g_{\mathrm{ETF}}(\epsilon)$, which implies the relations (2.201). A more formal proof is given in Ref. [255].

2.3.3 The extended Thomas-Fermi-plus-Strutinsky integral method

In order to overcome a certain ambiguity in the choice of the shell-model potential corresponding to a given liquid-drop-type macroscopic energy, Chu, Jennings, and Brack proposed the following procedure, halfway between self-consistent HF and the standard macroscopic-microscopic approach [187]. For a given Skyrme potential they solve the ETF equations for spherical nuclei self-consistently, using the generalized Fermi function (2.118) for a convenient representation of the proton and neutron densities. They obtain the total energy E_{ETF}, the densities $\rho_{\mathrm{ETF}}^{(q)}(r)$ and the kinetic energies $\tau_{\mathrm{ETF}}^{(q)}$ as functions of the density and of its gradients up to fourth order. With $\rho_{\mathrm{ETF}}^{(q)}(r)$ and $\tau_{\mathrm{ETF}}^{(q)}(r)$ the quantities $\tilde{V}_q(\mathbf{r})$, $\tilde{f}_q(\mathbf{r})$, and $\tilde{W}_q(\mathbf{r})$ are calculated according to Eqs. (2.109), (2.112), and (2.113). For the single-particle Hamiltonian

$$\tilde{h}_q = -\frac{\hbar^2}{2M_{\mathrm{nucl}}} \boldsymbol{\nabla} \tilde{f}_q(\mathbf{r}) \boldsymbol{\nabla} + \tilde{V}_q(\mathbf{r}) - i\tilde{W}_q(\mathbf{r}) \cdot (\boldsymbol{\nabla} \times \boldsymbol{\sigma})$$

the Schrödinger equation

$$\tilde{h}_q \phi_j^{(q)} = \varepsilon_j^{(q)} \phi_j^{(q)} \,,$$

is solved to get the eigenvalues $\varepsilon_i^{(q)}$. In terms of them the Strutinsky theorem can be written in the form

$$E_{\mathrm{HF}} = E_{\mathrm{ETF}} + \sum_q \left(\sum_{i \leq i_{\mathrm{Fermi}}} \varepsilon_i^{(q)} - \mathrm{tr}\, \tilde{h}_{\mathrm{ETF}}^{(q)} \tilde{\rho}_q \right) + \mathcal{O}(\delta\rho^2) \qquad (2.202)$$

with $\delta\rho = \rho_{\mathrm{HF}} - \tilde{\rho}$. Since it is less time-consuming to calculate self-consistent solutions in the ETF frame than in HF, this approach allows to calculate binding energies faster than in HF, but not so fast as with the standard macroscopic-microscopic approach.

It is the main advantage of this way of introducing shell corrections that no averaging over the single-particle level-density is needed. It can therefore be used up to the drip lines of the chart of nuclides (if, in addition, the pairing corrections are calculated appropriately). It may also appear to be an advantage that the smooth part of the binding energy, E_{ETF}, and the shell correction are derived from the same microscopic source, namely the underlying Skyrme potential. On the other hand, any errors caused by too narrow restrictions in the specific ansatz for the density $\rho(\mathbf{r})$ in the self-consistent solution of the ETF equations cannot be compensated as in the standard version of the macroscopic-microscopic approach, where the smooth part of the energy is fitted to the liquid drop or droplet model.

The method was called "Extended Thomas-Fermi plus Strutinsky Integral", ETFSI, by Dutta et al. [256–260] because the trace on the right-hand-side of Eq. (2.202) becomes an integral over $\mathbf{r}$ in space representation. These authors used the method in a version which was simplified for reasons of numerical convenience in two essential points: Skyrme potentials were restricted to those for which $M^*_{\mathrm{nucl}} = M_{\mathrm{nucl}}$ and the density $\rho(\mathbf{r})$ was restricted to a Fermi function. Presumably because of the latter restriction the authors found an intolerably large difference between the left and the right-hand-side of Eq. (2.202) when both sides were calculated independently with the same Skyrme potential. The authors then felt obliged to use different Skyrme parameters for HF and for ETFSI calculations.

2.4 Energy surfaces

There are two ways to construct energy surfaces: one may either solve a constrained HF or HFB problem, introducing appropriate constraining single-particle operators to control the shape of the resulting self-consistent density distribution or one adopts a priori a shape class, characterized by a number of shape parameters, for which the liquid-drop model yields the smooth part of the deformation energy to which shell and pairing corrections are added. These approaches will be discussed in the next two subsections. We will finally focus on the stationary points of the multidimensional energy surfaces and their dependence on the proton and neutron numbers of the corresponding nuclides.

2.4.1 Calculations with shape constraints

In this approach one looks for the lowest energy eigenvalue of the Schrödinger equation

$$(\hat{H}' - E)|\text{HF}\rangle = 0 \qquad (2.203)$$

in the set of N-particle HF states, where the Routhian $\hat{H}' = \hat{H} - \sum_{i=1}^{n} \lambda_i \hat{q}_i$ is defined in terms of n operators $\hat{q}_1 \ldots \hat{q}_n$ constraining the shape of the density distribution and corresponding Lagrange multipliers $\lambda_1 \ldots \lambda_n$. We restrict the discussion here to HF solutions, for HFB states analogous arguments hold. The solution of Eq. (2.203) yields a state $|\text{HF}(\lambda_1 \ldots \lambda_n)\rangle$ depending on the Lagrange multipliers λ_i. One introduces the expectation values $\overline{q}_i$ by the equations

$$\overline{q}_i(\lambda_1 \ldots \lambda_n) = \langle \text{HF}(\lambda_1 \ldots \lambda_n)|\hat{q}_i|\text{HF}(\lambda_1 \ldots \lambda_n)\rangle, \qquad i = 1, \ldots n, \quad (2.204)$$

which one may invert to obtain

$$\lambda_i = \lambda_i(\overline{q}_1 \ldots \overline{q}_n), \qquad i = 1, \ldots n. \qquad (2.205)$$

We will assume that this inversion exists and is unique in a certain area of the n-dimensional space of deformation parameters $\overline{q}_i$, where $\det(\partial_{\lambda_i} \overline{q}_j) \neq 0$. Inserting this result into the energy expectation-value

$$E(\lambda_1 \ldots \lambda_n) = \langle \text{HF}(\lambda_1 \ldots \lambda_n)|\hat{H}|\text{HF}(\lambda_1 \ldots \lambda_n)\rangle \,,$$

one obtains the energy $E(\overline{q}_1 \ldots \overline{q}_n)$ as function of the deformation parameters, which is referred to as the potential-energy surface. For a detailed discussion how these steps can be numerically implemented for HFB wave functions we refer to the work of Younes and Gogny [261].

There is a naive interpretation of the Lagrange multipliers in Eq. (2.203) that they are the components of the generalized force, needed to keep the system in equilibrium at the deformation $\overline{q}_1 \ldots \overline{q}_n$. In the context of a dynamical equation of motion in the collective variables $\overline{q}_i$ the force would be identified with the inertial force according to d'Alembert's principle. However, for a microsystem like a nucleus, one has to keep in mind that the variances $\langle \text{HF}|(\hat{q}_i - \overline{q}_i)^2|\text{HF}\rangle$ are in general not zero and cannot be neglected. The way in which the states $|\text{HF}(\lambda_1 \ldots \lambda_n)\rangle$ and the function $E(\overline{q}_1 \ldots \overline{q}_n)$ enter into equations of motion is therefore a more subtle subject, to which we will return in Chapter 5. Here we are interested in stationary points only, local minima and saddle points, for which all Lagrange multipliers vanish so that they are independent of constraining conditions.

Often used constraining operators are $\hat{Q}_{20} = r^2 Y_{20}(\theta, \phi)$ to control the elongation along the z-axis, $\hat{Q}_{22} + \hat{Q}_{2-2}$ to control ellipsoidal nonaxiality, $\hat{Q}_{30}$ to control deviations from reflection symmetry, and $\hat{Q}_{40}$ to influence neck-formation [263]. To manipulate the scission shape, the constraint $\hat{Q}_N = \exp(-[z/a]^2)$ was used for necked-in configurations [264] with some length parameter $a \approx 1$ fm.

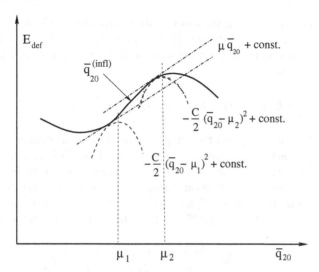

Fig. 2.14. Schematic plot of $E_{\text{def}}(\bar{q}_{20})$, thick line. Linear and quadratic constraints are shown with dash-dotted and dashed lines, respectively, at a point on the left and on the right side of the inflection point $\bar{q}_{20}^{(\text{infl})}$. The geometrical meaning of the parameter μ is also indicated (after Ref. [262]).

It might appear that one can obtain all minima and saddle points along the fission path by introducing only one elongation parameter, for instance $\bar{q}_{20}$. However, already for the energy surface of the liquid-drop model there are two different $\bar{q}_{20}$ values for the same λ as can be seen in Fig. 2.14, since $\partial_\lambda \bar{q}_{20} = 0$ for the inflection point $\bar{q}_{20}^{(\text{infl})}$. Sometimes a quadratic constraint $-(C/2)(\bar{q}_{20}-\mu)^2$ was introduced to overcome this problem [265]. The meaning of the parameter μ is shown in Fig. 2.14. The value of C is rather arbitrary, if $C > |d^2E/d\bar{q}_{20}^2|$. This constraint leads to an effective Lagrange parameter $\lambda_{\text{eff}} = C(\mu-\bar{q}_{20})$. For alternative methods to handle the multivalued functions (2.205) we refer to Sec. 7.6 of Ref. [108].

To calculate the energy surface around a saddle point with an unstable direction perpendicular to the elongation direction, a similar nonlinear constraint in that direction should be used. For instance, for a fissility x below the Businaro-Gallone point the fission saddle of the liquid-drop model is unstable with respect to reflection asymmetry. Therefore a quadratic constraint in $\bar{q}_{30}$ is required in this case. In order not to miss an extremum which has lower symmetry than the initial state of the iterative solution procedure of Eq. (2.203), it is necessary to include constraints in H' which have the same lower symmetry. It also happens that there are several solutions for the same $\bar{q}_{20}$. An example is given in Fig. 2.16. In such case an additional constraint is required – in the example a neck-size parameter – to distinguish between solutions. We refer to a discussion of this point by Berger, Girod, and Gogny [263] and to examples given in Ref. [261].

For the classical liquid-drop model with sharp-surface Coulomb, surface, and curvature energies Strutinsky [266] and Strutinsky, Lyashchenko, and Popov [267] proposed to look for a constrained, selfconsistent solution of the variational problem

$$\frac{\delta}{\delta\rho}[E_{\text{l.d.}}\{\rho(z)\} - \lambda_1 \text{vol}\{\rho(z)\} - \lambda_2 D\{\rho(z)\}] = 0,\qquad(2.206)$$

where the function $\rho(z)$ gives the shape of the nucleus in cylindrical coordinates (axial symmetry is assumed) and $E_{\text{l.d.}}\{\rho, \rho'\}$ is the liquid-drop energy, considered as functional of the shape function $\rho(z)$. The constraint $\lambda_1 \text{vol}\{\rho\}$ shall guaranty volume conservation during shape variation, and $D\{\rho\}$ is the distance between the centers of mass of the left and the right part of the figure. From Eqs. (2.148) and (2.153) follow the variational derivatives

$$\frac{\delta E_{\text{surf}}}{\delta\rho} = E_{\text{surf}}^{(0)}\frac{\delta B_{\text{surf}}}{\delta\rho} = \frac{E_{\text{surf}}^{(0)}}{2R_0^2}\frac{1 + \rho'^2 - \rho\rho''}{(1+\rho'^2)^{3/2}}\qquad(2.207)$$

$$\frac{\delta E_{\text{curv}}}{\delta\rho} = E_{\text{curv}}^{(0)}\frac{\delta B_{\text{curv}}}{\delta\rho} = -\frac{E_{\text{curv}}}{2R_0}\frac{\rho''}{(1+\rho'^2)^2}.\qquad(2.208)$$

The variation of the Coulomb energy when ρ changes by $\delta\rho$ in the interval dz, is

$$\delta E_{\text{Coul}} = \frac{3eZ}{4\pi R_0^3}2\pi\rho\,\delta\rho dz\,\Phi(\rho,z),$$

where $\Phi(\rho,z)$ is the Coulomb potential on the surface at the point ρ, z. Measuring Φ in units of eZ/R_0, this yields the functional derivative

$$\frac{\delta E_{\text{Coul}}}{\delta\rho} = \frac{3}{2}\left(\frac{eZ}{R_0^2}\right)^2\rho\Phi.\qquad(2.209)$$

Inserting the results (2.207)-(2.209) into Eq. (2.206) and dividing the equation by $E_{\text{surf}}^{(0)}/(2R_0^2)$ one obtains the Euler equation of the variational problem (2.206)

$$\frac{\rho\rho'' - 1 - \rho'^2}{(1+\rho'^2)^{3/2}} + \frac{E_{\text{curv}}^{(0)}}{E_{\text{surf}}^{(0)}}R_0\frac{\rho''}{(1+\rho'^2)^2} + 10x\frac{\rho}{R_0}\Phi - \lambda_1\frac{\rho}{R_0} - \lambda_2\frac{\rho|z|}{R_0^2} = 0.\quad(2.210)$$

In this equation x is the fissility and the potential Φ is given by

$$\Phi(z,\rho) = \frac{3}{4\pi R_0^2}\int_{\text{vol}}\frac{d^3r'}{|\mathbf{r}(\rho,z) - \mathbf{r}'|} = -\frac{3}{8\pi R_0^2}\int_{\text{surf}}\frac{(\mathbf{r}(\rho,z) - \mathbf{r}')\cdot d\mathbf{S}'}{|\mathbf{r}(\rho,z) - \mathbf{r}'|},$$

where the identity

$$\text{div}_{r'}\frac{\mathbf{r} - \mathbf{r}'}{|\mathbf{r} - \mathbf{r}'|} = -\frac{2}{|\mathbf{r} - \mathbf{r}'|}$$

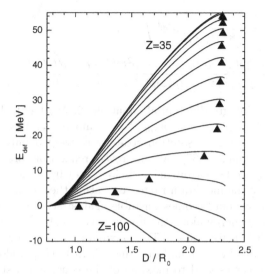

Fig. 2.15. Deformation energy along the fission valley as function of the distance D between the centers of mass of the two halves of the (reflection symmetric) shape for nuclei with $35 \leq Z \leq 100$. The neutron excess was taken to be $I = 0.26$. Liquid-drop parameters from Ref. [228]. The fission saddle points were indicated by black triangles (after Ref. [268]).

was used. In cylindrical coordinates one obtains

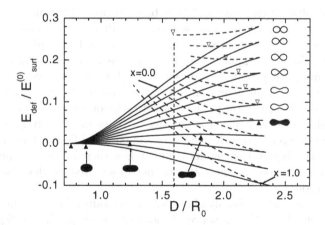

Fig. 2.16. Deformation energy along the fission valley as function of the distance D between the centers of mass of the two halves of the shape for nuclei with fissility $0 \leq x \leq 1$. Liquid-drop parameters of Ref. [228]. The fission saddle points with an unstable direction perpendicular to the elongation direction were indicated by open triangles (after Ref. [269]).

$$\Phi(\rho, z) = -\frac{3}{8\pi R_0^2} \int_0^{2\pi} d\phi' \int_{z_{min}}^{z_{max}} dz'$$

$$\times \frac{\rho(z)\rho(z')\cos(\phi - \phi') - \rho^2(z') - \rho(z)\rho'(z')(z - z')}{[\rho^2(z) + \rho^2(z') - 2\rho(z)\rho(z')\cos(\phi - \phi') + (z - z')^2]^{1/2}} \, .$$

The nonlinear integro-differential equation (2.210) has to be solved with boundary conditions $\rho(z_{min}) = \rho(z_{max}) = 0$ and $\rho'(z_{min}) = \infty$, $\rho'(z_{max}) = -\infty$. The original iterative solution strategy converged only in certain areas of the (x, D) plane [267]. An improved relaxation procedure allowed Ivanyuk [270] to map the bottom of the fission valley from the spherical ground state to the end of the mountain ridge, separating the fission from the fusion valley, where the system becomes unstable against scission at fixed distance D. In Fig. 2.15 the energy along the fission valley is shown as function of the distance D for various fissilities x. The maximum of these curves corresponds to the fission barrier in the liquid-drop model. In Fig. 2.16 also dashed lines are added, which correspond to maxima of the constrained stationary-value problem. These lines map the mountain ridge separating fission and fusion valleys. The minima of these lines, indicated by open triangles, correspond to saddle points for transitions between these valleys at fixed distance D. Unfortunately, the bottom of the fusion valley has not yet been mapped by this method.

2.4.2 Calculations based on shape parametrizations

By far most often energy surfaces have been calculated by the macroscopic-microscopic method. In this approach one has first to fix a class of shapes, which one expects to contain the actual fission pass. For shapes close to the ground-state shape an expansion in multipoles is a natural choice since any smooth, compact shape can be represented in this way. However, for strongly necked-in shapes not all points on the surface can be "seen" from the origin (the origin is no longer a star point). Close to scission the multipole expansion therefore fails. Moreover, it turns out that it takes more than 14 multipoles to achieve convergence even for the very compact saddle point of the lighter actinides in the liquid-drop model [180]. It was therefore also a question of economy to find shape parametrizations, better adapted to what one believes to be the fission path.

A first one-parameter family of shapes was introduced by Hill and Wheeler [87]. They considered the sequence of saddle-point shapes in the liquid-drop model, shown in Fig. 1.1 as function of the fissility x, or rather of the "deformation" parameter $y = 1 - x$. The shape class was therefore called y-family of shapes. For more detailed information on the y-family we refer to the monograph by Hasse and Myers [200].

Various other shape classes have been proposed for the same purpose. Their appropriateness has often been judged on their ability to describe fission saddle-points with a few deformation parameters only. However, one has to keep in mind that the fission path is very much influenced by mass and friction parameters and must not necessarily pass over the fission barrier, as will be shown in Chapter 5. In particular, beyond the saddle point, down to scission, there are no reasonable criteria for the choice of a restricted shape class, based on the statics of the energy surface alone.

Since there is no one-to-one mapping of one of these shape classes onto another, even dynamically calculated fission paths in one shape class cannot easily be compared with results obtained in another class. Unfortunately, one often finds plots of the first two "important" shape parameters versus those two parameters of another shape class, giving the erroneous impression that such mapping is possible in more than very limited areas of the shape variables. With this cautioning in mind we shall discuss some of the more frequently used shape classes and refer for a more comprehensive discussion to Hasse and Myers' book [200].

Nilsson's ε_n parametrization

In connection with deformed, axially symmetric oscillator potentials Nilsson [92] introduced stretched, dimensionless cylindrical coordinates ρ_t, z_t, related to normal cylindrical coordinates ρ, z by

$$\begin{aligned} \rho_t &= c\omega_\perp^{1/2}\rho, & z_t &= c\omega_z^{1/2}z, \\ \cos\theta_t &= z_t/\rho_t, & r_t^2 &= z_t^2 + \rho_t^2 \end{aligned} \tag{2.211}$$

with $c^2 = M_{\mathrm{nucl}}/\hbar$ and

$$\omega_\perp = \omega(\varepsilon)(1+\varepsilon/3), \quad \omega_z = \omega(\varepsilon)(1-2\varepsilon/3), \quad \omega(\varepsilon)/\omega_0 = (1-\varepsilon^2/3-2\varepsilon^3/27)^{-1/3}$$

in terms of the spheroidal deformation parameter ε and a scaling factor ω_0. The oscillator potential becomes in the stretched coordinates

$$V_{\mathrm{osci}}(\rho_t, z_t; \varepsilon; \omega_0) = \frac{1}{2}\hbar\omega(\varepsilon)r_t^2\left[1 - \frac{2}{3}\varepsilon P_2(\cos\theta_t)\right]. \tag{2.212}$$

The nuclear shape is identified with the equipotential surface

$$V_{\mathrm{osci}} = \epsilon_{\mathrm{Fermi}}. \tag{2.213}$$

Inserting Eqs. (2.211) into Eq. (2.212), one obtains from Eq. (2.213) an implicit equation for ρ as function of z and ε. Nilsson's shape parametrization was later generalized to an expansion around the spheroidal shape, involving additional shape parameters $\varepsilon_3, \ldots \varepsilon_N$ [93].

The Lawrence shape-class and its extensions

Lawrence proposed the quartic shapes

$$\rho^2 = az^4 + bz^2 + c\,, \tag{2.214}$$

where one of the three parameters a, b, c is determined by the requirement of volume conservation [271]. Hasse extended the ansatz to allow for mirror asymmetric shapes

$$\rho^2 = R_0^3 \lambda [z_0^2 - (z + z_s)^2][z_2|z_2| + (z + z_s - z_1)^2]\,, \tag{2.215}$$

where λ is determined as function of the other parameters to ensure volume conservation and z_s serves to bring the center of mass to the origin [272]. There are three deformation parameters: z_0 and z_1 controlling the elongation and the asymmetry, respectively and z_2, connected with constriction.

A further extension of the shape class (2.214) was introduced by Trentalange, Koonin, and Sierk [273]

$$\rho^2 = R_0^2 \sum_{n=0} a_n P_n(z/z_0) \tag{2.216}$$

in terms of Legendre polynomials $P_n(z/z_0)$. The sum was restricted to even multipoles in Ref. [273]. The length of the nuclear shape is $2z_0 = (4R_0)/(3a_0)$ and volume conservation yields $a_0 = -\sum_{n=2} a_n$. Further restrictions of the deformation-parameter space are required to ensure geometrically sensible values of ρ^2. To describe shapes without axial symmetry, Sierk extended the class (2.216) by the ansatz [274]

$$\rho(z, \phi) = \rho^{(0)}(z)\, \eta(\phi)/\lambda \tag{2.217}$$

with $\eta(\phi) = 1 + \alpha_1 P_2(\cos \phi) + \alpha_2 P_4(\cos \phi)$ and $\lambda = 1 + \alpha_1/4 + 9\alpha_2/64$ in terms of two additional deformation parameters α_1 and α_2.

Pomorski and Bartel [275] included also odd Legendre polynomials in the sum (2.216) to describe shapes without mirror symmetry

$$\rho^2(z) = R_0^2 \sum_{n=0}^{n_{max}} \alpha_n P_n \left(\frac{z - z_s}{z_0} \right)\,. \tag{2.218}$$

The first two α parameters should satisfy the relations

$$\alpha_0 = - \sum_{n=2,4,} \alpha_n\,, \qquad \alpha_1 = - \sum_{n=3,5,} \alpha_n$$

to yield reasonable shapes. Volume conservation requires $z_0 = 2R_0/(3\alpha_0)$ and $z_s = -2\alpha_1 R_0/(9\alpha_0^2)$ serves to keep the center of mass at the origin.

A modification of the shape class (2.215) was introduced by Brack et al. [103] to improve the description of deformed ground-state shapes

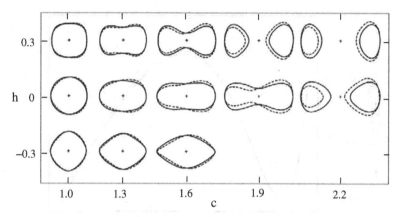

Fig. 2.17. Some shapes in the c, h, α parametrization. Full lines correspond to $\alpha = 0$, dotted ones to $\alpha = 0.2$ (after Ref. [103]).

$$\rho^2(z) = \begin{cases} (C^2 - z^2)[A + B(z/C)^2 + \alpha(z/C)] & \text{for} \quad B \geq 0 \\[2mm] (C^2 - z^2)[A + \alpha(z/C)]e^{BCz^2/R_0^3} & \text{for} \quad B < 0. \end{cases} \qquad (2.219)$$

Volume conservation implies one functional relation between A, B, α, and C/R_0. After elimination of A by this relation two independent, dimensionless deformation parameters

$$c = C/R_0$$
$$h = B/2 - (C/R_0 - 1)/4$$

are introduced besides the asymmetry variable α. The shape class (2.219) is shown in Fig. 2.17.

Still another modification of the class (2.215) is the so-called "Modified Funny-Hills" parametrization of Pomorski and Bartel [275] which allows to describe nonaxial shapes with an ellipsoidal cross section perpendicular to the z axis with eccentricity $\epsilon = \sqrt{2\eta/(1 + \eta)}$ – independent of z – in terms of one extra nonaxiality deformation-parameter η. In cylindrical coordinates (ρ, z, ϕ)

$$\rho^2(z, \phi) = \tilde{\rho}^2(z) \frac{\sqrt{1 - \eta^2}}{1 + \eta \cos 2\phi} \qquad (2.220)$$

with

$$\tilde{\rho}^2(z) = \frac{R_0^2}{c f(a, B)} (1 - u)(1 + \alpha u - B e^{-a^2 u^2}), \qquad (2.221)$$

where the function

$$f(a, B) = 1 - \frac{3B}{4a^2} \left[e^{-a^2} + \sqrt{\pi} \left(a - \frac{1}{2a} \right) \text{Erf}(a) \right]$$

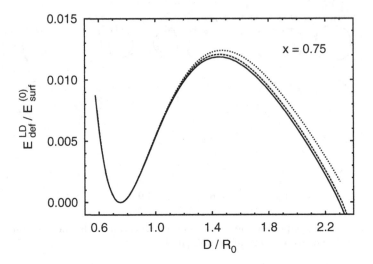

Fig. 2.18. Deformation energy along the fission valley as function of the distance D between the centers of mass of the two halves of the (reflection symmetric) shape for a nucleus with fissility $x = 0.75$, the region of lighter actinides. Liquid-drop parameters from Ref. [228]. The full line is the result of a variational calculation with Eq. (2.210). The dashed line was calculated within the shape class (2.220), where the parameter B was minimized out at fixed D (from Ref. [276]). The dotted line corresponds to the fission valley in the shape class (2.219), also with B being minimized out.

serves to keep the volume constant and $u = (z - z_s)/(cR_0)$. To fix the center of mass at the origin the shift $z_s = -4\alpha cR_0/[15f(a, B)]$ is introduced. The parameter a is taken to be 1 and instead of B the parameter $h = B - c + 1$ is introduced. This leaves four independent, dimensionless deformation parameters: c, h, α, and η. Fig. 2.18 shows that the liquid-drop deformation energy along the fission valley is very well reproduced for lighter actinides in this shape class.

The shape class of Cassinian ovaloids and its extension

Stavinsky, Rabotnov, and Seregin introduced the shape class of Cassinian ovaloids [277]

$$\rho^2(z) = \sqrt{a^4 + 4c^2z^2} - (c^2 + z^2) \; . \tag{2.222}$$

The volume-conservation condition allows to express the parameter a by $R_0 = r_0A^{1/3}$ so that a one-parameter family of shapes results. Fig. 2.19 shows this shape family as function of the dimensionless deformation parameter $u = c/a$. It yields a fairly good description of the fission valley in the liquid-drop model

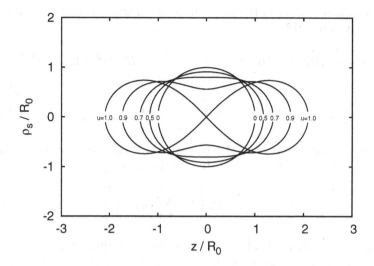

Fig. 2.19. The shape class of Cassinian ovaloids of constant volume (after Ref. [200]).

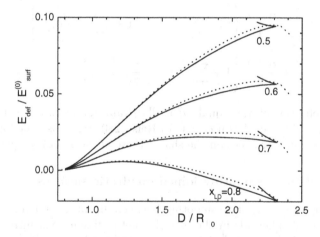

Fig. 2.20. Deformation energy along the fission valley as function of the distance D between the centers of mass of the two halves of the shape for nuclei with fissility $0.5 \leq x \leq 0.8$. Liquid-drop parameters of Ref. [228]. The full lines are calculated with the variational ansatz (2.206), (2.210), the dashed lines show the deformation energy in the shape class (2.222) (after Ref. [270]).

when compared with the constrained, self-consistent calculation using Eq. (2.210) as shown in Fig. 2.20.

An extension of the shape class (2.222) was proposed by Pashkevich [278]. He considers Cassinian oval coordinates [279], a system of orthogonal, curvilinear coordinates (R, x) in the plane, in terms of which cylinder coordinates (ρ, z) are given by [278]

$$\rho(R, x) = \frac{1}{\sqrt{2}} \{[R^4 + 2c^2 R^2 (2x^2 - 1) + c^4]^{1/2} - R^2 (2x^2 - 1) - c^2\}^{1/2}$$

(2.223)

$$z(R, x) = \frac{\text{sign}(x)}{\sqrt{2}} \{[R^4 + 2c^2 R^2 (2x^2 - 1) + c^4]^{1/2} + R^2 (2x^2 - 1) + c^2\}^{1/2}.$$

For $R = a = \text{const}$, elimination of x between these equations yields the shape class (2.222). Pashkevich extended the class by putting

$$R = R(x) = R_0 \left(1 + \sum_l \alpha_l P_l(x)\right)$$

(2.224)

with Legendre polynomials $P_l(x)$, and additional deformation parameters α_l. The expression (2.224) is to be inserted into Eqs. (2.223). Besides the α_l Pashkevich introduced instead of u a deformation parameter α, defined implicitly by the equation

$$u^2 = \frac{\alpha - 1}{4} \left\{ \left(1 + \sum_l \alpha_l\right)^2 + \left(1 + \sum_l (-1)^l \alpha_l\right)^2 \right\}$$

$$+ \frac{\alpha + 1}{2} \left\{ 1 + \sum_l (-1)^l \frac{(2l - 1)!!}{2^l l!} \alpha_{2l} \right\}^2.$$

(2.225)

With this definition of α, α equals u^2 if all α_l vanish, and for any values of the α_l the neck radius vanishes if $\alpha = 1$. In Refs. [280–282] α is called ϵ. Since α and α_2 have a similar effect on the shape, these authors set α_2 equal to zero.

The class of three smoothly joined quadratic surfaces

In an attempt to interpolate smoothly between an initial sphere or spheroid and two coaxial spheroids in the exit channel of fission, Nix introduced the shape class of three smoothly joined surfaces of revolution [283]

$$\rho^2(z) = \begin{cases} a_1^2 - (a_1/c_1)^2 (z - l_1)^2, & l_1 - c_1 \le z \le z_1 \\ a_2^2 - (a_2/c_2)^2 (z - l_2)^2, & z_2 \le z \le l_2 + c_2 \\ a_3^2 + (a_3/c_3)^2 (z - l_3)^2, & z_1 \le z \le z_2 \end{cases}$$

(2.226)

The meaning of the eleven parameters z_1, z_2, and a_i, c_i, l_i, $i = 1, 2, 3$ is described in Fig. 2.21 There are four relations between these parameters to ensure the smooth matching at z_1 and z_2, one relation to ensure volume conservation, and a sixth one to fix the center of mass at the origin. This leaves 5 independent deformation parameters, three of which are related with symmetric shapes. For ground-state shapes this parametrization is nevertheless

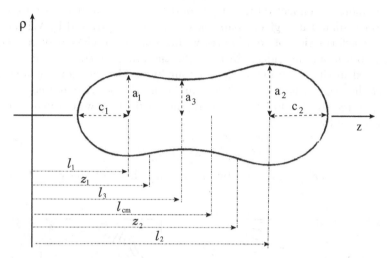

Fig. 2.21. The parameters of the three, smoothly joined quadratic surfaces (after Ref. [283]).

of inferior flexibility compared to a multipole expansion with the first three even multipoles. On the other hand none of the previously discussed shape classes contains two spheroids at given tip-to-tip distance, a configuration one expects in the fusion valley. Advantages of the shape class of three quadratic surfaces are therefore expected around and beyond scission. The shape class may be generalized to shapes with elliptic cross section perpendicular to the fission direction using the transformation (2.220).

In connection with dynamical calculations one is forced to restrict the dynamical variables to what are considered to be "essential" variables: elongation, constriction, and asymmetry. The shape class of three quadratic surfaces is for instance reduced by allowing only spheres for the two outer surfaces with radii R_1 and R_2. Błocki et al. [284–287] introduced the ρ, λ, Δ parametrization with

$$\rho = \frac{l_2 - l_1}{R_1 + R_2}, \quad \lambda = \frac{(z_1 - l_1) - (z_2 - l_2)}{R_1 + R - 2}, \quad \Delta = \frac{R_1 - R_2}{R_1 + R_2}. \qquad (2.227)$$

Other parametrizations of the same restricted shape class can also be found in the literature [288].

2.4.3 Survey of the geometry of stationary points

We have seen that in constrained selfconsistent calculations stationary points of the energy surface are in principle independent of the choice of the constraining operator with the caveat mentioned in Sec. 2.4.1. This is in general not true for such topographical features as valleys and mountain ridges. They are even more dependent on the choice of deformation parameters for energy surfaces calculated in a given shape class. As already stressed by Wilets [30] even a transformation of coordinates within one shape class may convert a valley into a mountain ridge. In Fig. 2.22 the function $f(x, y) = y^2 - x^2$ is represented in the left frame by its contour lines. The function has a saddle point at the origin where two valleys start along the positive and negative y-axis, separated by mountain ridges along the x-axis. In new coordinates x', y', related to x, y by

$$x = x'$$
$$y^2 = y'^3 - 3x'^2 y' + x'^2 \tag{2.228}$$

the function $f(x', y')$ is given by the contour lines of the right frame of

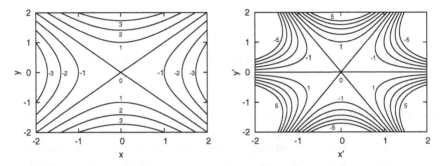

Fig. 2.22. Effect of the coordinate transformation (2.228) on the function f(x,y).

Fig. 2.22. The transformation leaves the saddle at the origin and the y-axis ($x = 0$) becomes the y'-axis ($x' = 0$). On the positive y'-axis is now a mountain ridge, instead of a valley. One should therefore not attach too much physical significance to valleys and ridge lines. In particular, one should not assume that in the fission process the fission valley is initially filled from the saddle with a rather sharp probability distribution in the phase space of the collective coordinates and that this probability distribution is subsequently running down the valley like honey. Note that only the equations of motion are covariant under coordinate transformations, but not topographical properties of the energy surface.

It is nevertheless useful to employ concepts like fission and fusion valleys to visualize the topological relation between various stationary points of the energy surface.

The energy surface in the liquid-drop model

The liquid-drop energy surface has one minimum which is the spherical ground state for nonrotating systems. For nuclei with finite angular momentum systematic investigations of the macroscopic energy surface are usually based on the assumption of a rigid rotation, irrespective of how that can be realized microscopically. We presented in the Introduction the result obtained by Cohen, Plasil, and Swiatecki, Ref. [44], for the pure liquid-drop model, consisting of a volume, a sharp-surface, and a Coulomb term. It is summarized in Figs. 1.3. and 1.4: below line y_I of the phase diagram of Fig. 1.3 the ground state is axially symmetric, oblate, and rotates around the symmetry axis, between

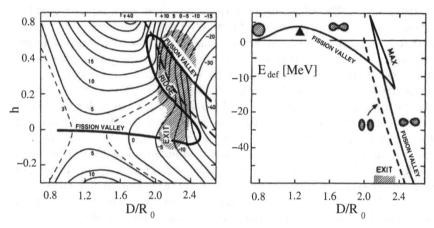

Fig. 2.23. Contour plot of the potential-energy surface of the liquid-drop model, calculated in the parametrization (2.219) and plotted vs. h and the centers-of-mass distance D for $\alpha = 0$. The fission and fusion valleys are indicated by thick lines, separated by a mountain ridge. Also shown by the thick, dashed line is the bottom of the fusion valley in a parametrization of two equal, coaxial spheroids. The right frame shows the energy of the valleys and the ridge line as function of D/R_0 (after Ref. [103]).

lines y_I and y_{II} the ground state is triaxial, rotating around the smallest axis, and beyond the line y_{II} there exists no minimum. The saddle-point shapes become increasingly more compact for increasing y at given fissility x. They are reflection symmetric between lines y_{II} and y_{III} and loose stability against mass-asymmetric distortions below line y_{III}, which is the continuation of the Businaro-Gallone point at $y = 0$ into the x, y plane.

A complete energy surface in the liquid-drop model for a nonrotating nucleus with $x = 0.8$ is shown in the left frame of Fig. 2.23. It was calculated in the shape class (2.219) and presented as function of the distance D between the centers-of-mass of the two nascent fragments and the neck parameter h. One sees the fission valley passing the saddle point in the lower half of the figure and the fusion valley leading upwards in the upper right corner of the figure. The two valleys are separated by a mountain ridge stretching from the upper end of the fusion valley at $D/R_0 \approx 1.9$ to the lower end of the fission valley at $D/R_0 \approx 2.6$. The shaded area was assumed by the authors of Ref. [103] to be the area where the neck rupture occurs. This can actually only be determined by a dynamical calculation.

Since the c, h, α parametrization is not very suitable to describe the fusion valley, the energy of two coaxial oblate spheroids is shown by a dashed line in Fig. 2.23. It leads to substantially lower energies for the same distance D than the shape class (2.219). With decreasing fissility x the mountain ridge becomes shorter and eventually the fission and the fusion valleys coalesce. In Fig. 2.24 we show as an example the fusion energy-surface for the system ^{40}Ca+^{40}Ca. The shape class (2.226) of three quadratic surfaces was used, better suited for the fusion valley. An early version of a finite-range liquid-drop model is employed to account for the proximity potential in the fusion entrance-channel. As shown in Fig. 2.15, independent of any parametrization, selfconsistent liquid-drop model (LDM) calculations for symmetric shapes yield the end point of the fission valley at $D_{\text{crit}}/R_0 \approx 2.23$, for all $x \leq 0.9$.

In the finite-range liquid-drop model (FRLDM) one can no longer absorb all model parameters in the dimensionless parameters x and y and collect all shape dependencies in the dimensionless functions B_{surf}, B_{Coul}, and B_{rot} of the shape parameters since there are additional independent lengths involved to characterize the diffusenesses of the nuclear energy density, of the charge density and the mass density, the latter entering the evaluation of the moments of inertia. Binding and deformation energies have therefore to be given as functions of A, Z, and the collective angular momentum L, besides the shape parameters, and are specific for each set of potential parameters. Results will

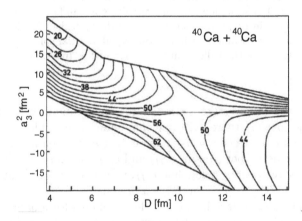

Fig. 2.24. Contour diagram of the potential energy for the system ^{40}Ca+^{40}Ca as function of the centers-of-mass distance D and the neck cross-section parameter a_3^2. The line $a_3 = 0$ is the scission line. The spherical compound state is in the upper left corner. The energy is minimized at each point with respect to the eccentricity of the outer spheroids. Potential parameters are taken from Ref. [225]. Energies on the contour lines are in MeV, counted from the configuration where the fragments are at infinite distance D (after Ref. [289]).

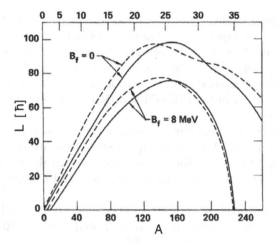

Fig. 2.25. Phase diagram of rotating nuclei in the L, A plane. The full lines are calculated in the FRLDM, the dashed lines in the LDM. The upper two lines are calculated for vanishing fission barrier, the two lower lines for a barrier height of 8 MeV (after Ref. [290])

be presented in the following for β stable nuclei, i.e. Z is given by Eq. (2.162) as function of A. It is not necessarily integer.

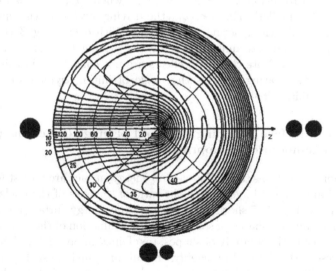

Fig. 2.26. Potential energy of the scission configuration of touching spheres in the FRLDM of Ref. [225]. The angular variable is the asymmetry ϕ, the radial variable is the charge number Z of the compound system. The compound nucleus, for $\phi = 180°$, is taken as the energy zero-point for each Z (after Ref. [289]).

For zero angular momentum the saddle-point shapes in the FRLDM have a little larger necks compared to the same nuclei in the pure LDM [274]. The main difference, however, is the barrier height, shown in Fig. 2.8. For strongly necked-in saddle-point shapes the LDM does not account for the proximity potential between juxtaposed surface areas of the nascent fragments, which is included automatically in the FRLDM and leads to a substantial lowering of the fission barrier.

For finite (collective) angular momentum L the phase-space diagram of the LDM, shown in Fig. 1.3, remains qualitatively the same in the FRLDM [274, 290]. In Fig. 2.25 the dashed lines correspond to the lines y_I and y_{II}, the full lines are calculated by Mustafa, Baisden, and Chandra [290] in the FRLDM.

There is no systematic investigation of the saddle points with an unstable direction in the neck degree of freedom, indicated in Fig. 2.16 by the open triangles, under rotation or in the FRLDM.

Sometimes the mass asymmetry is plotted as an angular variable [291], e.g. $\phi = \pi(a_1^3 - a_2^3)/(a_1^3 + a_2^3)$ in the shape class (2.226) [289]. In Fig. 2.26 we show as an example the potential energy in the fusion valley for touching spheres. The radial variable is the charge number of the compound state. The N/Z ratio is assumed to have equilibrated between the reacting nuclei and is further assumed in this figure to follow Green's formula for β-stable nuclei, Eq. (2.162). Changes in the mass asymmetry at fixed total mass (or Z) proceed in Fig. 2.26 on a circle. For instance for $Z = 100$ the driving force is seen to lead to fusion if $\pi \geq \phi > \phi_{\mathrm{crit}}(Z)$, where $\phi_{\mathrm{crit}}(100) \approx 2\pi/3$, "the larger eats the smaller". For $\phi < \phi_{\mathrm{crit}}$ the driving force is in the direction of making the size of the reacting nuclei equal. With decreasing Z the critical angle $\phi_{\mathrm{crit}}(Z)$ decreases and for $Z < 60$ the driving force is always in the direction of fusion. Note that shell effects are not considered in this picture. For collectively rotating systems the range of initial asymmetries favoring fusion is shrinking [289].

The energy surface including shell and pairing corrections (in the macroscopic-microscopic approach)

Microscopic corrections to the liquid-drop energy surface are largest for spherical nuclei and second largest when the ratio of the axes of the nuclear shape is roughly 1:2. Fig. 2.27 shows a cut through the energy surface for axial and reflection symmetric shapes of some actinides as function of the elongation parameter y within the y-family of shapes, (explained in Sec. 2.4.2). The dashed lines represent the liquid-drop energy, the full lines include shell-plus-pairing corrections. The first minimum corresponds to the ground state, the second minimum represents an isomeric state, and the two outer maxima are saddle points. Clearly it takes a representation of the energy in more dimensions and a more flexible shape parametrization to establish the character of these extrema.

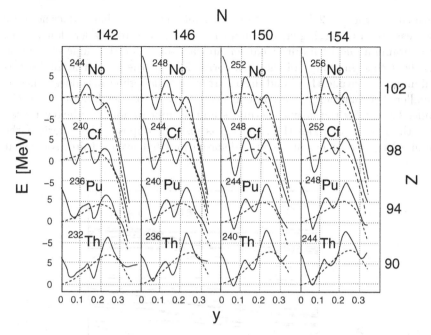

Fig. 2.27. Potential energy for some actinides within the y-family of shapes. The dashed lines are liquid-drop energies, the full lines include the microscopic corrections (after Ref. [96])

There are however two features which can be learned from this crude picture. The ground states of all nuclides of Fig. 2.27 are deformed. The reason is the following: All nuclei shown in the figure are mid-shell nuclei with respect to their protons as well as to their neutrons. In a spherical mean-field potential both types of nucleons would fill some of a bunch of levels lying closely together. By breaking the spherical symmetry of the potential some of these levels would go down, some would go up. Occupying mostly downsloping levels leads therefore to a lower Hartree-Fock ground-state energy. This lowering of the energy because of a spontaneous symmetry breaking of the mean field is called Jahn-Teller effect. It causes the shell correction for actinides to be positive for the spherical shape and to decrease with increasing elongation y, thus shifting the ground state to a finite deformation. The situation is opposite for the double magic nucleus ^{208}Pb and nuclides in its neighborhood where the shell correction is large and negative for the spherical ground state because no energy is gained by breaking the spherical symmetry. The Jahn-Teller effect is partly diminished by the pairing correction since the pairing correlation energy increases with increasing level density at the Fermi energy.

The other interesting feature in Fig. 2.27 is the position of the second minimum. It lies always at the same elongation of $y \approx 0.18$, corresponding roughly to an axes ratio of 1:2. The liquid-drop barrier on the other hand, does not only decrease with increasing charge number Z, but moves to smaller elongations. For uranium the liquid-drop barrier appears at about the same elongation as the second minimum of the shell correction, for the heavier actinides this minimum is beyond the liquid-drop barrier, for the lighter actinides it lies at smaller deformations. As a consequence the outer barrier is lower than the inner barrier for the heavier actinides and disappears for superheavy nuclei. For the lighter actinides the situation is just the opposite. For thorium the liquid-drop barrier lies at so large deformation that even a third, somewhat

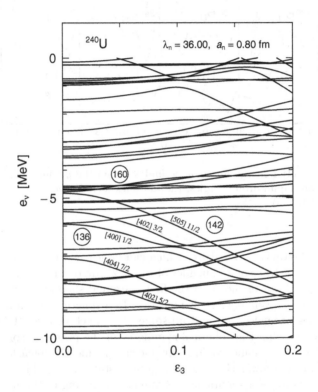

Fig. 2.28. Neutron single-particle levels in the folded Yukawa potential with fixed symmetric deformation parameters $\epsilon = 0.85$, $\epsilon_4 = 0.12$ as function of the octupole deformation parameter ϵ_3 with ϵ_5 being minimized out for fixed values of the other ϵ_n. Relevant couplings are between reflection-symmetric states [404 7/2], [402 5/2], [402 3/2], [400 1/2] and odd states [514 7/2], [512 5/2], [512 3/2], [510 1/2] (not labelled in the figure), respectively, (after Ref. [292]).

smaller, shell-correction minimum at axis ratio 1:3 has a chance to produce a further shallow, hyperdeformed minimum in the energy surface [293].

The latter and some other details are not visible in Fig. 2.27, but a more flexible shape parametrization than the y-family leads to a lowering of minima and saddle points. The most important modification is again due to a Jahn-Teller effect: Gustafsson, Möller, and Nilsson [294] first noticed that adding a parity-violating term $\sim \rho^2 P_3$ to the standard Nilsson potential leads to a strong coupling between positive-parity Nilsson levels with asymptotic quantum numbers $[40\Lambda\Omega]$ and negative-parity states $[51\Lambda\Omega]$ when the elongation corresponds to that of the second barrier. Some of the eigenstates in the octupole-deformed Nilsson potential are bending downwards and some go up

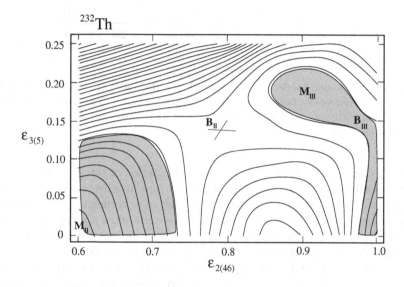

Fig. 2.29. Potential-energy surface for ^{232}Th in the region of the third minimum (M$_{\mathrm{III}}$) on the $(\varepsilon_2, \varepsilon_3)$ plane. The distance between contour lines is 0.5 MeV. The shaded area corresponds to lower energies (after [295]).

with increasing octupole deformation ϵ_3 as shown in Fig. 2.28. For actinides with neutron numbers between 132 and 160 the Nilsson states originating from reflection-symmetric $[40\Lambda\Omega]$ states are occupied, the levels coming from $[51\Lambda\Omega]$ states are unoccupied. In a simple single-particle picture such nuclei can therefore reduce their energy by breaking the mirror symmetry. Qualitatively the same situation as in the Nilsson potential occurs in finite-depth single-particle potentials.

In the region of the second minimum a similar energy gain is observed due to a breaking of the axial symmetry. This time it is a term $\sim \rho^2(Y_{22} + Y_{2-2})$, added to the standard Nilsson potential, which couples states $[Nn_z\Lambda\Omega]$ with states $[Nn_z\Lambda \pm 2\Omega \pm 2]$ as pointed out by Larsson, Möller, and Nilsson [296].

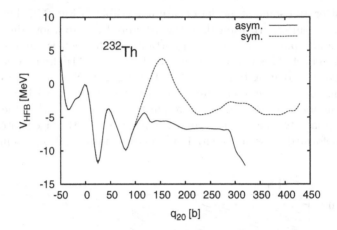

Fig. 2.30. The fission barrier of ^{232}Th evaluated in Ref. [304] within the HFB theory with the Gogny force as function of the quadrupole moment. The barrier evaluated without and with conserving of the reflection symmetry are marked by the solid and dashed lines, respectively.

For a discussion of the experimental determination of various saddle points and local minima of fissile nuclei we refer to review articles by Lynn [297] and Weigmann [298]. The latter contains also a list of experimentally determined barrier heights and energies of fission isomers. Stationary points of the energy hypersurfaces of fissile nuclei, calculated in various selfconsistent and macroscopic-microscopic schemes are compared in Refs. [299, 300].

There have been extensive experimental and theoretical studies of hyperdeformed minima. The existence of such hyperdeformed states was experimentally inferred from fission-fragment angular distributions together with an analysis of the microstructure in the excitation functions of the reactions ^{230,231,232}Th(n,f) [301] and of the (d,pf) reactions ^{229,230,231,232}Th(d,pf) [301], ^{233}U(d,pf) [302], and ^{231}Pa(d,pf) [303], measured with high energy resolution.

As an example for a calculated potential-energy surface in the region of the third minimum we show in Fig. 2.29 a two-dimensional plot for ^{232}Th. The energy is obtained in the macroscopic-microscopic model with a Nilsson single-particle potential [295].

Also selfconsistent calculations have predicted hyperdeformed shape isomers in nuclei close to ^{232}Th. The fission barrier of ^{232}Th, evaluated within the constrained HFB model with the Gogny force D1S, [304] is shown in Fig. 2.30. The barrier is plotted as function of the quadrupole moment. The solid line represents the energy, obtained with constraints on the quadrupole, hexadecapole, and octupole moments. The dashed line is calculated using only the constraining operators $\hat{q}_{20}$ and $\hat{q}_{40}$ and restricting the calculation to reflection-symmetric shapes. There is a sequence of an oblate, a prolate, a super-deformed, and a hyperdeformed minimum in the energy. The second

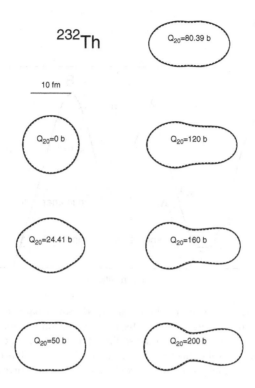

Fig. 2.31. The half-density contours of ^{232}Th for a spherical configuration, for the ground-state, at the top of the first barrier, in the second minimum, on top of the second symmetric and asymmetric barriers, as well as in the hyper-deformed minimum of the energy presented in Fig. 2.30. The neutron and proton shapes are marked by the solid and the dashed lines, respectively.

minimum at about 80 b is well developed. The height of the inner and outer barriers are of the order of 5 MeV, while the third super-deformed minimum is rather shallow with an outer barrier of only 0.5 MeV. Half-density shapes of ^{232}Th in the stationary points corresponding to subsequent extrema along the static fission path are presented in Fig. 2.31. The solid and dashed lines represent neutron and proton shapes, respectively. One can see that in the hyperdeformed minimum, which is located around q_{20}=200 b, the fission fragments are already well formed. The heavier fragment is almost spherical, while the lighter one has a significant prolate deformation. More detailed studies of the density distributions in both fragments reveal [305] that the heavier one corresponds to the isotope ^{132}Sn, while the light fragment is very close to ^{92}Zr. The remaining neutrons are mainly located in the neck region.

Fission isomers

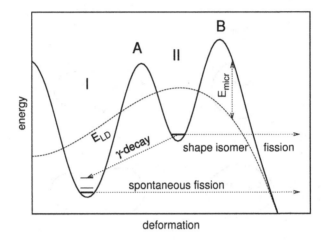

deformation

Fig. 2.32. Schematic plot of the double-humped fission barrier as function of the elongation. The ground state (I), first barrier (A), second minimum (II), and the second barrier (B) are marked. The dashed line corresponds to the macroscopic (LD-type) energy, while the solid line represents the whole energy. The difference between the both energies is equal to the microscopic, shell+pairing correction-energy.

If the potential pocket, corresponding to the secondary minimum, is deeper than about 1 MeV, the nucleus may be caught there. Traditionally states in the ground-state minimum are called class I states, those in the second minimum class II states. Decay modes of shape-isomeric states are shown schematically in Fig. 2.32. These states undergo either fission by tunneling through the outer barrier or decay by γ transitions into class I states [306]. Lifetimes for fission decay of shape-isomers are typically many orders of magnitude shorter than for spontaneous fission from the ground state because of the considerably larger barrier for the latter decay. At present around 40 shape-isomers between ^{232m}Th and ^{245m}Bk are found. Their fission half-lives lie between 14 ms and 6 ps.

Specht, Weber, Konecny, and Heunemann reported 1972 the first measurement of a rotational band in ^{240m}Pu by conversion-electron spectroscopy [307]. The moment of inertia corresponding to this band was more than two times larger than its value in the ground-state band. This was an indirect indication of the large deformation of fission isomers. Calculations of the moment of inertia by Sobiczewski, Bjørnholm, and Pomorski [308] in the framework of the cranking model with pairing confirmed this interpretation. In several uranium and plutonium isotopes rotational bands above the shape-isomeric state have been observed since by γ and by conversion-electron spectroscopy [302, 309].

Quadrupole deformations of shape isomeric states have been determined with optical spectroscopy and life-time measurements of the transitions in the rotational bands [309, 310]. They are in good agreement with calculations by Nerlo-Pomorska using the macroscopic-microscopic approach [311].

2.4.4 Effect of an odd-particle on the fission barrier

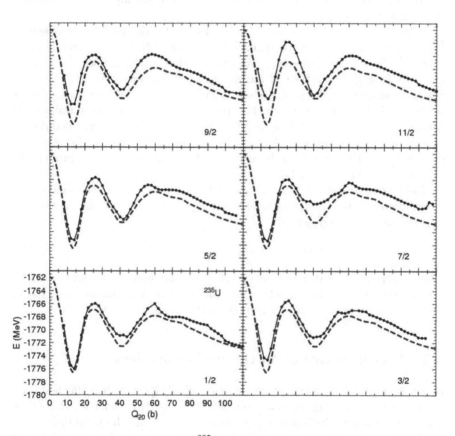

Fig. 2.33. The fission barriers for ^{235}U evaluated for different values of Ω (solid lines with points) in comparison with the corresponding barrier (dashed line) of an even-even system. The values of Ω of the odd particle are shown in the lower-right corners (private communication by L. M. Robledo).

The projection Ω of the angular momentum on the fission axis of the odd particle in an odd nucleus is a constant of motion along the fission path if the single-particle potential remains axially symmetric during fission. Since the quantum number Ω of the ground-state orbital of the odd particle is in general not the same as in the lowest single-particle state above the fission barriers of the neighboring even nuclei, the latter are typically lower than the barrier of the odd nucleus. This extra barrier energy of odd nuclei is called specialization energy [313]. A similar argument holds for the conservation of the parity of the odd particle as long as the nuclear shape is reflection-symmetric. However, since violations of the axial or reflection symmetry occur often along the fission path, a reduction of the actual specialization energy must be expected, compared to the case of a fully symmetric path. In addition, the blocking effect of the odd nucleon reduces the pairing energy compared to neighboring even nuclei.

The effect of an odd neutron on the fission barrier of ^{235}U is shown in Fig. 2.33 for various assumed quantum numbers Ω (the actual ground-state spin of ^{235}Pu is $1/2$). The energy was evaluated in the HFB frame [312] using the D1S Gogny interaction. For comparison the mean HFB energy of the neighboring even nuclei is also shown. The difference between both curves is the specialization energy, which corresponds roughly to the quasiparticle energy of the odd particle. It is seen that the specialization energy is different for different orbitals occupied by the odd neutron and varies significantly with deformation. The influence of this effect on the spontaneous fission probability will be discussed in Sec. 5.1.4.

Notice that the effect of an odd particle is not additive. In odd-odd nuclei only the sum of the angular momenta of the two odd nucleons and the product of their parities are conserved

$$\Omega = \Omega_p + \Omega_n \ , \qquad \pi = \pi_p \cdot \pi_n \tag{2.229}$$

This constraint is weaker than conserving angular momentum and parity of each particle separately. Therefore the modification of the fission barrier is in this case comparable to that of a single odd particle.

2.5 Effect of rotation on the fission barriers

It was already shown in the 1970's (cf. e.g. Ref. [314]) that shell effects do not disappear with growing angular momentum of a rotating nucleus. In fact, at high spin even new shells can appear. In order to obtain the potential energy of the rotating nucleus in mean-field approximation one has to satisfy the constraint

$$\langle \Psi | \hat{J}_x | \Psi \rangle = \sqrt{I(I+1)} \, . \tag{2.230}$$

In this equation the wave function $|\Psi\rangle$ is either $|\mathrm{HF}\rangle$ or $|\mathrm{HFB}\rangle$ and $\hat{J}_x$ is the sum over the x components of the single-particle angular-momentum operators. It is assumed that the collective rotation is around the x-axis, for which the moment of inertia shall be the largest and I is the total spin. Therefore an additional Lagrange term $-\omega \hat{J}_x$ has to be added in the energy-minimization conditions, Eqs. (2.9) or (2.26) in the HF or HFB scheme, respectively.

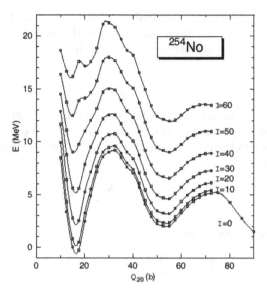

Fig. 2.34. Self-consistent fission barriers of ^{254}No, evaluated for various angular momenta using the Gogny potential D1S (from Ref. [315]).

An example of fission barriers evaluated selfconsistently for ^{254}No for a sequence of angular momenta is presented in Fig. 2.34 as function of the mass quadrupole moment Q_{20}. The curves were obtained by Egido and Robledo [315], using the Gogny potential D1S in a constrained HFB calculation. (No energy corrections for symmetry-violating HFB wave-functions are considered in Fig. 2.34.) It is seen that the double-humped barrier survives up to high angular momenta.

To formulate the macroscopic-microscopic approach for rotating nuclei [316], we consider the single-particle Schrödinger equation with the cranked shell-model Hamiltonian $\hat{h}' = \hat{h}^{\text{shell}} - \omega \hat{j}_x$

$$(\hat{h}^{\text{shell}} - \omega \hat{j}_x)|i\rangle = e_i^\omega |i\rangle , \qquad (2.231)$$

where e_i^ω and $|i\rangle$ are the eigenvalues and eigenstates in the rotating frame.

The shell-model energy of a rotating nucleus becomes

$$E(I_x) = \sum_i n_i^\omega e_i^\omega + \omega I_x , \qquad (2.232)$$

where the summation goes over all single-particle states and the angular momentum I_x is equal to

$$I_x = \sum_i n_i^\omega \langle i|\hat{j}_x|i\rangle . \qquad (2.233)$$

The occupation numbers n_i^ω in formulae (2.232) and (2.233) are 1 for the states with the Z (or N) lowest eigenvalues e_i^ω and zero else. The shell-model moment of inertia $\mathcal{J}_{\mathrm{SM}}$ is defined in analogy to classical mechanics by

$$\mathcal{J}_{\mathrm{SM}} = \frac{I_x}{\omega} . \tag{2.234}$$

To obtain the parameters of the reference nucleus in the Strutinsky procedure, one introduces averaged occupation numbers $\tilde{n}_i^{\tilde{\omega}}$ by Eq. (2.188) using ϵ_i^ω instead of ϵ_i. The two Lagrange parameters $\tilde{\lambda}$ and $\tilde{\omega}$ are determined implicitly by the two constraints

$$\sum_i \tilde{n}_i^{\tilde{\omega}} = Z(\text{ or } N) \quad \text{and} \quad \sum_i \tilde{n}_i^{\tilde{\omega}} \langle i|\hat{j}_x|i\rangle = I_x . \tag{2.235}$$

The energy of the reference nucleus is then

$$\widetilde{E}(I_x) = \sum_i \tilde{n}_i^{\tilde{\omega}} e_i^{\tilde{\omega}} + \tilde{\omega} I_x . \tag{2.236}$$

The smoothed moment of inertia is defined in analogy to Eq. (2.234) as

$$\widetilde{\mathcal{J}}_{\mathrm{SM}} = \frac{I_x}{\tilde{\omega}} . \tag{2.237}$$

One can show that for a realistic single-particle potential the smoothed moment of inertia is close to the moment of inertia of a rigid body [314] which is given by

$$\mathcal{J}_{\mathrm{rig}} = \frac{3AM_{\mathrm{nucl}}}{4\pi R_0^3} \int_V (y^2 + z^2)\, d^3r = \mathcal{J}_{\mathrm{sph}} \cdot B_{\mathrm{rot}}(\mathrm{def}) , \tag{2.238}$$

where

$$\mathcal{J}_{\mathrm{sph}} = \frac{2}{5} A M_{\mathrm{nucl}} R_0^2 = \frac{2}{5} M_{\mathrm{nucl}} r_0^2 A^{5/3}$$

is the rigid-body moment of inertia of a spherical nucleus and B_{rot} is a geometrical factor which depends on the deformation of the nucleus.

The shell-correction energy of a rotating nucleus is defined as

$$\delta E_{\mathrm{shell}}(I) = E(I) - \widetilde{E}(I) \tag{2.239}$$

and the total energy becomes in the macroscopic-microscopic approach

$$E_{\mathrm{tot}}(I, \mathrm{def}) = \left\{ E_{\mathrm{mac}}(\mathrm{def}) + \frac{I(I+1)}{2\mathcal{J}_{\mathrm{rig}}(\mathrm{def})} \right\} + \delta E_{\mathrm{shell}}(I, \mathrm{def}) . \tag{2.240}$$

As an example for the potential-energy surface of a rotating nucleus we show in Fig. 2.35 the energy of the superheavy nucleus ^{298}114. The liquid-drop model with parameters from Ref. [317] was used for the macroscopic part

240

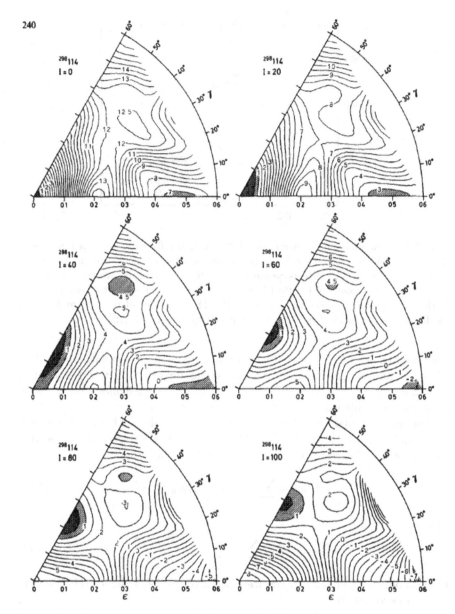

Fig. 2.35. The potential energy surfaces of the $^{298}114$ nucleus, projected on the (ε, γ) plane. The plots are made for $I=0$, 20, 40, 60, 80, and 100 in units of $\hbar$ (from Ref. [314]).

and in the Strutinsky shell-correction the Nilsson potential with deformation parameters ε and γ was employed. For their definition we refer to Eqs. (2.211) and (1.11), respectively. Fig. 2.35 shows that also a superheavy nucleus follows

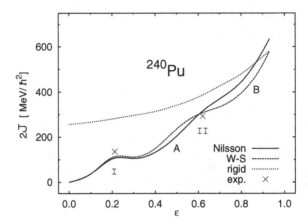

Fig. 2.36. The moment of inertia of ^{240}Pu along the fission path as a function of the quadrupole deformation ε. The microscopic moments (2.243) obtained with the Nilsson (solid line) and Woods-Saxon (dashed lines) potentials are compared with the rigid body moment of inertia (dotted line) and the experimental data (crosses) in the first (I) and the second minimum (II). The positions of the first and second saddle points are marked by A and B respectively (after Ref. [295]).

the general rule of Ref. [44] that with increasing angular momentum a nucleus becomes increasingly deformed in the oblate direction ($\gamma = 60^0$). Even a superheavy nucleus, whose fission barrier comes from shell effects only, is seen to be rather stable with respect to collective rotation.

For small angular velocities ω the eigenvalue problem of the single-particle Hamiltonian in a rotating frame (2.240) can be solved in second order perturbation theory. The result is Inglis' cranking formula [318]

$$E(\omega) = \sum_i n_i\, e_i + \frac{1}{2} \mathcal{J}_{\text{crank}}\, \omega^2 \ , \tag{2.241}$$

where

$$\mathcal{J}_{\text{crank}} = 2\hbar^2 \sum_{j>i_{\text{Fermi}}} \sum_{i\leq i_{\text{Fermi}}} \frac{|\langle i|\hat{j}_x|j\rangle|^2}{e_j - e_i} \tag{2.242}$$

is the moment of inertia. In the framework of the BCS theory the cranking formula becomes [97]

$$\mathcal{J}_{\text{crank}} = 2\hbar^2 \sum_{i,j>0} \frac{|\langle i|\hat{j}_x|j\rangle|^2}{E_j + E_i}(u_j v_i - u_i v_j)^2 \ , \tag{2.243}$$

where E_j and E_i are the quasiparticles energies (2.51), u_j and v_i the BCS occupation coefficients (2.46). The cranking moment of inertia, evaluated along the fission path, for ^{240}Pu with Nilsson and with Woods-Saxon potentials is compared in Fig. 2.36 with the rigid moment of inertia (2.238). It is seen that for large elongation, corresponding to the second saddle (B), both macroscopic-microscopic calculations yield results close to the rigid-body moment of inertia.

3

The Nuclear Level Density

To describe the compound state of a nucleus with proton number Z, neutron number N, excitation energy E^*, angular momentum J, and deformation $\{\beta_n\}$ one needs, besides its energy $E_0(Z, N, J, \beta)$ at zero temperature, the level density $\rho(Z, N, E^*, J, \beta)$. The latter is the derivative of the phase-space volume $\Phi(Z, N, E^*, J, \beta)$, measured in units of $h^{3(N+Z)}$, with respect to E^*

$$\rho = \frac{1}{h^{3(N+Z)}} \frac{d\Phi}{dE^*} .$$

Since the entropy S is related to this quantity by $S = \ln[\Phi(E^*)/h^{3(N+Z)}]$ (cf. §7 of Landau and Lifshitz' *Statistical Physics* [319]), one obtains the relation

$$\rho = e^S \frac{dS}{dE^*} = \frac{1}{T} e^S$$

between entropy and level density, where T is the temperature. Following common practice in nuclear physics, we will measure the temperature T in energy units. In terms of the conventional definition of the temperature T_{conv} in units of Kelvin one has $T = k_B T_{\text{conv}}$ with the Boltzmann constant k_B. From E_0 and S the total energy $E = E_0 + E^*$ and the free energy $F = E - ST$ are obtained. The latter is an important input for dynamical calculations to be discussed in Chap. 5.

The level density is usually calculated for an independent-particle model to which corrections for pairing correlations and collective vibrational and rotational states are later added. Only moderate excitation energies are considered in the following such that the compound state can be treated as a statistical equilibrium, which means that emission processes should happen on a much slower time scale than typical equilibration times within the nucleus.

From shell-model single-particle states level densities can be generated numerically by adding to the energies of particle-hole states in a certain energy interval all energies of two-particle two-hole states, three-particle three-hole states and so on that fall into the same energy interval. Powerful algorithms

H. J. Krappe and K. Pomorski, *Theory of Nuclear Fission*,
Lecture Notes in Physics 838, DOI: 10.1007/978-3-642-23515-3_3,
© Springer-Verlag Berlin Heidelberg 2012

have been developed to cope with the exponential increase of the level density with increasing excitation energy E^* [320, 321].

More popular for practical applications are analytic expressions for the level density. Their derivation for a system of N fermions starts from the partition sum of the grand canonical ensemble with temperature $T = 1/\beta$ and chemical potential μ (cf. Ref. [319], §35)

$$e^{-\beta\Omega(\mu,\beta)} = \sum_N \sum_i e^{-(E_i(N)-\mu N)\beta}, \tag{3.1}$$

were the second sum runs over all eigenvalues E_i of the N particle system. From the thermodynamic potential Ω the mean values of the particle number N, the energy E, and the entropy S are obtained by (cf. Ref. [319], §24)

$$N = -\left.\frac{\partial\Omega}{\partial\mu}\right|_\beta$$

$$E = \left.\frac{\partial(\beta\Omega)}{\partial\beta}\right|_\mu - \mu\left.\frac{\partial\Omega}{\partial\mu}\right|_\beta = \left.\frac{\partial(\beta\Omega)}{\partial\beta}\right|_{\alpha=\mu\beta} \tag{3.2}$$

$$S = \beta^2 \left.\frac{\partial\Omega}{\partial\beta}\right|_\mu .$$

From these equations follows trivially

$$\beta\Omega = \beta E - S - \mu\beta N. \tag{3.3}$$

The partition sum may be rewritten in terms of the level density $\rho(N,E)$ of a system with N particles and energy E

$$e^{-\beta\Omega} \approx \int dN \int dE\, \rho(N,E) e^{-(E-\mu N)\beta}, \tag{3.4}$$

where the level density is assumed to be a smooth function, averaged in N over one unit and in E over a typical level spacing δE. Equation (3.4) is therefore only valid for $N \gg 1$ and $E \gg \delta E$.

For an ideal Fermi gas the thermodynamic potential Ω can be expressed in terms of the single-particle level density $g(\epsilon)$ (cf. Ref. [319], §53)

$$\beta\Omega = -\sum_\nu \ln(1 + e^{(\mu-\epsilon_\nu)\beta}) \approx -\int g(\epsilon)\ln(1 + e^{(\mu-\epsilon)\beta})d\epsilon. \tag{3.5}$$

Inserting this result into Eq. (3.4), the level density $\rho(N,E)$ is obtained by inverting the double Laplace transform (3.4).

3.1 The Thomas-Fermi model

In Thomas-Fermi approximation the momentum distribution is assumed to be locally isotropic and the number of neutron or proton states in the volume element $d\tau$ is

$$dN = 2 \cdot \frac{4\pi}{3} p^3(\mathbf{r}) h^{-3} d\tau = \frac{p^3(\mathbf{r})}{3\pi^2 \hbar^3} d\tau, \tag{3.6}$$

where the contribution of the spin degree of freedom is included. In a leptodermous potential $V(\mathbf{r})$ we have $p^2 = 2M^*_{\text{nucl}}(\epsilon - V(\mathbf{r}))$, where the energy zero-point shall lie on the bottom of the potential and M^*_{nucl} is the effective mass of a nucleon in the interacting system. Using the result (3.6) the local number of neutron or proton states per unit volume becomes

$$z(\epsilon, \mathbf{r}) = \partial_\epsilon \frac{dN}{d\tau} = \frac{1}{3\pi^2} \left(\frac{2M^*_{\text{nucl}}}{\hbar^2} \right)^{3/2} \frac{d}{d\epsilon} (\epsilon - V(\mathbf{r}))^{3/2}.$$

With the abbreviation $C = (2M^*_{\text{nucl}} \hbar^{-2})^{3/2}/(2\pi^2)$ one has $z(\epsilon, \mathbf{r}) = C\sqrt{\epsilon - V(\mathbf{r})}$. In terms of this quantity the local density of the thermodynamic potential (3.5) is

$$\beta w(\mu, \beta; \mathbf{r}) = -\int z(\epsilon, \mathbf{r}) \ln(1 + e^{(\mu - \epsilon)\beta}) d\epsilon$$

and after partial integration,

$$w(\mu, \beta; \mathbf{r}) = -\frac{2}{3} C \int \frac{(\epsilon')^{3/2}}{e^{(\epsilon' - \mu')\beta} + 1} d\epsilon', \tag{3.7}$$

where $\epsilon'(\mathbf{r}) = \epsilon - V(\mathbf{r})$ and $\mu'(\mathbf{r}) = \mu - V(\mathbf{r})$ have been introduced. Applying the Sommerfeld expansion [185] (cf. Ref. [319], §58)

$$\int_0^\infty \frac{f(\epsilon)}{e^{(\epsilon - \mu)\beta} + 1} d\epsilon = \int_0^\mu f(\epsilon) d\epsilon + \frac{\pi^2}{6\beta^2} f'(\mu) + \mathcal{O}((\mu\beta)^{-4}) \tag{3.8}$$

to the integral in Eq. (3.7), one gets

$$w(\mu, \beta; \mathbf{r}) = -\frac{2}{3} C \left[\frac{2}{5} (\mu')^{5/2} + \frac{\pi^2}{4} \beta^{-2} (\mu')^{1/2} \right]. \tag{3.9}$$

The potential Ω is obtained by integrating $w(\mathbf{r})$ over that region of space where $\mu - V(\mathbf{r})$ is positive

$$\Omega(\mu, \beta) = \int w(\mu, \beta; \mathbf{r}) \Theta(\mu - V(\mathbf{r})) d\tau. \tag{3.10}$$

In order to write Ω as sum of a bulk and a surface term, $\Omega = \Omega^{(\text{bulk})} + \Omega^{(\text{surf})}$, we introduce a volume

$$v(\mu) = \int (1 - V(\mathbf{r})/\mu)^{3/2} \Theta(\mu - V(\mathbf{r})) d\tau \tag{3.11}$$

and parameters

$$q_\alpha(\mu) = \int [(1 - V(\mathbf{r})/\mu)^\alpha - (1 - V(\mathbf{r})/\mu)^{3/2}] \Theta(\mu - V(\mathbf{r})) d\tau \tag{3.12}$$

and obtain from Eqs. (3.9) and (3.10)

$$\Omega^{(\text{bulk})} = -\frac{4}{15}Cv\left(\mu^{5/2} + \frac{5}{8}\pi^2\mu^{1/2}\beta^{-2}\right), \tag{3.13}$$

$$\Omega^{(\text{surf})} = -\frac{4}{15}C\left(\mu^{5/2}q_{5/2} + \frac{5}{8}\pi^2\mu^{1/2}q_{1/2}\beta^{-2}\right). \tag{3.14}$$

The integrand in Eq. (3.12)

$$h_\alpha(\mu;\mathbf{r}) = [(1 - V(\mathbf{r})/\mu)^\alpha - (1 - V(\mathbf{r})/\mu)^{3/2}]\Theta(\mu - V(\mathbf{r}))$$

is a surface-peaked function. Introducing integration coordinates such that the z-coordinate has the direction of the normal to the nuclear surface, the other two being orthogonal surface coordinates, it is assumed that h_α depends only on the coordinate z. We write the volume element $d\tau = d\sigma(z)dz$ and expand $d\sigma(z) = (1 + \kappa z + \mathcal{O}(z^2))d\sigma(0)$, where κ is the local curvature of the surface [322]. With this expansion we get the (asymptotic) series

$$\int h_\alpha(z)d\tau = \int h_\alpha(z)\,dz \int d\sigma + \int h_\alpha(z)\,zdz \int \kappa\,d\sigma + \ldots, \tag{3.15}$$

where only the first two surface moments of h_α are kept. To determine the zero-point on the z axis we first consider spherical nuclei for which the equivalent sharp radius R_{eq} is defined by $(4\pi/3)R_{\text{eq}}^3 = v$ and we take $z = r - R_{\text{eq}}$. When the sphere is deformed (with constant saturation density $\rho(0)$ and particle number N) the equivalent sharp surface transforms into the new nonspherical equivalent surface on which the z-coordinate is assumed to be orthogonal and on which it has its origin.

Instead of using the moment expansion (3.15), the surface-peaked function $h_\alpha(\mu;\mathbf{r})$ can be represented by a folding product in analogy to the Yukawa-plus-exponential ansatz for the liquid-drop energy in Sec. 2.2.2. In this case the shape-dependence is characterized by a diffuseness parameter a and a generalized surface constant [323].

Using relations (3.2) one obtains for the mean values of the extensive variables N, E, and S from Eqs. (3.13) and (3.14) up to terms of order $(\mu\beta)^{-2}$

$$N = \frac{2}{3}C\mu^{3/2}(v + q_{5/2}) + \frac{\pi^2}{12}C\mu^{-1/2}\beta^{-2}(v + q_{1/2}) \tag{3.16}$$

$$E = \frac{2}{5}C\mu^{5/2}(v + q_{5/2}) + \frac{\pi^2}{4}C\mu^{1/2}\beta^{-2}(v + q_{1/2}) \tag{3.17}$$

$$S = \frac{\pi^2}{3}C\mu^{1/2}\beta^{-1}(v + q_{1/2}), \tag{3.18}$$

where the weak dependence of v and q_α on μ has been neglected in taking derivatives with respect to μ.

The Fermi energy μ_0 (at zero temperature) is defined implicitly by

$$N = \frac{2}{3}C\mu_0^{3/2}(v(\mu_0) + q_{5/2}(\mu_0)). \tag{3.19}$$

Subtracting this equation from Eq. (3.16) yields

$$\frac{2}{3}C(\mu^{3/2} - \mu_0^{3/2})(v + q_{5/2}) + \frac{\pi^2}{12}C\mu^{-1/2}\beta^{-2}(v + q_{1/2}) = 0.$$

In the lowest nonvanishing order of $(\mu_0\beta)^{-2}$ the relation

$$\mu - \mu_0 = -\frac{\pi^2}{12}\mu_0^{-1}\beta^{-2}\frac{v + q_{1/2}}{v + q_{5/2}} \tag{3.20}$$

follows, where v and q_α are to be taken at $\mu = \mu_0$. The energy at zero temperature is $E_0 = (2/5)C\mu_0^{5/2}(v + q_{5/2})$ and the excitation energy $E^* = E - E_0$ becomes, using Eqs. (3.17) and (3.20), again to leading order in $(\mu_0\beta)^{-1}$,

$$E^* = \frac{\pi^2}{6}C\mu_0^{1/2}\beta^{-2}(v + q_{1/2}) := a_N\beta^{-2}. \tag{3.21}$$

The expression

$$a_N = \frac{\pi^2}{6}C\mu_0^{1/2}(v + q_{1/2}) \tag{3.22}$$

is seen to consist of a volume term, proportional to v, and a surface and curvature term, proportional to $q_{1/2}$. (The result (3.22) does actually not depend on the approximation of v and q_α being independent of μ as can be shown by a straightforward, but somewhat lengthy calculation). Since the Thomas-Fermi approximation without gradient correction terms is not quantitatively reliable in the nuclear surface, it is not advisable to use Eqs. (3.11) and (3.12) to calculate v and $q_{1/2}$ in Eq. (3.22) from a given shell-model potential $V(\mathbf{r})$. Instead, only the form of the leptodermous expansion of a_N is used and the numerical values of the volume, surface, and curvature terms are determined empirically by fitting data as will be discussed in Sec. 3.3.7.

Eliminating β between Eqs. (3.18) and (3.21), yields

$$S = 2\sqrt{a_N E^*}.$$

The phase-space volume is $e^S = e^{2\sqrt{a_N E^*}}$ and therefore the level density $\rho(N, E)$ becomes

$$\rho(N, E) = \frac{\partial e^S}{\partial E^*} = \sqrt{\frac{a_N}{E^*}}e^{2\sqrt{a_N E^*}}. \tag{3.23}$$

The extensive quantities N, E, and S on the left-hand side of Eqs. (3.2) and (3.16)-(3.18) are average values in a grand-canonical ensemble. The level density ρ and its arguments N and E^*, however, refer to a single nucleus. The two sets of extensive quantities coincide only in the thermodynamic limit since their variances in the grand ensemble vanish in that limit. Bethe noticed [324] that for finite systems the microcanonical expression for $S(N, E^*)$

differs from the grand-canonical average of Eq. (3.20) by correction terms of the order $\ln aE^*$. Their neglect is in general an acceptable approximation for the thermodynamical potentials themselves. However, in Eq. (3.23) it would amount to the neglect of factors of powers of N and E^* which is less acceptable. Therefore Bethe gave the following derivation of the level-density formula retaining also terms of order $\ln N$ and $\ln E$ in the calculation of the entropy S. For the level density the ansatz

$$\rho(N, E) = \lambda(N, E^*)e^{S(N,E^*)}$$

is made with a correction function λ varying slowly with N and E. For S the approximate expression (3.18) is used. Inserting into Eq. (3.4) and making use of Eq. (3.3) yields

$$1 = \int dN' \int dE' \lambda(N', E')e^{f(N,N',E,E')} \tag{3.24}$$

with $f = \beta\mu(N'-N) - \beta(E'-E) + S(\beta', \mu') - S(\beta, \mu)$. Using Eqs. (3.16)-(3.20) Bethe showed that up to quadratic order in $(\mu_0\beta)^{-1}$ the exponent f can be rewritten as a quadratic function in $N - N'$ and $E^* - E^{*'}$ [324]

$$f = -\frac{\pi}{2}\left(\frac{N}{\mu_0 E^*}\right)^{1/2}\left[\frac{\mu_0}{3N}(N'-N)^2 + \frac{1}{4}\frac{(E^{*'}-E^*)^2}{E^*}\right].$$

Since f is sharply peaked at $N = N'$, $E^* = E^{*'}$, λ is taken out of the integral in Eq. (3.24) and the remaining Gauss integral is evaluated. One obtains $1 \approx \lambda(E^*)\sqrt{48E^*}$ and

$$\rho(N, E^*) = \frac{e^{2\sqrt{a_N E^*}}}{\sqrt{48E^*}}. \tag{3.25}$$

To account for protons as well as neutrons, one has to fold the level density for the protons into that of the neutrons

$$\rho(N, Z, E^*) = \int \frac{\exp(2\sqrt{a_N}[\sqrt{E_1^*} + \sqrt{E^* - E_1^*}])}{48E_1^*(E^* - E_1^*)}dE_1^*. \tag{3.26}$$

where it is assumed that the a parameter is equal for protons and neutrons. As long as the Fermi energies for both types of nucleons are not very different, this is a reasonable approximation. Otherwise the v and q_α parameters would be different for protons and neutrons. Expanding the exponent around its maximum at $E_1^* = E^*/2$

$$\sqrt{E_1^*} + \sqrt{E^* - E_1^*} \approx \sqrt{2E^*} - 2^{-1/2}(E^*)^{-3/2}(E_1^* - E^*/2)^2 + \cdots,$$

and taking the denominator out of the integral in Eq. (3.26) at $E_1^* = E^*/2$, the combined level density becomes

$$\rho(N, Z, E^*) = \frac{e^{2\sqrt{2a_N E^*}}}{12 E^{*2}} \int \exp\left\{-\sqrt{2a_N}(E^*)^{-3/2}\left(E_1^* - \frac{E^*}{2}\right)^2\right\} dE_1^*$$

$$= \frac{\sqrt{\pi}}{12} \frac{e^{2\sqrt{2a_N E^*}}}{(2a_N E^*)^{1/4} E^*}. \tag{3.27}$$

The quantity

$$a = 2a_N = \frac{\pi^2}{3} C \mu_0^{1/2}(v + q_{1/2})$$

is the usual level-density parameter, accounting for spin and isospin degeneracies. This equation can be rewritten

$$a = \frac{\pi^2}{6} g_{\text{total}}(\mu_0), \tag{3.28}$$

where $g_{\text{total}}(\epsilon) = 2 \int z(\epsilon, \mathbf{r}) d\tau$ is the combined single-particle level-density for protons and neutrons. Ericson [325] suggested an alternative way to obtain this result by applying the Laplace-transform inversion-formula several times to Eq. (3.4). However, assumptions about the behavior of the partition sum $\beta\Omega(\mu, \beta)$ for complex μ and β have to be made in this approach which need special justification [326], [327].

3.2 The picket-fence model

Although use of the specific form of the Thomas-Fermi single-particle level-density $g(\epsilon)$ was made to derive the result (3.27), only its value at the Fermi energy $g(\mu_0)$ enters Eqs. (3.27) and (3.28). One may therefore wonder whether another functional form of the single-particle level density with the same value at $\epsilon = \mu_0$ would lead to the same result. This can in fact be shown for a constant level density, the "picket-fence model",

$$g(\epsilon) = 1/\epsilon_0 = (6/\pi^2)a_N, \tag{3.29}$$

where, for simplicity, only neutrons are considered.

It is convenient to start in this case from the generating function for the canonical partition sum $Z(\beta, N)$ [326]

$$G(y) = \prod_{n=1}^{\infty}(1 + e^{-\beta\epsilon_n}y) = \sum_N Z(\beta, N)y^N \tag{3.30}$$

with single-particle energies ϵ_n. In the picket-fence model $\epsilon_n = n\epsilon_0$. Introducing the abbreviation $b = \exp(-\beta\epsilon_0)$ one obtains from $G(y) = \prod(1 + b^n y)$ the functional equation $G(y) = (1+by)G(by)$. Inserting the power series expansion in y (3.30) on both sides of this relation and equating coefficients of the same power in y, yields the recurrence relation $Z(N) = b^N(1 - b^N)^{-1}Z(N - 1)$. With the initial value $Z(\beta, 0) = 1$ one obtains

$$Z(\beta, N) = \frac{b^{N(N+1)/2}}{\prod_{n=1}^{N}(1 - b^n)} \tag{3.31}$$

and the free energy becomes

$$F(\beta) = -\beta^{-1} \ln Z = N(N+1)(\epsilon_0/2) + \beta^{-1} \sum_{n=1}^{N} \ln(1 - e^{-n\beta\epsilon_0}).$$

The denominator in Eq. (3.31) is the generator for the partition numbers $p_N(n)$

$$\frac{1}{\prod_{n=1}^{N}(1 - b^n)} = \sum_{n=0}^{\infty} p_N(n) b^n. \tag{3.32}$$

As shown in the elementary theory of numbers, $p_N(n)$ is the number of partitions of n into at most N integers $j \geq 0$ with repetition: $n = \sum_{j=1}^{N} n_j j$ with n_j counting the number of repetitions of j (including 0) [328]. In terms of the level density $\rho(N, E)$ the canonical partition sum is given by

$$Z(\beta, N) = \sum_i e^{-\beta E_i} \approx \int \rho(N, E) e^{-\beta E} dE. \tag{3.33}$$

From Eqs. (3.31) and (3.32) the partition sum $Z(\beta, N)$ is known analytically in terms of the partition numbers $p_N(n)$ and the Laplace transformation (3.33) can be inverted

$$\rho(N, E) = \frac{1}{2\pi i} \int_{\epsilon-i\infty}^{\epsilon+i\infty} Z(\beta, N) e^{\beta E} d\beta$$

$$= \frac{1}{2\pi i} \sum_n p_N(n) \int_{\epsilon-i\infty}^{\epsilon+i\infty} \exp\{[E - N(N+1)(\epsilon_0/2)]\beta - n\epsilon_0\beta\} d\beta,$$

and therefore

$$\rho(N, E^*) = \sum_{n=0}^{\infty} p_N(n) \delta(E^* - n\epsilon_0), \tag{3.34}$$

where the excitation energy $E^* = E - (N+1)(N\epsilon_0/2) = E - F(T = 0)$ has been introduced.

In order to get a smooth function $\rho(N; E^*)$, we consider the thermodynamic limit of the picket-fence model: $N \to \infty$, $\epsilon_0 = \mu_0/N$, and μ_0 fixed. With Eq. (3.29) one has

$$N/\mu_0 = 6a_N/\pi^2. \tag{3.35}$$

The level density (3.34) may be written in this limit

$$\rho(E^*) dE^* = \lim_{N \to \infty} \sum_{n=NE^*/\mu_0}^{N(E^*+dE^*)/\mu_0} p_N(n) = \lim_{N \to \infty} p_N(NE^*/\mu_0) N \frac{dE^*}{\mu_0}. \tag{3.36}$$

To take the limit, we use the asymptotic expression

$$p_N(n) \simeq \frac{1}{\sqrt{48n}} \exp\left(\pi\sqrt{2n/3}\right)$$

for the partition numbers $p_N(n)$ derived by Auluck and Kothari [329], valid for $N \gg n^{1/2}$. We insert $n = NE^*/\mu_0$; with Eq. (3.35) this is equal to $(6/\pi^2)a_N E^*$, and we obtain

$$\rho(N, E^*) \simeq \frac{1}{\sqrt{48E^*}} \exp 2\sqrt{a_N E^*}$$

in agreement with Eq. (3.25). This formula is valid for $a_N E^* \gg 1$. As discussed in Ref. [330] and [331] its range of validity can be extended to somewhat smaller excitation energies by adding the correction factor $(1 - [4aE^*]^{-1/2})$ to the result (3.25). This agrees with a variant of formula (3.25) derived by a different method [332]

$$\rho(N, E^*) \simeq \frac{e^{2\sqrt{a_N E^*}}}{\sqrt{48}(E^* + t/2)}$$

with the equation of state

$$E^* = a_N t^2 - t/2. \tag{3.37}$$

Note that the relation between the temperature $T = \beta^{-1}$ and the grand-canonical average of the excitation energy E^* is given by Eq. (3.21). However, in Eq. (3.37) E^* is meant to be the exact excitation energy of the micro-canonical ensemble. The quantity t in Eq. (3.37) appears in connection with asymptotic expansions, valid in the limit $a_N E^* \gg 1$. In this limit the mean value of the excitation energy and the microcanonical E^* become equal and $t \to T$ since

$$(a_N T)^2 = a_N E^* = a_N t\,(a_N t - 1/2) \to (at)^2. \tag{3.38}$$

More recently Ansaldo-Meneses introduced methods from the analytical theory of numbers to derive asymptotic expressions not only for the level density in the picket-fence model, but also for single-particle or quasiparticle spectra in some infinitely high potentials, like the infinite square well and the oscillator potentials [333, 334]. The same was earlier achieved by Ignatyuk et al. [335] by an analytic smoothing of the discrete single-particle level-density for large quantum numbers of a square well and a harmonic oscillator. Also for a Woods-Saxon potential a numerical result for the level density was reported by these authors.

For later use we also note the level density $\omega(p, h; E^*)$ of states with h holes and p particles in the picket-fence model, derived by Williams [336] in the limit $\epsilon_0 \beta \ll 2\pi/\max(p, h)$

$$\omega(p, h; E^*) = \frac{[E^*/\epsilon_0 - A(p, h)]^{p+h-1}}{\epsilon_0\, p! h! (p + h - 1)!} \tag{3.39}$$

with $A(p, h) = (p^2 + h^2)/4 + (p - h)/4 - h/2$. An extension of this expression to neutron and proton particle-hole excitations has been given by Fu [337]. The dependence of the particle-hole level-density on angular momentum and corrections for shell and pairing effects are discussed by Ignatyuk [338].

3.3 Angular momentum and odd-even dependence of the level density

As shown by Bloch [339] (cf. also Ref. [340]) a spherical nucleus with angular momentum J has in semiclassical approximation the rotational energy

$$E_{\rm rot}(J) = \hbar^2 J^2 / (2\mathcal{J}_{\rm TF})$$

with the rigid moment of inertia $\mathcal{J}_{\rm TF} = (2/5)M_{\rm nucl}AR^2$, where $M_{\rm nucl}$ is the mass of a nucleon and R is the equivalent sharp radius (disregarding surface-diffuseness effects). Spin degrees of freedom are neglected. This Thomas-Fermi expression for $E_{\rm rot}$ should not be confused with the rotational energy of the low-lying rotational states of deformed nuclei. Their effect on the level density will be discussed in a later section.

For a nucleus with excitation energy E^* (reckoned from the $J = 0$ ground state) and angular momentum J only the energy $E^* - E_{\rm rot}$ is available for "thermalization" since $E_{\rm rot}$ goes into ordered motion. The level density for a nucleus with excitation energy E^* and classical angular momentum vector $\mathbf{J}$ may therefore be written

$$\rho_{\rm class}(N, Z, E^*, \mathbf{J}) = c\rho(N, Z, E^* - E_{\rm rot}(\mathbf{J})) \qquad (3.40)$$

with the expression (3.27) for $\rho(N, Z, E^*)$ and $a = 2a_N$. The factor c follows from the requirement

$$\rho(N, Z, E^*) = \int \rho_{\rm class}(N, Z, E^*, \mathbf{J}) \, d^3J = c \int \rho(N, Z, E^* - E_{\rm rot}) \, d^3J. \quad (3.41)$$

In the limit $E^* \gg E_{\rm rot}$ the integral on the right-hand side can be evaluated by expanding the exponent of ρ

$$2\sqrt{a(E^* - E_{\rm rot})} \approx 2\sqrt{aE^*}(1 - (1/2)(E_{\rm rot}/E^*)) = 2\sqrt{aE^*} - E_{\rm rot}\beta, \quad (3.42)$$

(where $E^* = a\beta^{-2}$ was used), and neglecting in the denominator $E_{\rm rot}$ compared to E^*. In Cartesian coordinates one has

$$\frac{1}{c} = \int e^{-J^2/(2\sigma^2)} d^3J = (2\pi\sigma^2)^{3/2}, \qquad (3.43)$$

where the abbreviation

$$\sigma^2 = \frac{\mathcal{J}_{\rm TF}}{\hbar^2}\sqrt{\frac{E^*}{a}} \qquad (3.44)$$

was used. Conventionally the level density $\rho(N, Z, E^*, J)$ of a nucleus with nucleon numbers N and Z, excitation energy E^*, and angular momentum J is defined without counting the $(2J + 1)$-fold degeneracy with respect to the magnetic quantum number M. With the volume element $d^3 J = 4\pi J^2 \, dJ$ one therefore obtains

$$(2J + 1)\rho(N, Z, E^*, J) \, dJ = 4\pi J^2 \rho_{\text{class}}(N, Z, E^*, \mathbf{J}) \, dJ. \tag{3.45}$$

Replacing the classical angular-momentum expression $4J^2$ on the right-hand side of this equation by the quantal $(2J + 1)^2$, we finally get from Eqs. (3.27) and (3.41)-(3.45)

$$\rho(N, Z, E^*, J) = (2J + 1) \left(\frac{a\hbar^2}{2\mathcal{J}_{\text{TF}}} \right)^{3/2} \frac{e^{2\sqrt{a[E^* - \hbar^2 J^2/(2\mathcal{J}_{\text{TF}})]}}}{12a[E^* - \hbar^2 J^2/(2\mathcal{J}_{\text{TF}})]^2}. \tag{3.46}$$

This asymptotic equation is valid if $a(E^* - E_{\text{rot}}) \gg 1$. Since $\mathcal{J}_{\text{TF}}$ is only equal to the moment of inertia of rigid rotation in a rather rough, semiclassical approximation, this parameter is often just fitted to experimental data.

For too large excitation energies where a significant amount of particles are excited into the continuum, neither the picket-fence model, nor the Fermi gas model are adequate and formula (3.46) should not be used [341]. In such cases one has however to worry whether a description in terms of equilibrium thermodynamics makes sense.

To account for the significant differences of the level density of even and neighboring odd nuclei for low excitation energies, a special pair energy $-\Delta$ was associated with the last pair of nucleons in even nuclei, not contained in the single-particle Thomas-Fermi level-density $g(\epsilon)$. Its consideration in the partition function Ω leads to a "back-shifted" Bethe formula [342], where the excitation energy E^* in Eqs. (3.27) and (3.46) is replaced by $E^* - \Delta$. A convenient parametrization is $\Delta = 12\chi/\sqrt{A}$ MeV [343] with

$$\chi = \begin{cases} 0 & \text{for odd-odd,} \\ 1 & \text{for odd-even,} \\ 2 & \text{for even-even nuclei.} \end{cases} \tag{3.47}$$

Several variants of formula (3.46) are in use, which are derived in slightly different ways. In Ref. [343] the form

$$\rho(N, Z, E^*, J) = \frac{2J + 1}{2\sqrt{2\pi}\sigma^3} e^{-(J+1/2)^2/(2\sigma^2)} \rho(N, Z, E^* - \Delta) \tag{3.48}$$

was employed with σ from Eq. (3.44) and $\rho(N, Z, E^*)$ from Eq. (3.27). This form follows from formula (3.46) in the limit $E^* \gg E_{\text{rot}}$ using the expansion (3.42) in the exponent, neglecting E_{rot} in the denominator, and inserting the semiclassical expression $E_{\text{rot}}(J) = \hbar^2(J + 1/2)^2/(2\mathcal{J}_{\text{TF}})$.

The definition (3.1) is often generalized to describe systems with average neutron number N, proton number Z, angular momentum z-component M, and energy E [339]

$$e^{-\beta\Omega(\alpha_N,\alpha_Z,\alpha_M,\beta)} = \sum_i e^{\alpha_N N(i)+\alpha_Z Z(i)+\alpha_M M(i)-\beta E(i)} \qquad (3.49)$$

$$\approx \int \rho_M(N,Z,M,E)e^{\alpha_N N+\alpha_Z Z+\alpha_M M-\beta E}\,dN\,dZ\,dM\,dE. \qquad (3.50)$$

As usual, the chemical potentials for protons and neutrons μ_N and μ_Z are related to the α_i by $\alpha_N = \beta\mu_N$ and $\alpha_Z = \beta\mu_Z$. The sum in Eq. (3.49) is to be extended over all configurations i containing $N(i)$ neutrons, $Z(i)$ protons, angular momentum z-component $M(i)$, and energy $E(i)$. In the Fermi gas model the expression (3.5) generalizes to

$$\beta\Omega(\alpha_N,\alpha_P,\alpha_M,\beta) = -\sum_{\tau=N,Z}\sum_i \ln(1 + e^{\alpha_N+\alpha_Z+\alpha_M m_{\tau,i}-\beta\epsilon_{\tau,i}}). \qquad (3.51)$$

In terms of the mean occupation numbers $\overline{n}_{\tau,i} = [1+\exp(\beta\epsilon_{\tau,i}-\alpha_\tau-\alpha_M m_{\tau,i})]$ the equations of state become

$$N = \sum_i \overline{n}_{N,i}\,, \qquad\qquad Z = \sum_i \overline{n}_{Z,i}\,,$$

$$M = \sum_{\tau=N,Z}\sum_i m_{\tau,i}\overline{n}_{\tau,i}\,, \qquad E = \sum_{\tau=N,Z}\sum_i \epsilon_{\tau,i}\overline{n}_{\tau,i}\,,$$

$$S = \sum_{\tau=N,Z}\sum_i \{(\beta\epsilon_{\tau,i}-\alpha_\tau-\alpha_M m_{\tau,i})\overline{n}_{\tau,i} - \ln(1-\overline{n}_{\tau,i}). \qquad (3.52)$$

Inverting formally the Laplace integral in Eq. (3.50) yields the level density

$$\rho_M(N,Z,M,E) = \left(\frac{1}{2\pi i}\right)^4 \int_{c_0-i\infty}^{c_0+i\infty} d\beta \int_{c_1-i\infty}^{c_1+i\infty} d\alpha_M \int_{c_2-i\infty}^{c_2+i\infty} d\alpha_Z \int_{c_3-i\infty}^{c_3+i\infty} d\alpha_N$$

$$\times \exp\{-\alpha_N N - \alpha_Z Z - \alpha_M M + [E - \Omega(\alpha_N,\alpha_Z,\alpha_M,\beta)]\beta\} \qquad (3.53)$$

with $c_i > 0$, $i = 0,\ldots 3$. The inverse Laplace transformation is then evaluated in the saddle-point approximation [344]. The stationarity condition of the integrand in Eq. (3.53) yields the extensive variables

$$N = -\partial_{\alpha_N}(\beta\Omega), \quad Z = -\partial_{\alpha_Z}(\beta\Omega), \quad M = -\partial_{\alpha_M}(\beta\Omega), \quad E = \partial_\beta(\beta\Omega),$$

which define the parameters α_i and β implicitly as functions of N, Z, M, and E. Here the implicit assumption was made that there is only one saddle point in the complex β, α_ν planes, the one on the real axis. Evaluating Eqs. (3.52) in Thomas-Fermi approximation in the limit $\alpha_M m_{\tau,i} \ll 1$, it is shown by Gilbert and Cameron [345] that the four-dimensional version of the saddle-point approximation yields (neglecting the pairing backshift)

$$\rho_M(N,Z,M,E^*) = \frac{e^{2\sqrt{aE^*}-M^2/(2\sigma^2)}}{12\sqrt{2}\sigma[a(E^*-E_{\rm rot})]^{1/4}(E^*-E_{\rm rot})}. \qquad (3.54)$$

In view of the $2J + 1$ degeneracy of each J state the level density for fixed J is [324]

$$\rho(N, Z, E^*, J) = \rho_M(N, Z, E^*, M = J) - \rho_M(N, Z, E^*, M = J + 1)$$
$$\approx \partial_M \rho_M(N, Z, E^*, M)|_{M=J+1/2}.$$

Applying this rule to Eq. (3.54) yields Eq. (3.48).

Variants of this expression are obtained when all or some of the integrals in Eq. (3.53) are evaluated sequentially. E.g. Lang and Le Couteur [342] derived the equation of state $E^* - \Delta = at^2 - t$, which leads to the frequently used level density [346–349]

$$\rho(N, Z, E^*, J) = \frac{2J + 1}{24\sqrt{2}\sigma^3} \frac{e^{2\sqrt{a(E^* - \Delta)} - J(J+1)/(2\sigma^2)}}{[a(E^* - \Delta + t)]^{1/4}(E^* - \Delta + t)}. \tag{3.55}$$

About the meaning of the variable t the same remarks apply as in the case of Eqs. (3.37) and (3.38). With a different order of integrations in Eq. (3.53) the equation of state $E^* - E_{\rm rot} - \Delta = at^2 - 3t/2$ is obtained by Newton [350] and Lang [340]. This leads to the level density

$$\rho(N, Z, E^*, J) = \left. \frac{e^{2\sqrt{a(E^* - \Delta)} - M^2/(2\sigma^2)}}{12\sqrt{2}\sigma a^2 t^3} \right|_{M=J+1}^{M=J}, \tag{3.56}$$

which is used in Pühlhofer's CASCADE code [351] in the range of intermediate excitation energies. Except for very small excitation energies the variants (3.48) and (3.55) give practically the same result and so do (3.46) and (3.56). For $E_{\rm rot} \ll E^*$ and $S \gg 1$ all four variants coincide.

3.4 Shell corrections

The single-particle level-density in a shell-model potential at the Fermi energy differs significantly from the Thomas-Fermi level density used in Section 3.1. Empirically determined level-density parameters $a(N, Z)$ should therefore differ from the expression (3.28). Since Strutinsky's shell correction, discussed in the second chapter, is also constructed from the difference between the shell model and the smoothed, Thomas-Fermi single-particle level-density, it is not surprising that the empirical parameter a divided by the mass number A shows a similar dependence on A as the shell correction. This is shown in Fig. 3.1. For sufficiently large excitation energies the Thomas-Fermi description of the nucleus is expected to apply, i.e. shell effects should die out at high temperature. An empirical formula to describe the influence of shell effects on the level-density parameter a as function of the excitation energy was proposed by Ignatyuk, Smirenkin, and Tishin [352]

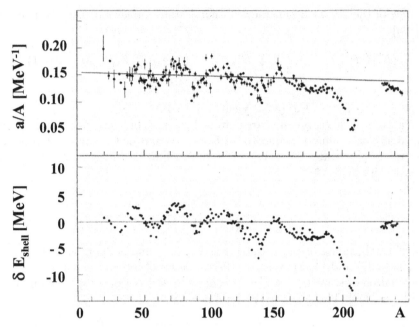

Fig. 3.1. Upper frame: the empirical level-density parameter a/A versus mass number, lower frame: Strutinsky shell correction for the same nuclei (after Ref. [352]).

$$a(E^*) = a_{\text{asympt}}[1 + f(E^*)\delta E_{\text{shell}}/E^*] \qquad (3.57)$$

with the correction function $f(E^*) = 1 - \exp(-E^*/E_D)$, the shell-correction energy δE_{shell}, and the parameter E_D characterizing the damping of the shell correction with increasing excitation energy. From Eq. (3.57) one finds

$$a(E^*)E^* = a_{\text{asympt}}[E^* + f(E^*)\delta E_{\text{shell}}]$$

$$\rightarrow \begin{cases} a_{\text{asympt}}(1 + \delta E_{\text{shell}}/E_D)E^* & \text{if } E^* \ll E_D \\ a_{\text{asympt}}(E^* + \delta E_{\text{shell}}) & \text{if } E^* \gg E_D. \end{cases} \qquad (3.58)$$

An often used value for E_D is 18.5 MeV [352]. For $E^* \gg E_D$ one has to subtract the shell correction δE_{shell} from the excitation energy E^* or, alternatively, reckon the excitation energy from the liquid-drop energy at zero temperature and use the Thomas-Fermi level-density parameter a_{asympt}. This is also sometimes called back-shifted Fermi-gas model [353], [354]. Kataria, Ramamurthy, and Kapoor [355] and Goriely [356] proposed alternative semi-empirical forms for the temperature dependence of the shell correction to the level-density parameter a.

The result of the combinatorial approach to the level density [320] has been used by Hilaire [357] to check the empirical formula (3.27) by fitting $a(E^*)$ for each excitation energy E^*. Sensible trends were obtained for $E^* \geq 15$ MeV from which Hilaire derived the more detailed formula

$$a(E^*) = a_{\text{asympt}}[1 - (f_{\text{neut}}(E^*)\delta E_{\text{shell}}^{(\text{neut})} + f_{\text{prot}}(E^*)\delta E_{\text{shell}}^{(\text{prot})})/E^*],$$

instead of Eq. (3.57). In this expression $f_\nu(E^*) = 1 - \exp(-E^*/E_D^{(\nu)})$, $\nu =$ neutron or proton with, in general different, damping parameters $E_D^{(\nu)}$ for protons and neutrons and $\delta E_{\text{shell}}^{(\nu)}$ are the shell corrections for protons and for neutrons.

It is very convenient in applications to use the simple formula (3.57). However, it is important to see also in a systematic way how the shell correction modifies the level density of the Thomas-Fermi approximation. Since nuclei are not a Fermi gas, but rather self-bound many-body systems, we shall base the following discussion on the self-consistent Hartree-Fock approximation and make sure that for $\beta^{-1} \to 0$ the total energy becomes the liquid-drop energy plus the Strutinsky shell-correction δE_{shell} [358].

The starting point shall therefore be the partition function Ω for neutrons in Hartree-Fock approximation [359] the generalization to two types of nucleons being rather straightforward

$$\beta\Omega^{\text{HF}} = -\sum_i \ln(1 + e^{\alpha - \beta\epsilon_i^{\text{HF}}}) - \frac{1}{2}\beta \sum_{i,j} n_i^{\text{HF}} n_j^{\text{HF}} V_{ij,ij} \qquad (3.59)$$

with Hartree-Fock energies ϵ_i^{HF} and average occupation numbers n_i^{HF} given in the selfconsistent Hartree-Fock basis $|i\rangle$ by

$$\epsilon_i^{\text{HF}} = t_{ii} + \sum_j n_j^{\text{HF}} V_{ij,ij}, \qquad\qquad n_i^{\text{HF}} = (1 + e^{\beta\epsilon_i^{\text{HF}} - \alpha})^{-1}, \qquad (3.60)$$

where t is the kinetic energy operator and $V_{ij,ij}$ the antisymmetrized matrix element of the two-particle interaction, which was denoted $\overline{V}_{ij,ij}$ in Sec. 2.1.1. The notation $\overline{V}$ will be used below with a different meaning. To make the following derivation of the shell correction to the level-density formula more transparent, we do not consider momentum and spin-dependent forces, which appear for instance in Skyrme functionals. We shall follow the arguments put forward in Refs. [358] by Gottschalk and Ledergerber and [360] by Junker, Hadermann, and Mukhopadhyay.

A smoothed occupation number

$$\overline{n}_i^{\text{HF}} = \int_{-\infty}^{\infty} f_M(\epsilon_i^{\text{HF}}, \epsilon) \frac{1}{1 + e^{(\epsilon - \mu)\beta}} d\epsilon \qquad (3.61)$$

is introduced in terms of the Strutinsky smoothing function $f_M(\epsilon_i, \epsilon)$ with the property that it transforms any polynomial $P_{2M+1}(\epsilon)$ of order $\leq 2M+1$ into itself

$$\int P_{2M+1}(\epsilon') f_M(\epsilon, \epsilon') d\epsilon' = P_{2M+1}(\epsilon). \qquad (3.62)$$

For large temperature, i.e. small β, the integrand in Eq. (3.61) may be expanded in a power series up to terms of order $2M + 1$

$$\overline{n}_i^{\mathrm{HF}}(\beta \to 0) = \int_{-\infty}^{\infty} f_M(\epsilon_i^{\mathrm{HF}}, \epsilon) \frac{1}{2} \left[1 - \frac{(\epsilon - \mu)\beta}{2} + \frac{(\epsilon - \mu)^3 \beta^3}{24} - \cdots \right]$$

That the polynomial in the bracket is mapped into itself by the smoothing function f_M means that $n_i^{\mathrm{HF}} \to \overline{n}_i^{\mathrm{HF}}$ for $\beta \to 0$.

In terms of the self-consistent Hartree-Fock single-particle potential $V_{ii} = \sum_j n_j^{\mathrm{HF}} V_{ij,ij}$ the energies are given by $\epsilon_i^{\mathrm{HF}} = (t + V^{\mathrm{HF}})_{ii}$. The smoothed single-particle potential $\overline{V}_{ii} = \sum_j \overline{n}_j^{\mathrm{HF}} V_{ij,ij}$ may be identified with a shell-model potential. Shell-model energies can then be defined as $\epsilon_i^s = (t + \overline{V})_{ii}$. Calculations with Skyrme functionals show that these quantities are rather independent of the temperature for moderate temperatures $\beta^{-1} < 4$ MeV [132] since $\overline{V}(\mathbf{r})$ keeps its shape in this temperature range. In terms of the differences

$$\delta n_i = n_i^{\mathrm{HF}} - \overline{n}_i^{\mathrm{HF}}(0) \,, \qquad\qquad \delta\epsilon_i = \sum_j \delta n_j V_{ij,ij} \,, \qquad (3.63)$$

where we use the notation $\overline{n}_i^{\mathrm{HF}}(0)$ for $\overline{n}_i^{\mathrm{HF}}(\beta^{-1} = 0)$, the partition sum (3.59) may be expanded to second order in δn_i

$$\beta\Omega^{\mathrm{HF}} = -\sum_i \ln(1 + e^{\alpha - \beta\epsilon_i^s}) - \frac{1}{2}\beta \sum_{i,j} \overline{n}_i^{\mathrm{HF}}(0)\overline{n}_j^{\mathrm{HF}}(0)V_{ij,ij} + R_1(\delta n) + R_2(\delta n^2)$$

$$(3.64)$$

where

$$R_1 = \beta \sum_i n_i^s \delta\epsilon_i - \beta \sum_{ij} n_i^{\mathrm{HF}}(0)V_{ij,ij}\delta n_j \qquad (3.65)$$

is the linear expansion term and R_2 contains second and higher order terms. Gottschalk and Ledergerber showed that they are negligibly small [358]. Insertion of expression (3.63) for $\delta\epsilon_i$ into Eq. (3.65) yields $R_1 = 0$, except for errors of the order $\delta n_i(n_i^s - n_i^{\mathrm{HF}}(0))$ which can be absorbed in R_2. For zero temperature this is the Strutinsky energy theorem. We now introduce a smoothed partition sum

$$\beta\overline{\Omega}^{\mathrm{HF}}(\alpha, \beta) = -\int \sum_i f_M(\epsilon_i^{\mathrm{HF}}, \epsilon) \ln(1 + e^{\alpha - \beta\epsilon}) d\epsilon - \frac{1}{2}\beta \sum_{i,j} \overline{n}_i^{\mathrm{HF}}(0)\overline{n}_j^{\mathrm{HF}}(0)V_{ij,ij}$$

$$(3.66)$$

and the difference

$$\delta\Omega(\alpha, \beta) = \Omega^{\mathrm{HF}}(\alpha, \beta) - \overline{\Omega}^{\mathrm{HF}}(\alpha, \beta)$$
$$= -\frac{1}{\beta}\int \sum_i [\delta(\epsilon - \epsilon_i^s) - f_M(\epsilon_i^{\mathrm{HF}}, \epsilon)] \ln(1 + e^{\alpha - \beta\epsilon}) d\epsilon.$$

Note that the difference contains only the single-particle contribution $\ln(1 + e^{\alpha - \beta \epsilon})$ to the partition sum. Parameters α and $\overline{\alpha}$ are given implicitly by the requirement

$$-\partial_\alpha \{\beta \Omega^{\mathrm{HF}}(\alpha, \beta)\} = -\partial_{\overline{\alpha}} \{\beta \overline{\Omega}^{\mathrm{HF}}(\overline{\alpha}, \beta)\} = N$$

and depend somewhat on the temperature.

Using standard procedures the dependence of the single-particle contribution to the smoothed partition sum $\beta \overline{\Omega}_{\mathrm{sp}}^{\mathrm{HF}}$ on $\overline{\alpha}$ and β can be written more explicitly. In terms of the smoothed level density $\overline{g}(\epsilon) = \sum_i f_M(\epsilon_i^{\mathrm{HF}}, \epsilon)$

$$\int_0^\infty \overline{g}(\epsilon) \ln(1 + e^{\overline{\alpha} - \beta \epsilon}) d\epsilon = \int_0^{\overline{\mu}} \overline{g}(\overline{\mu} - x) \ln(1 + e^{\beta x}) dx + \int_0^\infty \overline{g}(\overline{\mu} + x) \ln(1 + e^{-\beta x}) dx,$$

where $x = \overline{\mu} - \epsilon$ in the first integral and $x = \overline{\mu} + \epsilon$ in the second. Writing $1 + \exp(\beta x) = \exp(\beta x)[1 + \exp(-\beta x)]$ in the first integral, it can be rewritten as

$$\int_0^{\overline{\mu}} \overline{g}(\overline{\mu} - x) x \, dx + \int_0^{\overline{\mu}} \overline{g}(\overline{\mu} - x) \ln(1 + e^{-\beta x}) dx.$$

In the second of these two integrals the upper limit can be moved to infinity and the single-particle part of the smoothed partition sum becomes

$$-\beta \overline{\Omega}_{\mathrm{sp}}^{\mathrm{HF}} = \beta \int_0^{\overline{\mu}} x \overline{g}(\overline{\mu} - x) dx + \int_0^\infty [\overline{g}(\overline{\mu} - x) + \overline{g}(\overline{\mu} + x)] \ln(1 + e^{-\beta x}) dx \ ,$$

Expansion of $\overline{g}(\epsilon)$ in the second integral in a power series around $\epsilon = \overline{\mu}$ yields

$$-\beta \overline{\Omega}_{\mathrm{sp}}^{\mathrm{HF}} = \beta \int_0^{\overline{\mu}} (\overline{\mu} - \epsilon) \overline{g}(\epsilon) d\epsilon + 2 \overline{g}(\overline{\mu}) \int_0^\infty \ln(1 + e^{-\beta x}) dx + \mathcal{O}(\beta^{-3}) \quad (3.67)$$

$$= \overline{\alpha} N - \beta \overline{E}_{\mathrm{sp}}(0) + \overline{g}(\overline{\mu}) \frac{\pi^2}{6\beta}, \quad (3.68)$$

where $\overline{E}_{\mathrm{sp}}(0)$ is the smoothed shell-model energy at zero temperature.

The shell correction for the partition sum is conventionally defined as

$$\delta \Omega_{\mathrm{shell}}(\alpha, \beta) = \Omega^{\mathrm{HF}}(\alpha, \beta) - \overline{\Omega}^{\mathrm{HF}}(\overline{\alpha}, \beta)$$
$$\approx \Omega^{\mathrm{HF}}(\alpha) - \overline{\Omega}^{\mathrm{HF}}(\alpha) - \partial_{\overline{\alpha}} \overline{\Omega}^{\mathrm{HF}}(\overline{\alpha}, \beta)(\alpha - \overline{\alpha})$$
$$= \delta \Omega(\alpha, \beta) + N \beta^{-1}(\alpha - \overline{\alpha})$$

and according to Eq. (3.2) the shell correction for the total energy is

$$\delta E_{\text{shell}}(\beta) = \partial_\beta \{\beta \delta \Omega_{\text{shell}}(\alpha, \beta)\}|_\alpha, \tag{3.69}$$

where derivatives with respect to β are here understood at fixed α. The excitation energy $E^*(\alpha, \beta)$ is given in terms of Ω^{HF} or $\overline{\Omega}^{\text{HF}}$ and $\delta \Omega$ by

$$E^* = \partial_\beta \{\beta \Omega^{\text{HF}}(\alpha, \beta)\}|_{\beta=\infty}^\beta = \partial_\beta \{\beta \overline{\Omega}^{\text{HF}}(\overline{\alpha}, \beta) + \beta \delta \Omega_{\text{shell}}(\alpha, \beta)\}|_{\beta=\infty}^\beta.$$

According to definition (3.66) the two-particle part of Ω^{HF} is independent of the temperature in the approximation considered here. Therefore only the single-particle parts of Ω^{HF} and $\overline{\Omega}^{\text{HF}}$ contribute to the excitation energies E^* and $\overline{E}^*$, respectively. With Eq. (3.68) one obtains

$$\overline{E}^* = \partial_\beta \{\beta \overline{\Omega}_{\text{sp}}^{\text{HF}}(\overline{\alpha}, \beta)\}_{\beta=\infty}^\beta = \frac{\pi^2}{6\beta^2} \overline{g}(\overline{\mu}) = a_N \beta^{-2}.$$

With the small correction energy

$$E_{\text{corr}} = \overline{E}(\overline{\alpha}, T=0) - \overline{E}(\overline{\alpha}_0, T=0) \approx \partial_{\overline{\alpha}_0} \overline{E}(\overline{\alpha}_0, T=0)(\overline{\alpha} - \overline{\alpha}_0)$$

the excitation energy can therefore be written

$$E^* = a_N \beta^{-2} + E_{\text{corr}} + \delta E_{\text{shell}}(\beta) - \delta E_{\text{shell}}(T=0).$$

With the abbreviation $\Delta(\beta) = E_{\text{corr}} + \delta E_{\text{shell}}(\beta) - \delta E_{\text{shell}}(T=0)$

$$\beta^{-2} = (E^* - \Delta)/a_N. \tag{3.70}$$

The entropy is obtained from Eqs. (3.2), (3.68), and (3.69)

$$S(\mu, \beta) = \beta^2 \partial_\beta \Omega^{\text{HF}}(\mu, \beta)|_\mu = \beta^2 \partial_\beta (\overline{\Omega}^{\text{HF}}(\overline{\mu}, \beta) + \delta \Omega_{\text{shell}}(\mu, \beta))|_\mu$$
$$= 2a_N \beta^{-1} + (\delta E_{\text{shell}} - \delta \Omega_{\text{shell}})\beta.$$

Insertion of the result (3.70) yields

$$S(E^*) = 2\sqrt{a_N(E^* - \Delta(\beta))} \left[1 + \frac{\delta E_{\text{shell}}(\beta) - \delta \Omega(\beta)}{2(E^* - \Delta)}\right]. \tag{3.71}$$

To compare with the phenomenological ansatz (3.57), we consider

$$a(E^*)E^* = a_N(E^* - \Delta) \left[1 + \frac{\delta E_{\text{shell}}(\beta) - \delta \Omega(\beta)}{2(E^* - \Delta(\beta))}\right]^2$$

in the limit $E^* \gg \delta E_{\text{shell}}(0)$, where $\delta E_{\text{shell}}(\beta)$ and $\delta \Omega(\beta)$ vanish:

$$a(E^*)E^* \to a_N(E^* - E_{\text{corr}} + \delta E_{\text{shell}}(T=0)),$$

which agrees with the high-temperature limit (3.58) apart from the small correction term E_{corr}. Strictly speaking the expression (3.71) applies to the expectation value of the entropy in the grand canonical ensemble, which enters in the exponent of the level-density formula. We shall not consider shell corrections to the preexponential factor of the level density.

The smoothed partition sum $\overline{\Omega}^{\mathrm{HF}}$ is usually identified with the partition sum in the extended Thomas-Fermi approximation. Numerically good agreement between these quantities was observed. An analytical proof of this identity was however only given for the single-particle part of Ω and, in addition, only for an harmonic oscillator single-particle potential [108]. Both, the single-particle and the two-body contribution to the partition sum in Thomas-Fermi approximation admit a leptodermous expansion. For protons the same relations hold as for neutrons, except that the long-range Coulomb force cannot be included in the leptodermous expansion. Therefore the two-body part of the partition sum (3.59) gives rise to a temperature-dependent Coulomb energy in Thomas-Fermi approximation

$$\overline{E}_{\mathrm{Coul}} = \int d\mathbf{r} \int d\mathbf{r}' n^{\mathrm{TF}}(\mathbf{r}) V_{\mathrm{Coul}}(\mathbf{r} - \mathbf{r}') n^{\mathrm{TF}}(\mathbf{r}')$$

in terms of the temperature-dependent, Thomas-Fermi spatial density $n^{\mathrm{TF}}(\mathbf{r})$. To this expression a corresponding Coulomb exchange term is to be added.

3.5 Corrections for vibrational and rotational excitations

The lower end of the spectrum of many nuclei is dominated by states of rotational and vibrational character. Since they are of a collective nature, the Bethe formula, based on the independent particle model, can in principle not account for their contribution to the level density. Microscopic calculations show that these states consist of coherent particle-hole excitations, produced by the residual interaction. Some lie considerably lower in energy than the noninteracting particle-hole states from which they are formed. It has therefore been suggested that one adds artificially additional degrees of freedom to those of the Fermi gas, appropriate to describe rotations or vibrations. This leads obviously to an overcounting of states. One can try to eliminate the contribution to the entropy of those particle-hole excitations in the Fermi gas which get pulled down by the residual interaction. This strategy was pursued for the vibrational enhancement of the level density by Blokhin, Ignatyuk, and Shubin [361]. One can however also argue as Bjørnholm, Bohr, and Mottelson did [362]: The excitation energy of those particle-hole states which become involved in collective states is so high in the Fermi gas that at these energies there are many more states not related to collective dynamics. Therefore the contribution of those few to the entropy of the Fermi gas is altogether negligible and one does not have to consider their elimination.

With this reasoning in mind Bjørnholm, Bohr, and Mottelson proposed the following derivation of a rotational enhancement factor $K_{\mathrm{rot}}^{(0)}$ of the level density [362]: They start from the Fermi gas result

$$\rho(N, Z, E^*, M) = \frac{e^{2\sqrt{aE^*} - M^2/(2\sigma^2)}}{12\sqrt{2}\sigma(aE^*)^{1/4}E^*} \tag{3.72}$$

which follows from Eq. (3.54) by neglecting E_{rot} compared to E^* in the denominator. For a deformed nucleus with axial symmetry the natural quantization direction is the symmetry axis. In this case σ^2 becomes $\sigma_\perp^2 = \mathcal{J}_\perp T\hbar^{-2}$ in Eqs. (3.72), (3.46), and (3.54), where $\mathcal{J}_\perp$ is the moment of inertia perpendicular to the symmetry axis. Extra rotational degrees of freedom are introduced connected with a rotational energy E_{rot}'. For axially symmetric shapes $E_{\text{rot}}'(J, M) = \hbar^2[J(J+1) - M^2]/(2\mathcal{J}_\perp)$. Note that conceptually E_{rot}' is not identical with the quantity E_{rot} introduced in subsection 3.3.3. It is rather introduced as an additional rotational degree of freedom. The level density of the combined system of the rotor and the Fermi gas for fixed J is given by the folding expression

$$\rho_{\text{coll}}(N, Z, E^*, J) = \sum_{M=-J}^{J} \rho(N, Z, E^* - E_{\text{rot}}'(J, M), M). \tag{3.73}$$

With $E_{\text{rot}}' \ll E^*$ Eq. (3.73) yields

$$\rho_{\text{coll}}(N, Z, E^*, J) = \frac{2J+1}{12\sqrt{2}\sigma_\perp(aE^*)^{1/4}E^*} e^{2\sqrt{aE^*} - J(J+1)/(2\sigma_\perp^2)}.$$

Comparison with Eq. (3.48) (with $\Delta = 0$) and using Eq. (3.44), shows that

$$\rho_{\text{coll}} = 2\sigma_\perp^2 \rho(N, Z, E^*, J)$$

in the limit $E_{\text{rot}}' \ll E^*$. Therefore in this case $K_{\text{rot}}^{(0)}(T) = 2\sigma_\perp^2$. The introduction of collective rotations as additional degrees of freedom is in the spirit of the adiabatic rotor model [363], where each collective rotational state is shown to carry approximately the whole spectrum of intrinsic states.

We have assumed a rotor with axial symmetry, but without reflection symmetry since $|M|$ was assumed to take all even and odd values $\leq J$. With reflection symmetry the enhancement factor has half this size. For lower symmetries of the nucleus the following $K_{\text{rot}}^{(0)}$ factors were derived in an analogous way in Ref. [362].

$K_{\text{rot}}^{(0)}$	type of symmetry
$\sigma_\perp^2$	axial and R-symmetry
$2\sigma_\perp^2$	axial symmetry
$\sqrt{8\pi}\sigma_x\sigma_y\sigma_z$	no rotational, no parity or time-reversal symmetry
$\sqrt{2\pi}\sigma_x\sigma_y\sigma_z$	no rotational, but either parity or time-reversal symmetry.

With increasing excitation energy the rotational enhancement should decrease. Motivated by calculations in models with rather schematic separable two-particle interactions phenomenological damping factors have been proposed to account for this decrease. One often used form was suggested by Hansen and Jensen [364]

$$K_{\text{rot}}(E^*) = 1 + \frac{K_{\text{rot}}^{(0)}(E^*) - 1}{1 + \exp[(E^* - E_r)/d_r]} \qquad (3.74)$$

with empirical parameters $E_r = 120A^{1/3}\beta_2^2$ MeV and $d_r = 1400A^{-2/3}\beta_2^2$ MeV, where β_2 is the quadrupole deformation parameter.

To account for low-lying vibrational modes we consider first the canonical partition sum of a linear oscillator with characteristic energy $\epsilon = \hbar\omega$. It is given by $Z(\beta) = \sum_n \exp(-\beta\epsilon n) = (1 - e^{-\beta\epsilon})^{-1}$. A multipole oscillator with multipolarity λ is $(2\lambda + 1)$−fold degenerate. Its partition sum is therefore $Z_\lambda(\beta) = (1 - e^{-\beta\epsilon})^{-(2\lambda+1)}$. For a system of several independent oscillators with multipolarity λ and energies ϵ_λ the partition sum becomes

$$Z_{\text{osci}}(\beta) = \prod_\lambda (1 - e^{-\beta\epsilon_\lambda})^{-(2\lambda+1)}.$$

Typically only quadrupole ($\lambda = 2$) and octupole modes ($\lambda = 3$) are considered. The free energy of the system of oscillators is $F_{\text{osci}}(\beta) = -\beta^{-1}\ln Z_{\text{osci}}$. Ignatyuk writes the combined level density of the oscillators and the nuclear Fermi gas $\rho_{\text{coll}}(N, Z, E^*) = \rho(N, Z, E^* - F_{\text{osci}})$ in analogy to the rotational case. For $F_{\text{osci}} \ll E^*$ one obtains [338]

$$K_{\text{vib}}^{(0)} = \frac{\rho(N, Z, E^* - F_{\text{osci}})}{\rho(N, Z, E^*)} \approx e^{-\beta F_{\text{osci}}(\beta)} = Z_{\text{osci}}.$$

When the residual interaction, which is responsible for the collectivity of the vibrational states, is continuously switched off, the energies ϵ_λ will move to the shell-model single particle energies ϵ_λ^0. To avoid the double counting, mentioned above, the expression

$$K_{\text{vib}}^{(0)} = \prod_\lambda \left[\left| \frac{1 - e^{-\beta\epsilon_\lambda^0}}{1 - e^{-\beta\epsilon_\lambda}} \right| \right]^{(2\lambda+1)} \qquad (3.75)$$

was proposed in Ref. [361], which approaches 1 for vanishing residual interaction. For a systematic derivation on a microscopic basis in the RPA approximation we refer to a paper by Ignatyuk [365].

Vibrational modes are Landau-damped and increasingly so with growing temperature. To account for this effect the expression (3.75) was modified in Ref. [361] in a semiphenomenolgical way. The result is

$$K_{\text{vib}} = \prod_\lambda \left[\left| \frac{1 - e^{-\beta(\bar{\epsilon}_\lambda^0 + i\gamma_\lambda^0)}}{1 - e^{-\beta(\bar{\epsilon}_\lambda + i\gamma_\lambda)}} \right| \right]^{(2\lambda+1)}, \qquad (3.76)$$

where $\bar{\epsilon}_\lambda$ is the centroid of the strength function of the λth multipole, derived from the energy-weighted sum rule, and $\bar{\epsilon}_\lambda^0$ is the same quantity in the single particle model without residual interaction. From a schematic, two-level RPA model [366] for the collective vibrations Blokhin, Ignatyuk, and Shubin [361] obtained the approximate relation $\bar{\epsilon}_\lambda^2(T) = (\bar{\epsilon}_\lambda^0)^2 - \zeta_\lambda(T)[(\bar{\epsilon}_\lambda^0)^2 - (\epsilon_\lambda)_{\exp}^2]$ with a phenomenological function

$$\zeta_\lambda(T) = e^{-c_1 T^2/(\epsilon_\lambda)_{\exp}}.$$

The $(\epsilon_\lambda)_{\exp}$ are the experimental vibrational energies of multipolarity λ, $c_1 = 0.008$ MeV^{-1} is an empirical parameter and the averaged shell-model energies are approximated by

$$\bar{\epsilon}_\lambda^0 = \begin{cases} \hbar\omega_{\text{shell}}/2 \approx 20A^{-1/3} \\ \hbar\omega_{\text{shell}} \approx 40A^{-1/3} \end{cases} \quad \text{for} \quad \begin{matrix} \lambda^\pi = 2^+ \\ \lambda^\pi = 3^-. \end{matrix}$$

From Landau's theory of Fermi liquids follows the damping width

$$\gamma_\lambda = c_2(\bar{\epsilon}_\lambda^2 + 4\pi^2 T^2)$$

as function of the temperature with another empirical parameter $c_2 = 0.013$ MeV^{-1}.

Several variants for the temperature dependence and the structure of the enhancement factors have been proposed [338], [367], [368]. For deformed nuclei β and γ vibrations may be considered. However, in the context of the statistical theory of fission the β vibration is usually identified with the fission degree of freedom and shall therefore not contribute to the "intrinsic" level density of the fissioning nucleus.

3.6 Pairing correlations

The effect of pairing correlations on the level density has almost always been treated in the shell-model-plus-BCS approximation and mostly with constant pairing matrix-element. Therefore all caveats mentioned in Section 2.1.3 also apply in the present, temperature-dependent case [369]. We follow the presentation given by Ignatyuk [370], [338] and restrict it to neutrons; the generalization to protons as well as neutrons is obvious. We start with the BCS Hamiltonian

$$\hat{H} = \sum_k \epsilon_k(\hat{a}_k^+ \hat{a}_k + \hat{a}_{\bar{k}}^+ \hat{a}_{\bar{k}}) - G \sum_{k,k'}{}' \hat{a}_k^+ \hat{a}_{\bar{k}}^+ \hat{a}_{\bar{k}'} \hat{a}_{k'}, \qquad (3.77)$$

where ϵ_k are the energies of the neutron eigenstates $|k>$ in a given mean field (shell model or Hartree-Fock) which are supposed to be degenerate with respect to time reversal ($|\bar{k}>$) and temperature-independent. The prime on the second sum of Eq. (3.77) indicates a restriction to states $|k>$ in an energy band around the Fermi surface of width 2ω as discussed in Section 2.1.3. The operators of the neutron number and of the z component of the angular momentum are

$$\hat{N} = \sum_k (\hat{a}_k^+ \hat{a}_k + \hat{a}_{\bar{k}}^+ \hat{a}_{\bar{k}})$$

and

$$\hat{M} = \sum_k m_k (\hat{a}_k^+ \hat{a}_k - \hat{a}_{\bar{k}}^+ \hat{a}_{\bar{k}}) \,,$$

respectively.

The level density $\rho(N, E, M)$ is obtained by inverting a Laplace integral of type (3.50)

$$\rho(N, E, M) = \frac{1}{(2\pi i)^3} \int_{\gamma_1 - i\infty}^{\gamma_1 + i\infty} d\beta \int_{\gamma_2 - i\infty}^{\gamma_2 + i\infty} d\alpha_N$$

$$\times \int_{\gamma_3 - i\infty}^{\gamma_3 + i\infty} d\alpha_M e^{\beta E - \alpha_N N - \alpha_M M} \mathrm{tr}(e^{-\beta \hat{\Omega}}) \tag{3.78}$$

with

$$\beta \hat{\Omega} = \beta \hat{H} - \alpha_N \hat{N} - \alpha_M \hat{M}. \tag{3.79}$$

The chemical potential is $\lambda = \alpha_N / \beta$ and the corresponding quantity for the angular momentum constraint is $\mu = \alpha_M / \beta$. Evaluation of the triple integral in the saddle-point approximation yields

$$\rho(N, E, M) = (2\pi)^{-3/2} D^{-1/2}(\alpha_N^{(0)}, \alpha_M^{(0)}, \beta^{(0)}) e^{S(\alpha_N^{(0)}, \alpha_M^{(0)}, \beta^{(0)})} \tag{3.80}$$

with the entropy

$$S = \beta E - \alpha_N N - \alpha_M M - \ln \mathrm{tr}(e^{-\beta \hat{\Omega}}) \tag{3.81}$$

and the determinant D of the second derivatives of the exponent in the integrand of Eq. (3.78) with respect to β, α_N, and α_M

$$D = \begin{vmatrix} \langle \hat{H}^2 \rangle - \langle \hat{H} \rangle^2 & \langle \hat{H} \rangle \langle \hat{N} \rangle - \langle \hat{H} \hat{N} \rangle & \langle \hat{H} \rangle \langle \hat{M} \rangle - \langle \hat{H} \hat{M} \rangle \\ \langle \hat{H} \rangle \langle \hat{N} \rangle - \langle \hat{H} \hat{N} \rangle & \langle \hat{N}^2 \rangle - \langle \hat{N} \rangle^2 & \langle \hat{M} \rangle \langle \hat{N} \rangle - \langle \hat{N} \hat{M} \rangle \\ \langle \hat{H} \rangle \langle \hat{M} \rangle - \langle \hat{H} \hat{M} \rangle & \langle \hat{M} \rangle \langle \hat{N} \rangle - \langle \hat{N} \hat{M} \rangle & \langle \hat{M}^2 \rangle - \langle \hat{M} \rangle^2 \end{vmatrix}. \tag{3.82}$$

The brackets indicate grand canonical averages $\langle \hat{H} \rangle := \mathrm{tr}(e^{-\beta \hat{\Omega}} \hat{H}) / \mathrm{tr}(e^{-\beta \hat{\Omega}})$. The arguments $\beta^{(0)}, \alpha_N^{(0)}$, and $\alpha_M^{(0)}$ of S and D in Eq. (3.80) follow from solving the three saddle-point equations

$$E = \langle \hat{H} \rangle, \qquad\qquad N = \langle \hat{N} \rangle, \qquad\qquad M = \langle \hat{M} \rangle. \tag{3.83}$$

By a Bogolyubov transformation quasiparticle operators $\hat{\alpha}_k^+$ and $\hat{\alpha}_k$ are introduced

$$\hat{\alpha}_k^+ = u_k \hat{a}_{\bar{k}}^+ + v_k \hat{a}_k \quad \text{and} \quad \hat{\alpha}_{\bar{k}}^+ = u_k \hat{a}_k^+ - v_k \hat{a}_{\bar{k}}.$$

In this definition of quasiparticle operators, used in Ref. [338], the roles of quasiparticle creation and destruction operators are exchanged compared to our definition in Sec. 2.1.3. Unitarity of the transformation requires the relation

$$u_k^2 + v_k^2 = 1. \tag{3.84}$$

As approximation to the operator $\hat{\Omega}$ of Eq. (3.79) the independent quasiparticle operator

$$\hat{\Omega}_0 = U_0 + \sum_k (\eta_k \hat{\alpha}_k^+ \hat{\alpha}_k + \eta_{\bar{k}} \hat{\alpha}_{\bar{k}}^+ \hat{\alpha}_{\bar{k}})$$

is introduced in Eqs. (3.80) - (3.83). Averages of an operator $\hat{O}$ with respect to $\hat{\Omega}_0$, instead of $\hat{\Omega}$, will be written $\langle \hat{O} \rangle_0$. The thermal occupation probability of the quasiparticle state $|k>$ is given by $\overline{n}_k = \langle \hat{\alpha}_k^+ \hat{\alpha}_k \rangle_0 = [1 + \exp(\beta \eta_k)]^{-1}$. We also introduce the pair correlation function

$$\Delta = G \sum_k{}' u_k v_k (1 - \overline{n}_k - \overline{n}_{\bar{k}}). \tag{3.85}$$

To determine the quantities $U_0, \eta_k, \eta_{\bar{k}}, v_k$, and u_k, introduced above, one requires $\langle \hat{\Omega} \rangle_0 = \langle \hat{\Omega}_0 \rangle_0$ to obtain U_0 and

$$\partial \langle \hat{\Omega} \rangle_0 / \partial \overline{n}_k |_{v_k = \text{const}} = \partial \langle \hat{\Omega}_0 \rangle_0 / \partial \overline{n}_k |_{v_k = \text{const}},$$

which yields

$$\begin{aligned} \eta_k &= E_k + \mu m_k - (G/2)(\overline{n}_k - \overline{n}_{\bar{k}}) \\ \eta_{\bar{k}} &= E_k - \mu m_k - (G/2)(\overline{n}_k - \overline{n}_{\bar{k}}) \end{aligned} \tag{3.86}$$

with $E_k = (\mathcal{E}_k^2 + \Delta^2)^{1/2}$ and

$$\mathcal{E}_k = \epsilon_k - \lambda - (G/2)[1 + (v_k^2 - u_k^2)(1 - \overline{n}_k - \overline{n}_{\bar{k}})]. \tag{3.87}$$

Note that for $\mu m_k = 0$ one has $\eta_k = \eta_{\bar{k}}$ and $1 - \overline{n}_k - \overline{n}_{\bar{k}} = \tanh \beta \eta_k / 2$. Finally v_k and u_k follow from Eq. (3.84) and the optimization condition for the Bogolyubov transformation $\partial \langle \hat{\Omega} \rangle_0 / \partial v_k |_{\overline{n}_k, \overline{n}_{\bar{k}} = \text{const}} = 0$. One obtains

$$\left. \begin{aligned} u_k^2 \\ v_k^2 \end{aligned} \right\} = \frac{1}{2} \left(1 \pm \frac{\mathcal{E}_k}{E_k} \right).$$

Inserting this result into Eq. (3.85) yields the temperature-dependent gap equation

$$\frac{2}{G} = \sum_k{}' \frac{1 - \overline{n}_k - \overline{n}_{\bar{k}}}{E_k}. \tag{3.88}$$

Together with Eqs. (3.83) the gap equation allows to determine Δ, λ, and μ as functions of N, M, and E. For details of the calculation, in particular for explicit expressions of the variances of H and N in Eq. (3.82) in terms of u_k, v_k, $\bar{n}_k$, and $\bar{n}_{\bar{k}}$, we refer to Ignatyuk's paper [365].

The gap equation admits always the trivial solution

$$\begin{matrix} u_k = 1, \, v_k = 0 \\ u_k = 0, \, v_k = 1 \end{matrix} \quad \text{if} \quad \begin{matrix} \epsilon_k > \lambda \\ \epsilon_k \leq \lambda \end{matrix}$$

with $\Delta = 0$ and $E_k = |\epsilon_k - \lambda|$, which corresponds to the normal state. For sufficiently large single-particle level-density g at the Fermi energy and sufficiently small temperature and angular momentum there exists a solution with $\Delta \neq 0$ characterizing the superfluid state. The situation is most conveniently presented in the somewhat schematic uniform model [97]: When the single-particle level spacing is roughly constant within a band of width 2ω around the Fermi energy λ the sums in Eqs. (3.85) and (3.88) may be replaced by integrals $\sum_k' \to (\bar{g}/2) \int_{\lambda-\omega}^{\lambda+\omega} \ldots d\epsilon$ with the average level density $\bar{g}$. In addition, m_k is assumed to have a constant value $\bar{m}$ for all levels k in the 2ω-band. Sometimes also the terms proportional to $(G/2)$ in Eqs. (3.86) and (3.87) are dropped, or rather assumed to be absorbed in the mean field. In this model the angular-momentum dependence of the gap parameter at zero temperature, $\Delta_0(M)$, is given by the simple formula

$$\Delta_0(M) = \Delta_0(M = 0)\sqrt{1 - M/M_{\text{crit}}},$$

where $M_{\text{crit}} = \overline{gm}\Delta_0(T = 0)$ is the critical angular momentum beyond which the system is – at zero temperature – in its normal phase. A contour map of the gap parameter Δ for the neutrons is shown in Fig. 3.2 for the uniform model as function of the temperature T and the angular-momentum projection M. The outer line corresponds to $\Delta = 0$. Beyond that line the system is not superconducting. Note that for finite temperature the superconducting region stretches somewhat beyond M_{crit}. For an explanation of this temperature-driven pairing see Moretto [371].

In the normal region of the phase diagram the Fermi-gas level-density with shell corrections, (3.46) with (3.57), is appropriate if the excitation energy E^* is backshifted by the condensation energy E_{cond}, which is the difference between the Hartree-Fock and the BCS ground-state energies. In the uniform model one obtains [372]

$$E_{\text{cond}} = (g/4)\Delta_0^2 + \chi\Delta_0,$$

where the empirical value $\Delta_0 = 12A^{-1/2}$ MeV is often used and χ is defined in Eq. (3.47).

In the level-density formula

$$\rho_{\text{BCS}}(N, Z, M, E^*) = \frac{e^{S - M^2/(2\sigma^2)}}{(2\pi\sigma^2\Lambda)^{1/2}}$$

(cf. Eq. (3.54)) the recommended form for S, σ, and Λ is outside the superfluid phase [373], [374]

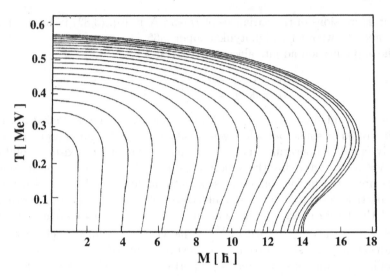

Fig. 3.2. Contour map of the gap parameter Δ for neutrons in the uniform model with parameters $\bar{g} = 7 \text{ MeV}^{-1}$, $\bar{m} = 2$, and $\Delta(T = 0, M = 0) = 1$ MeV. The spacing between successive contour lines is 0.05 MeV, starting with $\Delta = 1$ and ending with $\Delta = 0$ (after Ref. [371]).

$$S = 2\sqrt{aE'} = 2at, \quad \sigma^2 = (6/\pi^2)at\langle m^2 \rangle,$$

$$\Lambda = (144/\pi)a^3t^5, \quad a = a_{\text{asympt}}[1 - (1 - e^{-(E'/E_D)})\delta E_{\text{shell}}/E']$$

with $E' = E^* - E_{\text{rot}} - E_{\text{cond}}$ and $E_D = 18.5$ MeV. Disregarding the temperature-driven pairing [371] for $M > M_{\text{crit}}$ one has in the superfluid phase in terms of the function $\phi(E') = \sqrt{1 - E'/E_{\text{crit}}}$

$$S/S_{\text{crit}} = (t_{\text{crit}}/t)(1 - \phi^2), \quad \sigma^2/\sigma_{\text{crit}}^2 = (1 - \phi^2),$$

$$\Lambda/\Lambda_{\text{crit}} = (1 - \phi^2)(1 + \phi^2)^2, \quad a = a_{\text{crit}}$$

with

$$t_{\text{crit}} \approx (e^C/\pi)\Delta_0 = 0.567\Delta_0,$$

where $C = 0.577$ is the Euler constant and $E'_{\text{crit}} = a_{\text{crit}}t_{\text{crit}}^2$, $S_{\text{crit}} = 2a_{\text{crit}}t_{\text{crit}}$, $\sigma_{\text{crit}}^2 = (6/\pi^2)a_{\text{crit}}t_{\text{crit}}\langle m^2 \rangle$, $\Lambda_{\text{crit}} = (144/\pi)a_{\text{crit}}^3t_{\text{crit}}^5$. The parameter a_{crit} is determined by the equation

$$a_{\text{crit}} = a_{\text{asympt}}\{1 - [1 - \exp(a_{\text{crit}}t_{\text{crit}}^2/E_D)]\delta E_{\text{shell}}/(a_{\text{crit}}t_{\text{crit}}^2)\}.$$

The temperature t as function of E' follows from the equation $\phi = \tanh(\phi\, t_{\text{crit}}/t)$ [372]. For a discussion of the error in the entropy S, in particular around t_{crit}, because of the use of a grand canonical ensemble in the discussion above, instead of the more appropriate microcanonical ensemble, see Moretto [371].

3.7 Fits of level-density formulae and compilations of empirical level densities

We have seen that level densities contain a number of empirical parameters to be fitted with experimental data. With few exceptions we did not quote results of such fits because fits depend on the data basis and the specific variant of the level-density formula used, and both change with time. The Nuclear Data Section of the International Atomic Energy Agency (IAEA) has started a project to collect, evaluate and update the data basis for level densities and various level-density formulae, besides other quantities, relevant for fission and fusion research. These efforts resulted in the Reference Input Parameter Library (RIPL) [374], which is available through the internet, http://www-nds.iaea.org/RIPL-2/. The reader is referred to this library for specific, updated information.

In the following only three examples of fits will be discussed which are characteristic for a particular class of level-density formulae. We start with the work of Gilbert and Cameron [345]. Their input consisted of thermal neutron resonance-spacings, covering a range of excitation energies between 4.6 and 11.1 MeV for the sample of nuclei considered. The Fermi gas formula (3.48) is used for $E^* > E_x$ with $E_x = 2.5 + 150/A + \Delta$ [MeV] and $\Delta = P(Z) + P(N)$, where the pairing energies $P(Z)$ and $P(N)$ for protons and neutrons, respectively, are obtained from empirical odd-even nuclear mass differences. They are tabulated for each nuclid. To account for shell effects the level density parameter a of the Bethe formula is parametrized in the form

$$a = A[\alpha(S(Z) + S(N)) + Q(N, Z)], \qquad (3.89)$$

where $S(Z)$ and $S(N)$ are shell-correction energies obtained as the difference between the empirical mass and the liquid-drop mass for each nuclid. They were also tabulated. The parameters $\alpha = 0.00917$ MeV^{-2} and $Q = 0.142$ MeV^{-1} for spherical and $Q = 0.12$ MeV^{-1} for deformed nuclei were found to give reasonable fits of the level spacing for excitation energies around the neutron binding-energy. The spin cut-off parameter σ^2 of Eq. (3.48) was taken to be $\sigma^2 = 0.089\sqrt{aE^*}A^{2/3}$.

Since the Bethe formula does not account for the low-lying collective states, the level density below E_x was fitted by the purely phenomenological ansatz

$$\rho_1(N, Z, E^*) = \frac{1}{T_x}e^{(E^* - E_0)/T_x} \qquad (3.90)$$

with two parameters E_0 and T_x, which are determined by the requirement that ρ_1 joins ρ_2, the level density for all J, smoothly at $E^* = E_x$, where ρ_2 is defined as

$$\rho_2(N, Z, E^*) = \sum_J^\infty \rho(N, Z, E^*, J) = \frac{\exp(2\sqrt{a(E^* - \Delta)})}{12\sqrt{2}\sigma a^{1/4}(E^* - \Delta)^{5/4}}.$$

The level density (3.90) is referred to as constant-temperature formula.

More recent fits of the composite Gilbert-Cameron formula can be found in the Reference Input Parameter Library [374]. The formula (3.89) is not supposed to be used at higher energies.

A second example of a level-density fit to be discussed here is due to Reisdorf [375]. The input consists of level-spacing data at neutron binding energies and fission-evaporation branching ratios at excitation energies up to 11 MeV of the compound state. The discussion is based on the Fermi gas formula (3.27) with $E^* \rightarrow E^* - \Delta$ to account for pairing with $\Delta = (10.5 \pm 2)\chi/\sqrt{A}$ MeV and χ from Eq. (3.47). In the level-density parameter a shell corrections are considered according to Eq. (3.57) with $E_D = 18.5$ MeV and $\delta E_{\mathrm{shell}}$ from calculated Strutinsky shell corrections. The parameter a_{asympt} is expanded in volume, surface, and curvature terms. The leptodermous expansion is obtained from Eq. (3.22) by inserting Eq. (3.15). The result is

$$a_{\mathrm{asympt}} =$$
$$(0.04543 r_0^3 A + 0.1355 r_0^2 A^{2/3} B_{\mathrm{surf}} + 0.1426 r_0 A^{1/3} B_{\mathrm{curv}}) \ \mathrm{MeV}^{-1}, (3.91)$$

where B_{surf} is the ratio between the surface area of a (in general deformed) nucleus and a sphere of equal volume and B_{curv} is the ratio of the mean curvature κ integrated over the nuclear surface and the same integral over the surface of the sphere of equal volume. Fit of the parameter r_0 to the level-spacing data gave $r_0 = 1.153 \pm 0.01$ fm, somewhat smaller than the droplet-model value of $r_0 = 1.18$ fm corresponding to the saturation density of nuclear matter [212]. For finite nuclei along the β-stability line Myers gives empirical values for the r_0 parameter between 1.14 fm for $A \approx 60$ and 1.18 fm for very heavy nuclei [212]. No enhancement factors for collective states were considered in Reisdorf's analysis.

For excitation energies up to 12-15 MeV, in particular for thermal neutron induced reactions, neutron emission and fission are the only competing decay channels of importance of the compound state in a heavy nucleus. Measurement of the reaction and the fission cross sections, σ_R and σ_{fiss}, respectively yields the ratio of the neutron width to the fission width $\Gamma_n/\Gamma_{\mathrm{fiss}} = (\sigma_R/\sigma_{\mathrm{fiss}}) - 1$ (cf. Chap. 7 of Ref. [106]). Using Eq. (4.11) one can either determine the fission barrier B_{fiss} if the shape-dependence of the level-density parameter a is known or determine the latter when the barrier is known.

To test level-density formulae at higher energies experimental values for $\Gamma_{\mathrm{fiss}}/\Gamma_{\mathrm{reaction}}$ are being used. Unfortunately, the relation of this quantity to the level density is not as direct as for neutron-resonance spacings. Models are needed to determine the distribution of angular momenta and excitation energies in the compound nuclei formed in a given fusion reaction. From such initial distribution a cascade calculation leads to the observables. It is in the cascade codes, to be discussed in the next section, that the level densities enter. In Ref. [375] Reisdorf claimed that the level density fitted to low-energy

level spacings is also reproducing (α, f) reactions in the actinide region satisfactorily for moderate excitation energies.

As third example we consider the fit by Ignatyuk, Itkis, Okolovich, and
Smirenkin [335]. The input consisted again of neutron-resonance spacings at
neutron binding energies and $\Gamma_{\text{fiss}}/\Gamma_n$ values for some pre-actinide nuclei. The
level-density formula used is the same as in the previous case, but a rotational
enhancement factor of the form

$$K_{\text{rot}}^0 = \sigma_\perp^2 \frac{\sqrt{2\pi}\sigma_0}{2J+1} \operatorname{erf}\left(\frac{J+1/2}{\sqrt{2}\sigma_0}\right)$$

was used with $\sigma_0^{-2} = \hbar^2(1/\mathcal{J}_\parallel - 1/\mathcal{J}_\perp)/T$. Fitting the level-density parameter
a gave

$$a_{\text{asympt}} = (0.114A + 0.162A^{2/3})\ \text{MeV}^{-1} \qquad (3.92)$$

and the thermal damping parameter in the shell correction was $E_D = 18.5$
MeV.

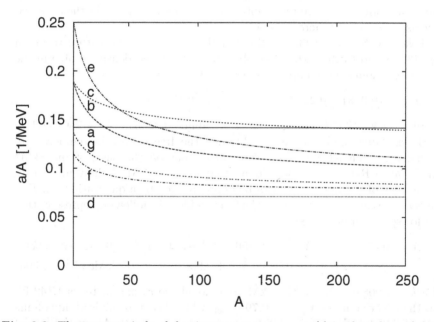

Fig. 3.3. The asymptotic level-density parameter a_{asympt}/A is plotted vs. A for
spherical nuclei along the line of β-stability. Line a corresponds to Gilbert and
Cameron's result $a/A = Q = 0.142\ \text{MeV}^{-1}$, curves b and c represent Eqs.
(3.91) (Reisdorf [375]) and (3.92) (Ignatyuk et al. [335]), respectively. The simplest
Thomas-Fermi estimate $a/A = 0.071\ \text{MeV}^{-1}$ yields line d and Eqs. (3.94) and (3.95)
are given by curves e (Töke [322]) and f (Guet et al. [376]), respectively. Curve g
shows the result (3.96) (Bartel et al. [377]).

In Fig. 3.3 the results of the three fits for a_{asympt}/A are plotted vs. A for spherical nuclei. Standard empirical fits of this quantity (without corrections for collective states) range between $1/7 = 0.143$ for medium-size nuclei and $1/10 = 0.1$ for heavy nuclei. The bold lines (a - c) are seen to fall into this margin although Reisdorf's result (b) is significantly lower than most other fits.

For comparison also theoretical estimates of a_{asympt}/A for spherical nuclei along the line of β-stability are presented. The simplest estimate is obtained from Eqs. (3.19) and (3.22) neglecting the surface terms $q_{1/2}$ and $q_{5/2}$ and writing the volume $v = (4\pi/3)r_0^3 A$. For β-stable nuclei the Fermi energy μ_0 for protons and neutrons should be the same in Thomas-Fermi approximation. Adding to Eqs. (3.22), and (3.19) similar equations for the protons, one obtains $A = (16\pi/9)C\mu_0^{3/2}r_0^3 A$ and $a = (4\pi^3/9)C\mu_0^{1/2}r_0^3 A$. From these equations follows

$$a/A = (2\pi/9)(9\pi)^{1/3}M^*_{\text{nucl}}\hbar^{-2}r_0^2. \tag{3.93}$$

With $r_0 = 1.18$ fm and assuming that the effective mass M^*_{nucl} is equal to the mass of a free nucleon, $M^*_{\text{nucl}} = M_{\text{nucl}}$, one gets $a/A = 0.071$ MeV^{-1}. Taking also surface and curvature terms into account, one would obtain the observed decrease of a/A with increasing A.

Töke and Swiatecki [322] estimated the surface and curvature terms in Eqs. (3.22) and (3.19) rather roughly from an assumed density profile in the surface in Thomas-Fermi approximation. They gave

$$a/A = (0.0684 + 0.213A^{-1/3} + 0.385A^{-2/3})(1 - I^2/9)\ \text{MeV},^{-1} \tag{3.94}$$

where $I = (N - Z)/A$. For the line of β-stability $I \approx 0.4A/(A + 200)$ [231].

Brack, Guet, and Håkansson obtained the free energy in temperature-dependent, extended Thomas-Fermi calculations on the basis of the SkM* functional in Ref. [132]. It was found that for $T < 4$ MeV the temperature dependence of the free energy can be approximated by a quadratic form $F = E_0 - aT^2$. Fitting the curvature of the form yielded the following representation of a for spherical nuclei [376]

$$a/A = \{0.053 + 0.095A^{-1/3} + 0.056A^{-2/3} + 0.098A^{-1} + (0.031 + 0.384k)$$
$$\times(1 + k)^{-2}I^2 + (0.0169A^{-2} + 0.00056A^{-4/3})Z^2\}\ \text{MeV}^{-1} \tag{3.95}$$

with $k = 1.908A^{-1/3}$. For the SkM* functional the r_0 parameter is 1.142 fm and the effective mass $M^*_{\text{nucl}} = 0.79M_{\text{nucl}}$, which leads to the leading volume term of 0.053 MeV in Eq. (3.95) according to Eq. (3.93). Along the line of β-stability the charge number is $Z \approx A(100 + 0.3A)/(A + 200)$. Note that the representation of the free energy by a quadratic form in T is only valid along the positive, real T axis and does not amount to an expansion in a power series in T. In fact, the thermodynamic potentials have an essential singularity at $T = 0$.

Bartel, Pomorski, and Nerlo-Pomorska made temperature-dependent Hartree-Fock calculations in a relativistic mean field approach [377]. They obtained the free energy from which the shell correction was subtracted to get the smooth background. Representation of the latter by a quadratic form in T gave the following result

$$a/A = \{0.032(1 - 3.265I^2) + 0.22A^{-1/3}(1 + 5.644I^2)$$
$$+0.0021Z^2/A^{4/3}\} \text{ MeV}^{-1}. \quad (3.96)$$

There are considerable differences between the theoretical predictions (e - g) and between fits to data (a - c). None of these curves has even approximately reached the asymptotic value for $A \to \infty$. The surface and curvature terms are therefore seen to shift the level-density parameter significantly upwards in the interesting range of mass numbers $50 \leq A \leq 250$. It is obviously difficult to obtain independently values for the volume term and the surface term by fitting a sample of data from this limited mass range only. Additional information from $\Gamma_{\text{fiss}}/\Gamma_n$ is needed. Theoretical approaches based on phenomenological energy-density functionals suffer from a large variation of the effective nucleon mass M^*_{nucl} in infinite nuclear matter from one functional to the other.

4

Statistical Decay Models

If there are points on the fission path where the fissioning system can be reasonably assumed to be in statistical equilibrium with respect to all or almost all of its degrees of freedom, a number of observables can be predicted. Only the deformation energy and the level density at such equilibrium point must be given. In particular, one does not have to know collective transport parameters like the mass and friction tensors for collective motion since one must not know how the system has reached the equilibrium, provided the equilibrium is actually reached. There are two points along the fission path where the hypothesis of statistical equilibrium has been used: At the saddle point and at scission. In this chapter we shall describe the consequences of the statistical assumption at each of these two points.

We assume that the nuclear temperature T lies in the range

$$B_{\text{fiss}} \geq T \gg 20/A \text{ MeV}. \tag{4.1}$$

The first inequality in (4.1) shall ensure that it is the fission barrier B_{fiss} which controls the fission decay. The second inequality in (4.1) shall ensure the approximate validity of the concept of a nuclear temperature: Since the compound nucleus is a closed system, it should be represented by a microcanonical ensemble. Its total energy is a sharp quantity in the thermodynamical sense and its temperature has a probability distribution. Substituting this microcanonical ensemble approximately by the more convenient canonical one, requires at least that the fluctuation of the excitation energy in the canonical ensemble is much smaller than the average excitation energy

$$\langle E^{*2} \rangle - \langle E^* \rangle^2 \ll \langle E^* \rangle^2.$$

Using the relation

$$\langle E^{*2} \rangle - \langle E^* \rangle^2 = T^2 \partial_T \langle E^* \rangle,$$

which follows directly from the definition of canonical ensemble averages

$$\langle E^* \rangle = \frac{\sum_n E_n \exp(-E_n/T)}{\sum_n \exp(-E_n/T)},$$

H. J. Krappe and K. Pomorski, *Theory of Nuclear Fission*,
Lecture Notes in Physics 838, DOI: 10.1007/978-3-642-23515-3_4,
© Springer-Verlag Berlin Heidelberg 2012

and the Fermi gas relation $\langle E^* \rangle = aT^2$ with the level density parameter $a \approx A/10 \text{ MeV}^{-1}$, the second inequality in (4.1) is obtained.

4.1 Transition-state theory

4.1.1 Excitation energy well above the barrier

Both, the rates for the emission of light particles and for fission can be obtained in the transition-state picture [26]: The reaction rate is taken to be the ratio between the phase space of those events that pass the reaction barrier per unit time and the total phase space. In order to calculate the former phase space volume, the degrees of freedom connected with the passage over the barrier are separated from the remaining "intrinsic" degrees of freedom and the phase-space density is obtained by folding the intrinsic level density of the system at the saddle point into that part of the phase space of the translational motion which leads to a decay. Two cases should be distinguished: either the saddle point lies inside the scission point, as sketched in Fig. 4.1 or the saddle lies beyond the scission configuration, as is the case for light particle emission.

In the latter case there are three translational degrees of freedom. Assuming that the only further degrees of freedom of the ejectile are the $2s + 1$ orientations of its spin s in space, the rate can therefore be written in the transition-state theory

$$r(E^*)dt = \frac{2s + 1}{h^3 \rho(E^*)} \int d^3p \, \rho_{\text{sd}}(E^* - B - \epsilon) \int d\Omega_r \int_{R(\Omega_r) - v \cos \theta_p \, dt}^{R(\Omega_r)} dr r^2 \, \mathcal{T}(\epsilon, \Omega_r, \theta_p),$$

(4.2)

where $\rho(E^*)$ and $\rho_{\text{sd}}(E^*)$ are the level densities at excitation energy E^* for the compound nucleus and for the intrinsic system at the saddle point, respectively. For neutron emission ρ_{sd} is the level density of the residual nucleus. In Eq. (4.2) $\mathcal{T}$ is the transmission coefficient.

One has to remember that in high-dimensional convex bodies as the $6A$-dimensional phase space of an A-body system almost all the volume is located in a thin surface layer. Therefore the phase space volume is taken to be proportional to the level density. The contribution of the translational degrees of freedom to the phase space is represented by the integrals over $\mathbf{p}$ and $\mathbf{r}$ space, the latter consisting of the volume of those particles which reach the surface in the time interval dt and can therefore contribute to the decay. The $\mathbf{p}$ integration is to be extended over the half sphere of the outward pointing momenta $\mathbf{p}$. In Eq. (4.2) B is for charged emitted particles the barrier height and $R(\Omega_r)$ is the barrier distance. For neutrons B is the separation energy B_n and R is the radius of the single-particle potential. The translational kinetic energy is $\epsilon = p^2/2m$, where m is the reduced mass, θ_p is the emission angle with respect to the local surface normal of the parent nucleus, and $\mathcal{T}(\epsilon, \Omega_r, \theta_p)$

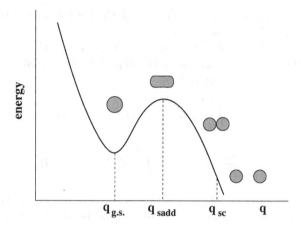

Fig. 4.1. Schematic picture of the potential energy along the fission path (after Ref. [281]).

the transmission coefficient at energy ϵ and emission angle θ_p of the emitting surface element between polar angles Ω_r and $\Omega_r + \delta\Omega_r$. The transmission coefficient is assumed to be independent of the spin s of the ejectile.

For a spherical emitter R and $\mathcal{T}$ do not depend on the polar angles Ω_r and the spatial integrations in Eq. (4.2) become trivial

$$r(E^*) = (2s+1)\frac{4\pi R^2}{h^3 m\rho(E^*)} \int d^3p \, \rho_{\text{sd}}(E^* - B - \epsilon) \, p\cos\theta_p \, \mathcal{T}(\epsilon, \theta_p). \quad (4.3)$$

Transmission coefficients are usually not calculated as functions of the emission angle θ_p, but of the conjugate variable l, the quantum number of the orbital angular momentum. It is convenient in this case to consider first the contribution of a single partial wave to the emission rate. This amounts to a one-dimensional emission problem [378]. Similar to Eq. (4.2) one can write

$$r_l(E^*)dt = \frac{2s+1}{h\rho(E^*)} \int dp_r \int_{R-(p_r/m)dt}^{R} dr \rho_{\text{sd}}(E - B_l - \epsilon_r) \, \mathcal{T}_l(\epsilon)$$

$$= \frac{2s+1}{h\rho(E^*)} \int_0^{E^*-B_l} d\epsilon_r \, \rho_{\text{sd}}(E^* - B_l - \epsilon_r) \, \mathcal{T}_l(\epsilon) \, dt$$

in terms of the radial momentum p_r, the radial energy $\epsilon_r = p_r^2/(2m)$, with $\epsilon = \epsilon_r + \hbar^2 l^2/(2mR^2)$ and the barrier for the lth partial wave $B_l = B + \hbar^2 l^2/(2mR^2)$. Changing the integration variable from ϵ_r to ϵ and summing over all angular momenta, including their orientation in space, gives

$$r(E^*) = \frac{2s+1}{h\rho(E^*)} \sum_l (2l+1) \int_0^{E^*-B} d\epsilon \, \rho_{\text{sd}}(E^* - B - \epsilon) \, \mathcal{T}_l(\epsilon). \quad (4.4)$$

The level density ρ_{sd} is usually identified with the level density of the daughter nucleus, cf. Ref. [378] for further details.

The result (4.4) can of course also be derived directly from Eq. (4.3). Performing the trivial integration with respect to the polar angle ϕ_p, the integral in Eq. (4.3) can be rewritten

$$\frac{\pi}{2} \int_0^{p_{\max}} d(p^2)p^2 \int_0^1 d(\cos\theta_p)^2 \rho_{\mathrm{sd}}(E^* - B - \epsilon)T(\epsilon,\theta_p)$$

$$= \frac{\pi m \hbar^2}{R^2} \int_0^{E^*-B} d\epsilon \int_0^\infty dl\, 2l\, \rho_{\mathrm{sd}}(E^* - B - \epsilon)T_l(\epsilon),$$

where use has been made of the relation $p^2 = 2m\epsilon$ and of the fact that the tangential momentum component at the saddle point R is $p\sin\theta_p = \hbar l/R$ and therefore $-p^2\, d(\cos\theta_p)^2 = (\hbar/R)^2 2l\, dl$. Replacing the integration over continuous angular momenta $\int 2l\, dl$ by the corresponding quantal expression $\sum_l(2l+1)$ yields the rate (4.4).

When the scission point lies beyond the saddle point and $E^* > B_{\mathrm{fiss}}$, the reaction is usually treated in the statistical model as one-dimensional classical motion in the fission direction. The fission rate is therefore given in the transition-state picture by the Bohr-Wheeler formula [24]

$$r_{\mathrm{BW}}(E^*)dt = \frac{1}{h\rho(E^*)} \int_0^{p_{\max}} dp \int_{-vdt}^0 dq\, \rho_{\mathrm{sd}}(E^* - B_{\mathrm{fiss}} - \epsilon)$$

and with $d\epsilon = v dp$

$$r_{\mathrm{BW}} = \frac{1}{h\rho(E^*)} \int_0^{E^*-B_{\mathrm{fiss}}} d\epsilon \rho_{\mathrm{sd}}(E^* - B_{\mathrm{fiss}} - \epsilon). \tag{4.5}$$

For excitation energies well above the fission barrier the relevant level densities are conveniently calculated in the Fermi-gas approximation. In the following we shall first consider this case. Using the relations

$$S = 2aT = 2\sqrt{aE^*} \tag{4.6}$$

for the entropy S of an ideal Fermi gas, which are valid in nuclear matter for moderate temperatures, $T \leq 4$ MeV, [132], the level density becomes $\rho = \exp(S)\partial S/\partial E^* = \exp(S)/T$, where all finite-size effects have been neglected. For the fission rate (4.5) we then get

$$r_{\mathrm{BW}} = \frac{T}{h} \frac{e^{S_{\mathrm{sd}}} - 1}{e^S}, \tag{4.7}$$

where S_{sd} denotes the entropy at the saddle point. If $S_{\mathrm{sd}} \gg 1$ and if the shape-dependence of the level-density parameter a is neglected, one obtains

$$r_{\mathrm{BW}} \approx (T/h)e^{\{2\sqrt{a(E^*-B_{\mathrm{fiss}})}-2\sqrt{aE^*}\}}$$

and for $E^* \gg B_{\text{fiss}}$ with Eq. (4.6)

$$2\sqrt{a(E^* - B_{\text{fiss}})} - 2\sqrt{aE^*} \approx -B_{\text{fiss}}/T. \tag{4.8}$$

Therefore an Arrhenius-type rate-formula

$$r_{\text{BW}} \approx (T/h)e^{-B_{\text{fiss}}/T} \tag{4.9}$$

follows in the lowest nonvanishing order of B_{fiss}/E^*. The derivation of this formula shows that the simple expression (4.9) for the fission rate follows from the transition-state theory only if the conditions $T_{\text{sd}} \gg 20/A$ MeV and $E^* \gg B_{\text{fiss}}$ are satisfied. The latter condition implies that the temperature is approximately constant during the reaction.

In 1928 Polanyi and Wigner derived the Arrhenius formula (4.9) under the same condition for unimolecular reactions [379], though on a somewhat different way [380]. For the preexponential factor T/h in Eq. (4.9) of the dimension of a frequency, $\tilde{\omega}/(2\pi) = T/h$, Polanyi and Wigner gave the following interpretation: The Planck formula for the average occupation number $\bar{n}_\omega = [\exp(\hbar\omega/T) - 1]^{-1}$ of an oscillator with frequency ω in a Bose gas with temperature T (cf. e.g. Landau, Lifshitz, Ref. [319], §63) shows that for $\omega = \tilde{\omega}$ the exponent in this expression is just unity. One can therefore take $\tilde{\omega}$ as an estimate of the highest frequency of the vibrational eigenmodes of a molecule which are not frozen-in at temperature T. Polanyi and Wigner therefore argue that $\tilde{\omega}$ sets the time scale on which energy is fed into the unstable dissociation mode to excite it up to the barrier.

Formula (4.9) was derived with the simple Fermi-gas expression for the level density. Consideration of shell and pairing effects, to be discussed in later sections, leads to a modification of the expansion (4.8) and consequently of the rate formula (4.9). There are numerous examples in the literature where the rate formula (4.9) is used although the conditions for its validity are not all fulfilled.

Often the ratio of the neutron width Γ_{n} to the fission width Γ_{fiss}, $\Gamma_{\text{n}}/\Gamma_{\text{fiss}} = r_{\text{neut}}/r_{\text{BW}}$ is used for comparison with experimental data. From Eq. (4.3) one obtains for spherical nuclei with the classical value for the transmission coefficient, $\mathcal{T}(\epsilon) = 0$ for $\epsilon < B_{\text{n}}$ and $\mathcal{T}(\epsilon) = 1$ for $\epsilon > B_{\text{n}}$,

$$r_{\text{neut}} = \frac{4mR^2}{2\pi\hbar^3\rho(E^*)} \int_0^{E^*-B_{\text{n}}} \rho_{A-1}(E^* - B_{\text{n}} - \epsilon)\epsilon d\epsilon \tag{4.10}$$

with the neutron binding energy B_{n}. The level density ρ_{A-1} in the integral refers to the daughter nucleus after neutron emission. Together with the Bohr-Wheeler formula (4.5) one obtains

$$\frac{\Gamma_{\text{n}}}{\Gamma_{\text{fiss}}} = \frac{4mR^2 \int_0^{E^*-B_{\text{n}}} \rho_{A-1}(E^* - B_{\text{n}} - \epsilon)\epsilon d\epsilon}{\hbar^2 \int_0^{E^*-B_{\text{fiss}}} \rho_{\text{sd}}(E^* - B_{\text{fiss}} - \epsilon)d\epsilon}. \tag{4.11}$$

It is in the spirit of the transition-state model that the quantities B_n and B_fiss should be taken at zero temperature [381]. This is to be distinguished from the temperature-dependent barrier height in the Helmholtz free energy in Kramers' rate formula to be discussed in the fifth chapter.

Formula (4.9) is valid for compound nuclei with zero angular momentum. For finite angular momentum the rotational energy at the saddle point can be written for axially symmetric shapes in terms of the total angular momentum J and its projection K onto the symmetry axis

$$E_\mathrm{rot}(J,K) = (\hbar^2/2\mathcal{J}_\perp)(J^2 - K^2) + (\hbar^2/2\mathcal{J}_\parallel)K^2$$
$$= (\hbar^2 J^2/2\mathcal{J}_\perp) + (K^2/2K_0^2)$$

with

$$K_0^{-2} = \hbar^2(1/\mathcal{J}_\parallel - 1/\mathcal{J}_\perp), \qquad (4.12)$$

where $\mathcal{J}_\perp$ and $\mathcal{J}_\parallel$ are the components of the inertial tensor at the saddle point perpendicular and parallel to the symmetry axis, respectively. They are treated as empirical parameters, not necessarily equal to the values for rigid rotation. In many cases they are in fact closer to the smaller values of an irrotational liquid drop. To the potential barrier B_fiss in Eq. (4.9) the rotational energy $E_\mathrm{rot}(J,K)$ is to be added to yield $B_\mathrm{fiss}(J,K)$ and instead of Eq. (4.9) one obtains for the fission rate of a compound state with angular momentum J

$$r_\mathrm{fiss}(J) = \frac{T}{h}\sum_K P_J(K)e^{-B_\mathrm{fiss}(J,K)/T}, \qquad (4.13)$$

where $P_J(K)$ is the probability that the angular momentum J has the projection K onto the symmetry axis of the fissioning system.

It is consistent with the assumptions of the statistical model to require that the orientation of the axis of the nucleus with respect to the spin direction in the lab system is in thermal equilibrium with the intrinsic degrees of freedom of the system, in particular at the saddle point, and $P_J(K)$ therefore becomes

$$P_J(K) = P_J\frac{\exp[-B_\mathrm{fiss}(J,K)/T]}{\sum_K \exp[-B_\mathrm{fiss}(J,K)/T]}, \qquad (4.14)$$

with P_J independent of K. For the construction of P_J from the entrance-channel angular momenta of a compound reaction we refer to Vandenbosch and Huizenga, Ref. [106] Section VI B. In heavy-ion fusion-fission reaction studies, in which one is dealing with particularly large angular momenta, it is however often assumed that the angular momentum is not only perpendicular to the reaction plane, but also to the fission direction. Therefore the probability

$$P_J(K) = \begin{cases} P_J & \text{if } K = 0 \\ 0 & \text{if } K \neq 0 \end{cases} \qquad (4.15)$$

is used in Eq. (4.13). As pointed out by Lestone [382], this assumption means that in the average the angular momentum component perpendicular to the symmetry axis is larger than with the distribution (4.14) and since it is connected with the larger moment of inertia, the barrier $B_{\mathrm{fiss}}(J, K)$ is lower and fission appears enhanced in this case compared to the use of the probability (4.14). The multiplicity of evaporated light particles per fission event is therefore reduced when (4.15) is used instead of (4.14).

For rather low fission barriers $B_{\mathrm{fiss}}(I)$ Ramamurthy and Kapoor argued that the fission time may be comparable to the time of thermal equilibration of the K quantum number [383]. Therefore a situation evolving from Eq. (4.15) to Eq. (4.14) may be more realistic. Rather schematically they split the total probability between (4.14) and a second component having a K value with a narrow Gaussian distribution around a most probable value, corresponding to (4.15). A less schematic, smooth time evolution of the width of an initial Gauss distribution was later used by Vorkapić and Ivanišević [384].

In particle-induced fission experiments one is interested in the angular distribution of fission fragments with respect to the beam axis. Halpern and Strutinsky [385] assumed that the fission saddle is axially symmetric and that the fissioning system retains its axial symmetry on the way from saddle to scission and that furthermore the angular momentum component K remains constant between saddle and scission. The angular distribution of the fragments is then given by the angular distribution of the symmetry axis with respect to the beam axis at the saddle point. Treating the saddle-point configuration as a spinning symmetric top, the probability distribution with respect to the cosine of the angle θ between beam and symmetry axis for given J with component M on the beam axis is

$$W_{JM}(\theta) = \frac{2J+1}{2} \sum_{K=-J}^{J} P_J(K)|d_{M,K}^J(\theta)|^2 \qquad (4.16)$$

in terms of the eigenfunctions of the top $d_{M,K}^J$ (after integration over the other two Euler angles) [83]. The normalization of the d functions is

$$\frac{2J+1}{2} \int_{-1}^{1} |d_{M,K}^J(\theta)|^2 d\cos(\theta) = 1.$$

In Eq. (4.9) B_{fiss} is the height of the saddle point of the potential energy surface with respect to all shape degrees of freedom. Moretto [386] also introduced the conditional saddle-point energies $B_{\mathrm{fiss}}(\eta_0)$ for fixed mass asymmetry $\eta_0 = (A_1 - A_2)/(A_1 + A_2)$, where A_1 and A_2 are the mass numbers of the fission fragments. For nuclei and mass splits where the scission point is very close to the conditional saddle point, a fission rate for decay with mass split η was calculated [387] by using $B_{\mathrm{fiss}}(\eta)$ in Eq. (4.5) instead of the unconditional barrier.

4.1.2 Excitation energy close to the fission barrier

To investigate the fission excitation function for energies close to the fission barrier, photoabsorption, transfer reactions, e.g. (d,pf) or (t,pf) reactions, and slow neutron absorption measurements are used. The gross structure of the excitation function shows an exponential increase below the fission barrier (in case of several barriers, below the largest one) and a slower increase for higher energies. Superimposed on this threshold behavior is often an intermediate structure of a few tens of keV width. At the energy resolution, reached with slow neutrons from white neutron sources and the time-of-flight technique, a fine structure on a few eV scale becomes visible, reflecting individual compound resonances and their coupling to fission channels.

The interpretation of these measurements requires a discussion of several quantal effects, not considered in the preceding subsection: Shell corrections of the liquid-drop energy surface become important, leading to multiply-humped potentials along the fission path of fissile nuclei, and the transmission coefficient in the fission degree of freedom have to account for tunneling effects. We will postpone a discussion of the latter to the fifth chapter, since their evaluation requires the introduction of fission dynamics. As first pointed out by A. Bohr [388] there are only a few, in general collective rotational or vibrational states above the fission barrier of even nuclei. If the excitation energy of such a nucleus is just a few hundred keV above its barrier, the fission reaction has to pass through these fission-channel states, characterized by quantum numbers J, K, and the parity π.

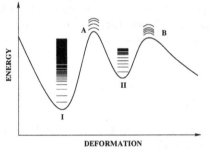
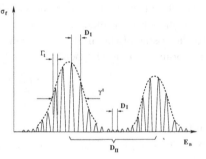

Figure 4.2: Intrinsic excitations in the two potential wells and at the two saddle points (after Ref. [389]).

Figure 4.3: Schematic picture of the resonant (n,f) excitation function with subbarrier neutrons (after Ref. [389]).

In the case of a double-humped potential the fission excitation function for energies below the higher of the two barriers can be interpreted qualitatively in the following way: When the transmission coefficients of both barriers are much smaller than 1, one speaks of class I states in the ground-state potential minimum and of class II states in the second minimum. The situation is sketched in Fig. 4.2. When the excitation energy is close to the bottom of the second well, vibrational states around this minimum can serve as doorway states towards fission, leading to sharp resonance maxima in the fission excitation function at the energy of these states [390].

At higher excitation energies vibrational states become increasingly damped in both wells. Nevertheless an intermediate structure may arise. The level spacing of class I states is often denoted by D_I, that of class II states by D_{II}, Γ_t is the total width of the compound state and the width of class two states is called γ^t. If one assumes for example that $D_{II} > \gamma^t > D_I > \Gamma_t$, the fission excitation function sketched in Fig. 4.3 arises. For a more detailed analysis of such excitation functions we refer to Section IVF of Vandenbosch and Huizenga's book [106] and to the review papers by Bjørnholm and Lynn [391] and Weigmann, Knitter, and Hambsch [392].

The Halpern-Strutinsky formula (4.16) for the angular distribution of the fission fragments requires some additional qualifications for multiple-humped potentials. We refer to the review article by Kapoor and Ramamurthy [393] for details.

4.1.3 Parity violation in fission

A special feature of fission by thermal neutrons from high-flux reactors is the larger incoming flux, compared to fission induced by light, charged particles or gamma absorption and the high energy resolution of up to $\delta E \approx 0.03$ eV. This resolution is small compared to the width Γ of individual compound resonances in fissile nuclei at an excitation energy of approximately 6 MeV as obtained by the absorption of a thermal neutron. An interesting effect which has been studied with polarized, slow neutrons is the parity violation in fission. It was first observed by Danilyan and his collaborators [394] in the reaction ^{239}Pu$(\vec{n}_{\text{therm}}, \text{f})$. For the dependence of the $(\vec{n}, \text{f})$ reaction cross-section on the angle β between the unit vector $\hat{\mathbf{p}}$ in the emission direction of the light fission fragment and the unit vector $\hat{\sigma}$ of the neutron spin one can make the ansatz $W(\beta) \propto 1 + \alpha_{\text{PNC}} \, \hat{\mathbf{p}}\hat{\sigma}$ and would expect that the parity-non-conservation coefficient α_{PNC} vanishes if the compound states of ^{240}Pu would have well-defined parity since the scalar product $\hat{\mathbf{p}}\hat{\sigma}$ is parity-odd and would therefore have a vanishing expectation value. The authors of Ref. [394] measured the quantity

$$\alpha_{\text{PNC}} = \frac{(\hat{\mathbf{p}}\hat{\sigma})_{\text{parallel}} - (\hat{\mathbf{p}}\hat{\sigma})_{\text{antiparallel}}}{(\hat{\mathbf{p}}\hat{\sigma})_{\text{parallel}} + (\hat{\mathbf{p}}\hat{\sigma})_{\text{antiparallel}}}$$

and found the coefficient α_{PNC} to be of the order of 10^{-4}. The same parity-violating effect, and of the same order of magnitude, was also observed in the reactions ^{235}U$(\vec{n}_{\text{therm}}, \text{f})$ [395] and ^{233}U$(\vec{n}_{\text{therm}}, \text{f})$ [396]. How the weak interaction can produce parity violating compound states which lead to such a surprisingly large effect, was explained by Sushkov and Flambaum in a series of papers [397–399]. We follow essentially the presentation given in Ref. [399].

Shell-model single-nucleon states $|JMl\pi\rangle$ with angular momentum quantum-numbers J, M, l and parity π get an admixture of states $|JMl' - \pi\rangle$ due to the weak interaction. Writing the single-particle Hamiltonian $H = H_0 + H_w$

with the shell-model Hamiltonian H_0 and an effective, parity-violating weak Hamiltonian H_w, the amplitude of admixture α_{sp} becomes in first order perturbation theory

$$\alpha_{sp} = \frac{\langle JMl' - \pi|H_w|JMl\pi\rangle}{E_{JMl\pi} - E_{JMl'-\pi}}. \tag{4.17}$$

An estimate for the parity violating, but time-reversal conserving contribution of the weak interaction to the single-nucleon potential was derived by Michel [400]: $H_w \sim G(\boldsymbol{\sigma}\mathbf{p}_{sp})\rho_{nucl}/(2cM^*_{nucl})$, where $G = (\hbar c)^3(1.166 \cdot 10^{-11}\ \mathrm{MeV}^{-2})$ is the Fermi constant of the weak interaction, $\boldsymbol{\sigma}$, $\mathbf{p}_{sp}$, M^*_{nucl} are the spin, Fermi momentum and effective mass of a nucleon, respectively, and ρ_{nucl} is the nuclear saturation density. We do not derive this result here. For a more detailed discussion we refer to the review by Adelberger and Haxton [401]. With $p^2_{sp}/(2M^*_{nucl}) = 40$ MeV, $\rho_{nucl} = 0.18\ \mathrm{fm}^{-3}$, $cp_{sp} = 270$ MeV, and taking the energy denominator in Eq. (4.17) equal to $\hbar\omega = 41\,A^{-1/3}$ MeV≈ 6.63 MeV for ^{238}U one obtains an admixture coefficient α_{sp} of p-states to s-states in fissile nuclei of the order of 10^{-7}. One has therefore to explain an amplification factor of several orders of magnitude for the parity violation in fission and one has to understand the transfer of the effect from the compound state to the macroscopic observable $\hat{\mathbf{p}}$, the emission direction of the light fragment.

We expand the compound state, formed by the absorption of the neutron in terms of antisymmetrized products ϕ_i of single-particle wave functions,

$$\Psi = \sum_{i=1}^{N} a_i\phi_i. \tag{4.18}$$

The number N of terms in the sum is of the order $N \sim \Delta E\,\rho$, where ΔE is the order of magnitude of the residual interaction and ρ is the level density of a fissile nucleus after absorption of a thermal neutron. With $\Delta E \sim 1$ MeV and $\rho \sim 1$ eV^{-1} follows $N \sim 10^6$. Taking the coefficients a_i to be all of the same order of magnitude, the normalization of Ψ leads to the estimate $|a_i| \sim N^{-1/2}$. For the sake of the present order-of-magnitude estimate we assume that the dominant part of H_w is an effective single-particle potential V_w and therefore the matrix element $\langle i|H_w|j\rangle$ vanishes unless the states i and j contain the same single-particle wave functions, except for one single-particle state which must have the same quantum numbers J and M but opposite parity in i and j. Assuming in this case $\langle i|H_w|j\rangle \sim V_w\delta_{i(\pi)j(-\pi)}$ the mixing matrix element C becomes

$$C = \left\langle \sum_i a_i\phi_i|H_w|\sum_j b_j\phi_j \right\rangle = \sum_{i,j} a_i^* b_j\langle i|H_w|j\rangle \sim \sum_{i=1}^{N} V_w\,a_i^* b_i. \tag{4.19}$$

Because of the chaotic character of the compound states the coefficients a_i and b_i are treated as random numbers with root-mean-square values $\sim N^{-1/2}$. Therefore

$$\overline{C^*C} \sim \overline{|V_w|}^2 \sum_{i,j=1}^{N} a_i a_j^* b_i^* b_j \sim \overline{|V_w|}^2 \sum_{i=1}^{N} |a_i|^2 |b_i|^2 \sim \overline{|V_w|}^2 / N \ .$$

Using this result and Eq. (4.17), the admixture coefficient α_{comp} of parity $(-\pi)$-states to compound states with dominant parity π becomes

$$|\alpha_{\text{comp}}| \sim |C| \rho \sim \alpha_{\text{sp}} \sqrt{N},$$

where we assumed in this order-of-magnitude estimate that $\Delta E \sim E_{JMl\pi} - E_{JMl'-\pi}$. The enhancement of the odd-parity admixture to even-parity compound states by a factor $\sqrt{N} \sim 10^3$ in fissile nuclei compared to single-particle shell-model states is called dynamical enhancement.

In Eq. (4.19) we considered for a given compound state (4.18) the admixture of just one compound state of opposite parity. In general there are several compound states $\Psi_\nu^{(-\pi)}$ of opposite parity and the same total angular momentum with energies $E_\nu^{(-\pi)}$ close to the neutron absorption-energy E, contributing to the parity-breaking effect. Taking into account that the compound states Ψ are not bound states, but have a finite decay width Γ, the admixture coefficient of the νth state becomes

$$\alpha_{\text{comp}}(\nu) = \frac{\langle \Psi_\nu^{(-\pi)} | H_w | \Psi^{(\pi)} \rangle}{E - E_\nu^{(-\pi)} + i\Gamma_\nu^{(-\pi)}/2}. \tag{4.20}$$

We have now to explain how the parity violation in the compound state is transferred to the observable $(\hat{\mathbf{p}}\hat{\boldsymbol{\sigma}})$ involving the collective variable $\hat{\mathbf{p}}$. The transition matrix element between the initial compound state $|\text{in}\rangle$ and the final state of two outgoing fragments $|\text{out}\rangle$ may be written as a sum over a complete set of intermediate states $|a\rangle$

$$\langle \text{out}|\text{in} \rangle = \sum_a \langle \text{out}|a \rangle \langle a|\text{in} \rangle.$$

Choosing the states $|a\rangle$ to be the fission channels above the fission barrier, there are only very few terms in the sum over a when the excitation energy of the compound state is only a little above the barrier. For the sake of simplicity we shall assume that there is only one open channel with quantum numbers J, M, and K. For a nucleus with a strongly asymmetric saddle-point shape each fission channel actually consists of a parity doublet. We further assume that the polarized neutron is absorbed at an s-wave resonance with width $\Gamma < \rho^{-1}$.

First neglecting parity non-conservation effects, we expand the initial compound state with parity π in terms of products of quasistationary single-quasiparticle wave functions ϕ_i and single out the one open fission-channel state $|a\rangle$

$$|\text{in}\rangle^{(\pi)} = \sum_{i=1}^{N} a_i \phi_i^{(\pi)} + \gamma^{(\pi)} |a\rangle^{(\pi)},$$

where $\gamma^{(\pi)} = \langle a|\text{in}\rangle$ is the transition amplitude to the fission channel. Similarly compound states of the opposite parity are given by

$$|\text{in}, \nu\rangle^{(-\pi)} = \sum_{j=1}^{N} b_{j\nu}\phi_j^{(-\pi)} + \gamma_\nu^{(-\pi)}|a\rangle^{(-\pi)}.$$

Under the action of H_w mixed-parity compound states (with dominant parity π) are formed

$$|\widetilde{\text{in}}\rangle^{(\pi)} = |\text{in}\rangle^{(\pi)} + \sum_\nu \alpha_{\text{comp}}(\nu)|\text{in}, \nu\rangle^{(-\pi)}$$

$$= \sum_i a_i\phi_i^{(\pi)} + \sum_\nu \alpha_{\text{comp}}(\nu)\sum_j b_{j\nu}\phi_{j\nu}^{(-\pi)} + \gamma^{(\pi)}|\widetilde{aK}\rangle^{(\pi)}$$

with

$$|\widetilde{aK}\rangle^{(\pi)} = |aK\rangle^{(\pi)} + \left(\sum_\nu \alpha_{\text{comp}}(\nu)\frac{\gamma_\nu^{(-\pi)}}{\gamma^{(\pi)}}\right)|a\,K\rangle^{(-\pi)}.$$

The mirror-asymmetry of the fission saddle-point may be represented by an octupole deformation parameter, in the simplest case described by β_{30}. In the collective energy surface around the saddle point there are two minima with respect to β_{30} at $\pm\beta_{30}^0$. In the adiabatic model the fission-channel state $|aK\rangle$ is characterized by the angular momentum quantum numbers J and M and the intrinsic quantum numbers K and the sign p of β_{30}^0. They refer to the intrinsic state $|K\,p\rangle$ which is a product of single-particle wave functions χ_K in the octupole-deformed single-particle potential and an oscillator ground-state wave-function $\zeta(\beta_{30})$ in the collective octupole variable β_{30} in either the left or the right minimum $|\beta_{30}^0|$ of the octupole potential [402]. The intrinsic state is therefore given by $|K\,p\rangle = \zeta(\beta_{30} - p|\beta_{30}^0|)\,\chi_K$. In terms of the parity operator $\hat{P}$ the total saddle-point state becomes

$$|aK\rangle^{(\pi)} = \sqrt{(2J+1)/(8\pi)}(1 + \pi\hat{P})\mathcal{D}_{MK}^J(\alpha, \beta, 0)|Kp\rangle.$$

We use the relation $\hat{P}\mathcal{D}_{MK}^J|K\,p\rangle = (-1)^{J+\delta}\mathcal{D}_{M-K}^J|-K-p\rangle$, where the phase $(-1)^\delta$ depends on the specific structure of χ_K [403], and get

$$|aK\rangle^{(\pi)} = \sqrt{\frac{2J+1}{8\pi}}\left[\mathcal{D}_{MK}^J(\alpha, \beta, 0)|Kp\rangle\right.$$
$$\left. +(-1)^{J+\delta}\pi\mathcal{D}_{M-K}^J(\alpha, \beta, 0)|-K-p\rangle\right]$$

and

$$|\widetilde{aK}\rangle^{(\pi)} = \sqrt{\frac{2J+1}{8\pi}}\left[(1 + \overline{\alpha}_{\text{comp}})\,\mathcal{D}_{MK}^J(\alpha, \beta, 0)|Kp\rangle\right.$$
$$\left. +(-1)^{J+\delta}\pi\,(1 - \overline{\alpha}_{\text{comp}})\,\mathcal{D}_{M-K}^J(\alpha, \beta, 0)|-K-p\rangle\right] \quad (4.21)$$

with $\bar{\alpha}_{\text{comp}} = \sum_\nu \alpha_{\text{comp}}(\nu)\gamma_\nu^{(-\pi)}/\gamma^{(\pi)}$. If $K = 0$, the fission axis is perpendicular to the total spin $\mathbf{J}$. Therefore $\mathbf{J}$ defines only a plane in which the fission axis lies, but not a direction along the fission axis. In the following we therefore consider only the case $K \neq 0$.

Assuming, as in the derivation of Eq. (4.16), that the fission axis is not bent on the way from saddle to scission and that K is conserved, the orientation of the fission axis in the saddle-point state $|aK\rangle$ gives the emission direction of the fragments at scission. Taking the expectation value $^{(\pi)}\langle\widetilde{aK}|\widetilde{aK}\rangle^{(\pi)}$ of Eq. (4.21) with respect to the intrinsic variables one obtains for the angular distribution of the light fragment

$$W_{JM}(\beta) \propto |\mathcal{D}_{MK}^J(\beta)|^2(1 + \alpha') + |\mathcal{D}_{M\,-K}^J(\beta)|^2(1 - \alpha') \tag{4.22}$$

with $\alpha' = 2\Re\,\bar{\alpha}_{\text{comp}}$. Terms of order α'^2 are neglected in deriving Eq. (4.22). Writing the amplitudes of the saddle-point states $\gamma^{(\pi)} = \sqrt{\Gamma}e^{i\psi(E)}$ and $\gamma_\nu^{(-\pi)} = \sqrt{\Gamma_\nu}e^{i\psi_\nu}$ and using Eq. (4.20) one gets

$$\alpha' = 2\Re\left(\sum_\nu \sqrt{\frac{\Gamma_\nu}{\Gamma}} \frac{\langle\Psi_\nu^{(-\pi)}|H_w|\Psi^{(\pi)}\rangle}{E - E_\nu + i\Gamma_\nu^{(-\pi)}/2} e^{i(\psi(E)-\psi_\nu)}\right),$$

where the ratios Γ_ν/Γ are of order 1. Only because $\Gamma_\nu \neq 0$, i.e. only because the states $\Psi_\nu^{(-\pi)}$ have a finite life time, α' does not vanish. (Note that the matrix element of the operator $H_w \propto \boldsymbol{\sigma}\mathbf{p}_{\text{sp}}$ between an $s_{1/2}$ and a $p_{1/2}$ state is purely imaginary). The phase $\psi(E)$ increases by π when the incoming beam energy E increases through a Breit-Wigner compound resonance [404, 405], with the consequence that α' changes sign.

In experiments with polarized neutrons and an unpolarized target with ground-state spin I the angular distribution becomes in terms of the Clebsch-Gordon coefficient $(I\,M', \frac{1}{2}\frac{1}{2}|JM) = (I + 1/2 \pm M)^{1/2}(2I + 1)^{-1/2}$ for $J = I \pm 1/2$ and $M' = M - 1/2$

$$W(\beta) \propto \sum_{MM'} (I\,M', \frac{1}{2}\,\frac{1}{2}|JM)^2 W_{JM}(\beta)$$

$$\propto 1 + (-1)^{J-I-1/2}\alpha' \frac{K}{I + 1/2}\cos\beta. \tag{4.23}$$

The sum rules $\sum_{M=-J}^J |\mathcal{D}_{MK}^J|^2 = 1$ (unitarity of the $\mathcal{D}$ matrices) and $\sum_{M=-J}^J M|\mathcal{D}_{MK}^J|^2 = KP_1(\cos\beta) = K\cos\beta$ have been used to obtain this result. The latter sum rule can be derived with the help of formula 4.3.4 of Ref. [83] by putting $j = 1$, $m = m' = 0$ in this formula. Kötzle et al. [406] measured the angular dependence of the correlation function $W(\beta)$ and confirmed the $\cos\beta$-dependence of Eq. (4.23). For the correlation function $W(\beta)$ in the case of overlapping resonances we refer to Ref. [399], where Sushkov and Flambaum also discuss the interesting case of a neutron p-wave resonance overlapping with an s-wave resonance. There is an interference between contributions from $J = I + 1/2$ and $J = I - 1/2$ compound states feeding the same intrinsic state $|Kp\rangle$ at the saddle point. The interference is shown to lead to a nonvanishing angular dependence of $W(\beta)$ even for $K = 0$ fission channels.

The preceding discussion was based on the assumption that the parity violation in fission is "decided" at the fission saddle-point, rather than on the way from saddle to scission. There is experimental evidence that this assumption is correct: The parity-violation coefficient $\alpha_{\mathrm{PNC}} = (-1)^{J-I-1/2} \alpha' \frac{K}{I+1/2}$ was found to be independent of the mass ratio A_L/A_H and the total kinetic energy of the fission products [406–408]. Both quantities were determined for actinides at and beyond the saddle point as will be discussed below. It was also found [409] that binary and the corresponding ternary fission reactions lead to the same asymmetry coefficient α_{PNC}, although they correspond to different scission configurations. But they were fed from the same saddle point.

4.1.4 Transmission coefficients

Since we treated the motion in the fission degree of freedom in the framework of classical mechanics, only the fission barrier B_{fiss} appears in Eq. (4.5) and not the transmission coefficient. Formally one can introduce the classical transmission coefficients in this equation

$$\mathcal{T}(E^*) = \begin{cases} 1 & \text{if } E^* \geq B_{\mathrm{fiss}} \\ 0 & \text{if } E^* < B_{\mathrm{fiss}} \end{cases} \tag{4.24}$$

to stress the similarity with Eq. (4.3). For the emission of light particles however, the motion in the separation degrees of freedom has always to be treated as quantized. From a given interaction potential $V(r)$ the transmission coefficient $\mathcal{T}_l(\epsilon)$ for spherical nuclei in Eq. (4.4) is obtained by numerically integrating the radial Schrödinger equation for the lth partial wave.

It might appear that the transmission of particles from the compound state is the time-reverse of the corresponding fusion reaction and that interaction potentials used for fusion might also be suitable to calculate transmission coefficients for decay. However, apart from the fact that fusion experiments are made on targets in their ground state, whereas the daughter nucleus after particle emission is usually not left in its ground state, fission and decay paths follow different shape sequences for heavier systems. For lighter systems a reasonably good description of fusion excitation functions is achieved for $Z_1 Z_2 < 80$ and $Z_1, Z_2 > 2$ [410] with the proximity force from Ref. [203]. In these cases the proximity potential may then be used to obtain transmission coefficients. Still it is not clear which boundary condition one should use inside the barrier: the compound nucleus is not to be modeled as a box in which the ejectiles are moving freely.

In practice the integration of the Schrödinger equation is rather time-consuming and therefore the approximate, but closed-form Hill-Wheeler expression [87] $T_l = [1 + \exp(2W(\epsilon, l))]^{-1}$ is used, where

$$W(\epsilon, l) = \int_{r_1}^{r_2} \left[\frac{2m}{\hbar^2} (V(r) - \epsilon) + \frac{l(l+1)}{r^2} \right]^{1/2} dr \qquad (4.25)$$

is the action integral of the WKBJ approximation across the barrier [411]. The integration limits r_1 and r_2 in this equation are the classical turning points. To avoid the numerical integration, the even further approximated expression

$$W = \pi(B_{\mathrm{barr}}(l) - \epsilon)/\hbar\omega_{\mathrm{barr}} \qquad (4.26)$$

is used, where the potential, including the centrifugal potential, is expanded around its peak $B_{\mathrm{barr}}(l)$ to second order in $r - r_{\mathrm{barr}}(l)$

$$V(r) + \frac{\hbar^2}{2m} \frac{l(l+1)}{r^2} = B_{\mathrm{barr}}(l) - \frac{1}{2} \omega_{\mathrm{barr}}^2(l)\, m\, (r - r_{\mathrm{barr}})^2. \qquad (4.27)$$

The expression (4.27) is valid as long as the barrier is well approximated by a parabola over a distance of the order of magnitude of $\hbar\omega_{\mathrm{barr}}$. This is typically not the case for very shallow potential pockets, where a modelling of the barrier by an inverse hyperbolic cosine $V_0 \cosh^{-2}((r - r_{\mathrm{barr}})/a)$ (Eckart potential [412]) is more appropriate. There is an analytic expression for the transmission coefficient through this barrier (see e.g. Landau, Lifshitz, Ref. [413]), p. 76.

For light particles transmission coefficients are often generated from the optical potential $V_{\mathrm{optical}}(r)$, fitted to measured elastic cross sections

$$V_{\mathrm{optical}}(r) = V_{\mathrm{Coul}} - V_r f(X_r) + \left(\frac{\hbar}{m_\pi c} \right)^2 V_{\mathrm{so}}(\boldsymbol{\sigma} \cdot \boldsymbol{l}) \frac{df(X_{\mathrm{so}})}{r\, dr}$$

$$- i[V_i f(X_i) - 4V_D f'(X_D)]$$

with

$$V_{\mathrm{Coul}} = \frac{Z_1 Z_2 e^2}{r} \begin{cases} 1 & \text{if } r \geq R_c \\ (3 - r^2/R_c^2)r/(2R_c) & \text{if } r \leq R_c, \end{cases}$$

$$f(X_\nu) = \frac{1}{1 + e^{X_\nu}}, \qquad \nu = r, i, \mathrm{so}, \text{ or } D,$$

$R_c = r_c A^{1/3}$, $X_\nu = (r - r_\nu A^{1/3})/a_\nu$, and $\hbar^2/(m_\pi c)^2 = 2$ fm^2. Standard parameter sets for emitted light particles are given in table 4.1. In this approach it is assumed that the spectroscopic factor of the emitted particle is one. For composite particles it is however considerably smaller [419]. This is partly compensated by larger barriers to be tunneled.

Clearly the optical model is even less suitable to be time-inverted. The imaginary part would change sign! Also the proposal by Alexander, Magda,

Table 4.1. Woods-Saxon parameters for the scattering of light particles on target nuclei with charge Z, mass number A, and neutron excess I. For neutrons, protons, and deuterons the entries refer to pure surface absorption, $V_i = 0$, for the other projectiles to pure volume absorption, $V_D = 0$, and ϵ is the lab energy in MeV.

projectile	V_r[MeV]	r_r[fm]	a_r[fm]
n	$47.01 - 0.267\epsilon - 0.00118\epsilon^2$	$1.322 - 7.6 \cdot 10^{-4}A + 4 \cdot 10^{-6}A^2 - 8 \cdot 10^{-9}A^3$	0.66
p	$53.3 - 0.55\epsilon + 27I + 0.4Z/A^{1/3}$	1.25	0.65
d	$91.13 + 2.2Z/A^{1/3}$	1.05	0.80
t	$165 - 0.17\epsilon - 6.4I$	1.20	0.72
^{3}He	$151.9 - 0.17\epsilon + 50I$	1.20	0.72
α	50	$1.17 + 1.77A^{-1/3}$	0.576

projectile	$V_{i/D}$[MeV]	$r_{i/D}$[fm]	$a_{i/D}$[fm]
n	$9.52 - 0.53\epsilon$	$1.266 - 3.7 \cdot 10^{-4}A + 2 \cdot 10^{-6}A^2 - 4 \cdot 10^{-9}A^3$	0.48
p	$3A^{1/3}$	1.25	0.47
d	$218/A^{2/3}$	1.43	$0.5 + 0.013A^{2/3}$
t	$46 - 0.33\epsilon - 110I$	1.40	0.84
^{3}He	$41.7 - 0.33\epsilon + 44I$	1.40	0.88
α	$5.87A^{1/3} - 12.01$	$1.17 + 1.77A^{-1/3}$	0.576

projectile	V_{so}[MeV]	r_{so}[fm]	a_{so}[fm]	r_c[fm]	Ref.
n	–	–	–	–	[414]
p	7.5	1.25	0.47	1.25	[415]
d	7.0	0.75	0.50	1.30	[416]
t	2.5	1.20	0.72	1.30	[417]
^{3}He	2.5	1.20	0.72	1.30	[417]
α	–	–	–	1.17	[418]

and Landowne [378] to fit elastic scattering data with a real potential and ingoing wave boundary-conditions [420] can only be successful if direct reactions are an insignificant fraction of the total reaction cross section. One may therefore try to express the decay rate in terms of the fusion excitation function without the intermediate step of a potential. In cases where the emission of ejectiles can reasonably be considered as the time inverse of the corresponding fusion or absorption reaction, one finds for the evaporation rate from the principle of detailed balance [421]

$$r(\epsilon, J_R, s; E^*, J) = \frac{(2s + 1)(2J_R + 1)}{2J + 1} \frac{m\epsilon}{\pi^2 \hbar^3}$$
$$\times \frac{\rho_R(E^* - B - \epsilon; J_R)}{\rho(E^*, Z, A, J)} \sigma_{\text{fus}}(J_R, s, \epsilon; J, E^*). \quad (4.28)$$

The first factor on the right hand side accounts for the spin multiplicity of the ejectile spin s, residual nucleus spin J_R, and compound nucleus spin J.

Note that the evaporation rate r of Eq. (4.28) refers to fixed angular momenta J_R, s, and J, rather than fixed orbital angular momentum l as in Eq. (4.4). The level densities of the residual nucleus and the compound nucleus are ρ_R and ρ, respectively and $\sigma_{\text{fus}}(J_R, s, \epsilon; J, E^*)$ is the fusion cross section of two nuclei with angular momenta J_R and s and with channel energy ϵ forming a compound state with angular momentum J and excitation energy E^*. Finally m is the reduced mass. The conservation laws require

$$E^* = \epsilon + E_R + Q_{\text{reaction}}$$
$$J = J_R + s + l,$$

where E_R is the excitation energy of the daughter nucleus and Q_{reaction} is the Q-value of the emission reaction. No internal degrees of freedom of the ejectile other than its spin are assumed to contribute to the level density of the final system.

Experimental fusion excitation functions σ_{fus} are of course only available for $E_R = 0$ and $J_R =$ ground-state spin. Moreover, the empirical σ_{fus} are often substituted by the empirically fitted functions

$$\sigma_{\text{fus}} = \alpha(1 + \beta/\epsilon)\pi R^2$$

for neutrons [422] with $\alpha = 0.76 + 1.93A^{-1/3}$, $\alpha\beta = 1.66A^{-2/3} - 0.05$, and $R = 1.70A^{1/3}$ fm and for charged particles by Weisskopf's classical expression [423]

$$\sigma_{\text{fus}}(\epsilon) = \pi R^2(1 - V/\epsilon) \qquad (4.29)$$

with $R = 1.42A^{1/3} + R_i$ and $V = ZZ_ie^2/(r_eA^{1/3} + R_i)$, where for protons $Z_i = 1$, $R_i = 1.44$ fm, $r_e = 1.81$ fm and for ^{4}He $Z_i = 2$, $R_i = 2.53$ fm, $r_e = 2.452 - 0.408\log_{10}(2Z)$ fm [424]. In these expressions A is the mass number of the daughter nucleus.

As to be expected the Richardson law for the thermionic electron current from a heated metal surface (see e.g. Ashcroft and Mermin, Ref. [206], Chapter 2) can be derived from the rate (4.28) using the relations (4.8) for the infinitely extended system of conduction electrons. In this case $s = 1/2$, $J = J_R = 0$, m is the electron mass, and Q the work function W. With the expression (4.29) for σ_{fus} with $V = W$, integration of Eq. (4.28) with respect to ϵ yields for the number of fermions emitted per unit time

$$\dot{N} = \frac{2R^2m}{\pi\hbar^3}e^{-\sqrt{4aE^*}}\int_0^{E^*-W}(\epsilon - W)e^{\sqrt{4a(E^*-W-\epsilon)}}d\epsilon. \qquad (4.30)$$

Introducing the new integration variable $x^2 = 4a(E^* - W - \epsilon)$ and the abreviation $s^2 = 4a(E^* - W)$, the integral becomes

$$\frac{1}{8a^2}\int_0^s(s^2 - 4aW - x^2)xe^x dx \approx \frac{s^2}{4a^2}e^s,$$

where $s \gg 1$ and $s^2 \gg 4aW$ has been used. Inserting this result into Eq. (4.30) yields

$$\dot{N} = \frac{2R^2 m}{\pi \hbar^3} \frac{E^* - W}{a} e^{2\sqrt{a(E^* - W)} - 2\sqrt{aE^*}} \approx \frac{2R^2 m T^2}{\pi \hbar^3} e^{-W/T}$$

in lowest nonvanishing order of W/E^*. A connection between the time rate of change of the charge $e\dot{N}$ and the electric current density j is obtained by integrating the continuity equation over a sphere with radius R, i.e. $4\pi R^2 j = e\dot{N}$ and therefore one gets the Richardson law

$$j = \frac{emT^2}{2\pi^2 \hbar^3} e^{-W/T}.$$

Several approximate schemes have been proposed to generalize the result (4.28) to the case of non-spherical emitting nuclei. In Ref. [425] transmission coefficients are calculated for a spherical nucleus, however with an angle-dependent nuclear radius parameter $R(\theta)$ in the optical potential. The resulting, angle-dependent transmission coefficients are averaged over the angle, thus neglecting the nonsphericity of the Coulomb field. Iwamoto and Herrmann [426] used Eq. (4.3) with the classical expression for the transmission coefficient $T(\epsilon, \theta) = 1 - V_{\text{barr}}(\theta)/\epsilon$ [423]. In Ref. [427] Hill-Wheeler transmission coefficients in the approximation (4.26) are used with $V(r)$ in Eq. (4.27) replaced by $V(\mathbf{r}, \omega)$, where the angular dependence of the nuclear and Coulomb single-particle potentials and the rotation of the potential well with angular frequency ω are taken into account in the framework of semiclassical dynamics. For α decay a more careful treatment of the coupling between the angular momenta of the orbital motion of the ejectile and the deformed daughter nucleus in the noncentral potential is considered by Rasmussen and Segal [428] and by Fröman [429]. However, the contribution of the nuclear proximity force to the potential is not considered in these calculations.

4.1.5 Cascade calculations

A compound nucleus with excitation energy E^* and angular momentum J may decay by particle emission according to the rate formula (4.4) or fission with the rate (4.13), provided E^* exceeds the respective barriers. In addition, decay through γ-emission might occur. For the γ emission rate a formula analogous to Eq. (4.4) holds [430]

$$r_\gamma(\epsilon_\gamma, l_\gamma; Z, A, E^*, J) = \frac{1}{\hbar \rho(E^*, Z, A, J)} \sum_{J_R = |J-l|}^{J+l} \rho_R(E^* - \epsilon_\gamma, J_R) f_l(\epsilon),$$

where $f_l(\epsilon)$ is often given as an empirical multiple of the Weisskopf single-particle unit [431], which is for electric multipole transitions [432]

$$r_W(\epsilon, l \to 0) = \frac{2(l+1)}{l[(2l+1)!!]^2} \left(\frac{3}{l+3}\right)^2 \frac{e^2}{\hbar c} \left(\frac{\omega R}{c}\right)^{2l} \hbar\omega\delta(\hbar\omega - \epsilon).$$

Sometimes only giant dipole γ-emission is considered. In this case

$$f_l(\epsilon) = \frac{4}{3\pi} \frac{1+\kappa}{M_{\text{nucl}}c^2} \frac{e^2}{\hbar c} \frac{NZ}{A} \frac{\Gamma_D\epsilon^4}{(\Gamma_D\epsilon)^2 + (\epsilon^2 - E_D^2)^2},$$

where Γ_D and E_D are the width and energy of the dipole giant resonance, respectively, and κ is the relative increase of the dipole sum-rule because of nonlocal parts in the particle-hole residual interaction in the dipole channel, cf. Chap. 7 of Ref. [404].

If the excitation energy E^* of the compound nucleus is sufficiently large, there may be enough energy left after gamma or particle emission to overcome the fission barrier ("second chance fission"). For still higher initial excitation energy even two or more generations of light particles may be emitted prior to fission ("multiple chance fission"). Such decay cascades can proceed until either an evaporation residue nucleus emerges, which has no longer sufficient excitation energy to decay by fission, gamma or particle emission or fission has occurred. In the latter case the fission fragments will deexcite by further particle emission. Observables are, at least in principle, the energy distributions of the light particles and gammas in coincidence with the distribution of evaporation residues over mass and charge and/or these distributions for the deexcited fission fragments.

Two methods have been implemented in cascade codes to calculate some or all of these observables. In the first method one starts with the original compound state and calculates with the rate equations for the various decay channels the probability distribution for the decay products of the first generation. For those products which can still further decay the probability distributions over energy, mass, and charge of the second generation of decay products are determined and so forth until the excitation energy of all remaining nuclei is below all particle-decay thresholds. In the end the deexcited nuclei from all generations are collected in mass, charge, and kinetic energy bins. This strategy is used in the codes CASCADE [351], ALERT [433], and GEMINI [387]. In another approach the probability of each particular decay line from the initial compound state through all generations is followed. A Monte-Carlo sampling strategy is used to give the lines with large probability preference compared to less probable ones. An implementation of this strategy resulted in the codes JULIAN [434], PACE [435], and JOANNE [436].

We have so far assumed that the deexcitation cascade starts from a thermally equilibrated compound nucleus. When the compound nucleus results from bombardment of a heavy target nucleus with light particles (n, p, α) of sufficiently high energy or from heavy-ion fusion reactions with more than 10 MeV/A kinetic energy, the energy spectrum of emitted particles shows high-energy tails beyond the standard thermal distribution. The corresponding events have been associated with particle emission before thermal equilibrium of the emitter was reached. The spectrum of these precompound decays was quite successfully described by Griffin's exciton model [437].

Assume the reaction is induced by the absorption of a nucleon with sufficient energy by a larger target nucleus. It is then kept in the potential well of the latter in a continuum state and can either leave the target without energy loss or collide with a nucleon of the target producing a particle-hole excitation. In the latter case collisions of the two particles above the Fermi surface may produce another particle-hole excitation and so on. Eventually the energy of the incoming particle is distributed among emitted precompound nucleons and a complicated, hopefully thermalized configuration of the target residue. The particle-hole excitations are called excitons. In each step of the deexcitation process the state of the precompound system is characterized by the number of particles p, the number of holes h, and the excitation energy E^*. The sum $n = p + h$ is called the exciton number. If the residual interaction between particles above the Fermi energy and between particles and holes is assumed to be a two-body force, the exciton number can either change in each reaction step by two or not at all, $\Delta n = 0, 2$. This picture has been generalized by allowing more than one particle in the initial state, thus providing for the description of precompound effects in reactions induced by composite projectiles. Also the emission of (light) composite precompound particles has been included in this picture by introducing a semiempirical preformation probability of such particles in a p particle exciton-state.

The evolution of the exciton cascade is to be described by the probability for the transition from one exciton generation to the next and the emission probability of a particle from a given exciton generation (during its lifetime). Invoking Fermi's Golden Rule the transition probability from an exciton generation n with energy E^* to an exciton generation with $n + \Delta n$ is written

$$\lambda(n, \Delta n, E^*) = (2\pi/\hbar)\overline{\langle |M|\rangle^2}\, w(n, \Delta n, E^*),$$

where $\Delta n = 0, \pm 2$. The squared transition matrix element $\langle |M|\rangle^2$ is averaged over the many possible members of the initial and final exciton generation and is treated as another semiempirical parameter in the exciton model. The last factor, $w(n, \Delta n, E^*)$, is the level density of the final exciton state. It can be expressed in terms of the density $\omega(p, h, E^*)$ of states with p particles, h holes, and excitation energy E^*: For $\Delta n = 2$ either one particle is destroyed and two particles and one hole are created or one hole is destroyed and one particle and two holes are created. One therefore has to multiply the probability of having $p - 1$ particles, h holes, at energy $E^* - \epsilon$ with the number of two-particle one-hole states at energy ϵ and integrate over the energy ϵ. A similar term for the creation of two-hole one-particle states is to be added to obtain $w(n, 2, E^*)$

$$w(n, 2, E^*) = \frac{1}{\omega(p, h, E^*)} \int_0^{E^*} [\omega(p-1, h, E^* - \epsilon)\omega(2, 1, \epsilon)$$

$$+\omega(p, h-1, E^* - \epsilon)\omega(1, 2, \epsilon)]d\epsilon. \tag{4.31}$$

Obložinský, Ribanský, and Běták have given analogous expressions for $\Delta n = 0, -2$ [438]. For the particle-hole level density $\omega(p, h, E^*)$ in the picket-fence model (3.39) analytic formulae for the $w(n, \Delta n, E^*)$ are also given in Ref. [438].

For the emission width of a particle of species α the expression

$$\Gamma_\alpha(n, E^*) = \frac{2s_\alpha + 1}{\pi^2 \hbar^2} m_\alpha \int_0^{E^* - B_\alpha} \epsilon \sigma_{\text{fus}}^\alpha(\epsilon) \frac{\omega(p - p_\alpha, h, E^* - B_\alpha - \epsilon)}{\omega(p, h, E^*)}$$

$$\times N_\alpha \gamma_\alpha(p) R_\alpha(p) d\epsilon \tag{4.32}$$

is used, where the first part is essentially the integral over ϵ of Eq. (4.28) with the level densities restricted to the exciton degrees of freedom, B_α is the binding energy of particle α and m_α its reduced mass. The factor N_α counts the internal number of states of the emitted particle α, γ_α is the preformation probability of the species α and $R_\alpha(p)$ accounts for the probability that p_α nucleons of the total number p have the right isospin, the latter three factors being 1 for nucleon emission.

With these ingredients the evolution of the exciton cascade is described as a Markov process by the master equation

$$\dot{P}(n) = \lambda(n - 2, 2, E^*)P(n-2) + \lambda(n + 2, -2, E^*)P(n+2)$$

$$- \left[\lambda(n, 2, E^*) + \lambda(n, -2, E^*) + \sum_\alpha \Gamma(n, E^*) \right] P(n)$$

for the probability $P(n, t)$ to find n excitons in the nucleus at time t.

There are a number of refinements and variants of the exciton model. The reader is referred to the review articles by Blann [439] and by Seidel, Seeliger, Reif, and Toneev [440] and to the monograph by Gadioli and Hodgson [441] for details. For kinetic energies E_{kin} of the projectile nucleons such that the de Broglie wave-length $\lambda = E_{\text{kin}}/(\hbar c)$ is smaller than the nuclear distance parameter r_0 and the range of the effective nucleon-nucleon interaction a, the motion of the projectile nucleons in the mean potential of the target may be described in the ray-optic approximation and the scattering events by classical scattering theory. The resulting intra-nuclear cascade models shall not be discussed in this book. Again we refer to the literature [442].

4.2 Scission-point models

In the standard definition of the scission point along the fission path it is the point where the two nascent fragments have just one point in common.

For an n-dimensional parametrization of the fission landscape the scission configuration thus defines in general an $(n-1)$-dimensional hypersurface in the space of deformation parameters. Since the sharp-surface characterization of a nuclear shape is to be understood as being the equivalent sharp surface of a leptodermous nuclear density distribution with a diffuse surface, this definition appears to be rather formal. However, it has the advantage of being invariant under continuous transformations of the shape coordinates.

In another, more physically motivated, definition of scission it is the point on the fission path where the Fermi surface hits the single-particle potential barrier, emerging between the nascent fragments, thus blocking the exchange of particles between fragments. Since the equivalent sharp radius of the density distribution is about 0.7 fm smaller than the radius of the potential [443], this second definition of scission leads to a somewhat larger distance between the centers of mass of the two fragments at scission than the first definition. However, taking tunneling into account and many-body correlations, also the second definition of scission overemphasizes the sharpness of the transition. In any case the shape of the scission hypersurface depends on the particular choice of a set of deformation parameters and on the shape class to which they belong.

Passing through the scission region, all observables should change continuously. This does however not always happen in approximate models. If, for example, the leptodermous expansion of the potential energy is carried beyond the mean curvature term, the next term, proportional to the Gaussian curvature, jumps at scission because the curvature jumps by 2π according to the Gauß-Bonnet theorem [201]. This is indeed a reminder of the fact that the leptodermous expansion is only of semiconvergent nature and such terms should therefore not be kept. Note that the Yukawa-plus-exponential form of the generalized surface energy does not yield a discontinuity at scission.

There is also an apparent discontinuity in the Strutinsky prescription for the shell correction to the potential energy: In the internal region one would construct one Fermi energy $\tilde{\lambda}$ for the averaged level density of the total system. Beyond scission there are two, in general different, Fermi energies $\tilde{\lambda}_i$ resulting from different averaged level densities in each fragment. Because of the different treatment of the reference energy before and after scission also the shell correction comes out slightly different before and after scission. One may group the levels all along the fission path into two classes according to whether they correlate to the lighter or the heavier fragment beyond scission. The Strutinsky shell correction for the total system may then be written as the sum of the shell corrections calculated for each class of states separately.

A similar discontinuity appears in the standard expression for the pairing energy in monopole-pairing BCS approximation because one considers one single chemical potential in the inside region, but two, in general different, chemical potentials for each fragment beyond scission. The deeper reason for this problem is a change in the gauge symmetry of the system around scission: Before scission only the total number of particles is a conserved quantity, corresponding to a one-dimensional gauge symmetry. Beyond scission the particle number in each fragment is conserved, corresponding to a two-dimensional gauge symmetry. Therefore there is a phase transition around scission, which is known to be very poorly described in terms of single particle or quasiparticle mean-field approaches. For attempts to overcome these problems by introducing the two classes of states, mentioned above, we refer to the literature [444, 445].

4.2.1 Neck rupture

Scission is usually assumed to be caused by a Rayleigh instability of the neck connecting the nascent fragments. Rayleigh considered an infinitely long cylinder of an incompressible liquid. A sinusoidal disturbance of the surface of the form $\epsilon \cos(kz) \cos(m\phi)$ leads in cylindrical coordinates (ρ, ϕ, z) to a radial coordinate $\rho_s(\epsilon; z, \phi)$ of the surface of the deformed cylinder

$$\rho_s(\epsilon; z, \phi) = \overline{\rho}_s(\epsilon) + \epsilon \cos kz \cos m\phi. \tag{4.33}$$

Volume conservation yields $\overline{\rho}_s(\epsilon) = \rho_s(0)[1 - (1/2)(1 + \delta_{0m})(\epsilon/2\rho_s(0))^2]$ up to second order in $\epsilon/\rho_s(0)$, where $\rho_s(0)$ is the radius of the unperturbed cylinder. Inserting this result into Eq. (4.33) one obtains for the surface area per unit length in leading order of $\epsilon/\rho_s(0)$

$$S = \lim_{L \to \infty} (1/L) \int_{-L/2}^{L/2} dz \int_0^{2\pi} d\phi \sqrt{[1 + (\partial_z \rho_s)^2][(\partial_\phi \rho_s)^2 + \rho_s^2]}$$

$$= 2\pi\rho_s(0)[1 + (1/2)(1 + \delta_{m0})(\epsilon/2\rho_s(0))^2(k^2\rho_s^2(0) + m^2 - 1)]. \tag{4.34}$$

The restoring force from the surface tension decreases with increasing ϵ, i.e. the disturbance grows exponentially in time, if the factor $(k\rho_s(0))^2 + m^2 - 1$ in Eq. (4.34) becomes negative. This is only possible if $m = 0$ and $k^2 < (\rho_s(0))^{-2}$. The cylinder is therefore unstable against axially symmetric disturbances whose wavelength $\lambda = 2\pi/k$ is larger than the circumference $2\pi\rho_s(0)$.

If the finite surface diffuseness is also taken into account in terms of the Yukawa-plus-exponential expression for the surface energy, the condition for instability becomes in terms of the diffuseness parameter a (up to correction terms of order $(a/\rho_s(0))^2$) [446]

$$2\pi\rho_s(0)/\lambda - (3a)^2/(2\rho_s(0))^2 < 1. \tag{4.35}$$

For a cylinder of finite length L one finds the condition [40]

$$(2\pi\rho_s(0)/L)^2 + 2\rho_s(0)/L < 1 \tag{4.36}$$

for instability. Note that full dynamical calculations of the fission of heavy nuclei show indeed that beyond the saddle point rather cylinder-like shapes

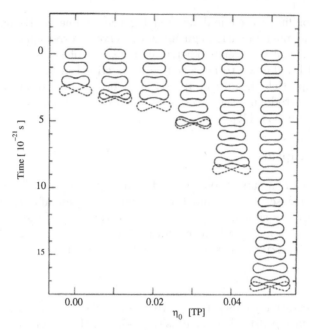

Fig. 4.4. Sequence of shapes of the fissioning nucleus ^{236}U, starting with 1 MeV kinetic energy at the saddle point in the shape parametrization of three quadratic surfaces as function of the size of the viscosity constant η_0. Only symmetric paths are considered (after Ref. [82]).

with an increasing necking-in of a roughly sinusoidal form occur, as seen, for example, in Fig. 4.4.

Less schematic nuclear shapes were used in stability investigations by Strutinsky, Lyashchenko, and Popov [267]. They determined shape functions $\rho(z)$ for axially and mirror symmetric drops by minimizing the liquid-drop energy $E_{\mathrm{l.d.}}\{\rho(z)\}$, with respect to the shape function $\rho(z)$, with the constraints of a given volume, given total charge or fissility x, and given distance D between the centers of mass of the two halves of the drop. As shown in Fig. 2.14 the fission valley ends for symmetric shapes at a critical distance $D_{\mathrm{crit}}/R_0 \approx 2.32$ for all $x \leq 0.9$ (R_0 denotes the radius of a sphere with equal volume). The neck radius at the critical distance D_{crit} is almost independent of the fissility x: it varies between $r_{\mathrm{neck}}/R = 0.24$ ($x = 0$) and 0.27 ($x = 0.9$). The approach has been generalized to asymmetric shapes by Ivanyuk [270]. However, as shown in Fig. 4.5 the critical distance D_{crit} depends very much on the asymmetry $(V_R - V_L)/(V_R + V_L)$, where V_R and V_L are the volumes of the right and left portions of the fissioning nucleus. To understand the overall topography of the liquid-drop energy surface with its various valleys and mountain ridges separating them, we refer to Fig. 2.24 and the associated discussion in Chapter 2.

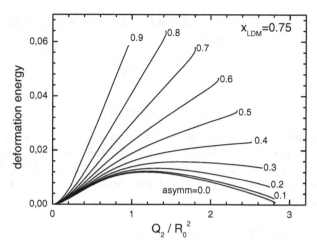

Fig. 4.5. The liquid-drop deformation energy calculated with selfconsistent shapes as function of the quadrupole moment Q_2 for fixed values the of mass asymmetry (indicated in the figure). The end points of the lines lie at the onset of instability. The value of the fissility parameter is $x_{LDM} = 0.75$ (after Ref. [270]).

In all preceding stability considerations nuclear matter was treated as incompressible. Since the actual incompressibility is large, but finite, one may suspect that the strain in the neck produced by the Coulomb repulsion of the nascent fragments becomes probably larger than the nuclear tensil strength. Neck rupture would in that case not arise from an instability of a surface mode, but of a sound mode. However, for a homogeneously charged cylinder of finite length Martinot and Gaudin [447] showed that the disruptive Coulomb force is almost completely balanced by the cohesive surface tension so that the total energy E_{total} of a cylinder with the charge and the volume of an uranium nucleus is almost independent of its length. The strain in the neck $\partial_L E_{\text{total}}(L)$ is therefore small and sound modes should not play an important role in neck-rupture dynamics.

Purely static stability considerations do not fully account for the physics of scission as a dynamical process. In the hydrodynamical model of fission one considers the velocity field $\mathbf{v}(\mathbf{r})$ of the liquid, induced by a change in the surface shape, assuming ϵ in Eq. (4.33) to be time-dependent. The field $\mathbf{v}(\mathbf{r})$ is proportional to the time derivative $\dot{\epsilon}$. The kinetic energy T of the flow in a cylinder of length L, which is a quadratic functional of $\mathbf{v}(\mathbf{r})$, is therefore quadratic in $\dot{\epsilon}$, $T = (1/2)M\dot{\epsilon}^2$, with the inertia $M = \pi \rho_m \rho_s(0)^2 L f_0(\rho_s(0)/L)$ for the mode (4.33) with $m = 0$ [40]. The mass density is denoted by ρ_m and the function $f_0(x)$ is given in terms of the modified Bessel function I_0 [238] by $f_0(x) = I_0(2\pi x)/[2\pi x I_0'(2\pi x)]$. When the potential energy for the Rayleigh mode (4.33) is written $V(\epsilon) = (1/2)C\epsilon^2$ in terms of the stiffness $C = (\sigma/2)L\partial_\epsilon^2 S$, where σ is the surface tension and S the surface area per unit length (4.34), the characteristic time for the unstable mode becomes

$\tau = (-M/C)^{1/2}$. Since M and C depend on L, so does τ. Lord Rayleigh [38] showed that τ has a minimum for

$$L_0 = 1.435 \times 2\pi\rho_s(0). \tag{4.37}$$

and the smallest τ is

$$\tau_0 = [8.493\, \rho_m\rho_s^3(0)/\sigma]^{1/2}. \tag{4.38}$$

In a viscous liquid the rupture is slowed down and the fastest mode shifts to larger L. Großmann and Müller [448] solved the linearized Navier-Stokes equations with appropriate boundary conditions for an infinite cylinder and showed that for a moderate dynamical viscosity constant η_0, more precisely for $\eta_0^2/(\rho_m\sigma\rho_s(0)) \ll 1$,

$$\tau(\eta_0)/\tau_0 = 1 + 0.743\, \eta_0\tau_0/(\rho_m\rho_s^2(0)) + \mathcal{O}(\eta_0^{3/2}). \tag{4.39}$$

and [40]

$$\frac{L(\eta_0)}{L_0} = 1 + 1.164 \left(\frac{\eta_0^2}{\rho_m\sigma\rho_s(0)}\right)^{1/2} + \mathcal{O}(\eta_0^{3/2}) \tag{4.40}$$

Assuming that the fastest unstable mode determines the onset of rupture and taking the corrections (4.35) and (4.36) to Rayleigh's result $L_{\text{crit}} = 2\pi\rho_s(0)$ for the critical length L_{crit} into account, Brosa, Großmann, and Müller [40] conclude that rupture occurs in the liquid-drop model for $L_{\text{crit}} \approx 11\rho_s(0)$.

One can estimate a typical rupture time with Eqs. (4.38) and (4.39). Using a standard value for the nuclear hydrodynamical viscosity of $\eta_0 = 0.015$ terapoise (TP) [96], the nuclear radius parameter $r_0 = 1.16$ fm, and the surface tension $\sigma = c_2/(4\pi r_0^2)$ with a surface-energy term c_2 from the Myers-Swiatecki mass formula [224] $c_2 = 17.944[1 - 1.783(N-Z)^2/A^2]$ MeV, one finds for $^{238}_{92}$U a rupture time of

$$\tau = \{1.19\, (\rho_s(0)/\text{fm})^{3/2} + 0.61\, (\rho_s(0)/\text{fm})\} \cdot 10^{-22}\text{sec}.$$

For a typical value of the neck radius $\rho_{\text{neck}} = \rho_s(0) = 1$ fm the rupture time becomes $1.8 \cdot 10^{-22}$ sec. This is short compared to the time scale on which the elongation of the fissioning nucleus proceeds as, for example, Fig. 4.6 shows. The reason is that not much mass has to be shifted to produce a cleavage, which implies small inertia, whereas the inertia connected with a change of the center-of-mass distance D between the nascent fragments is of the order of the reduced mass. The rupture time is however still an order of magnitude larger than the time t_{sp} a single nucleon with a Fermi energy of about 40 MeV needs to cross the neck with radius ρ_{neck}, $t_{\text{sp}} \approx 1.14\, (2\rho_{\text{neck}}/\text{fm}) \cdot 10^{-23}$ sec. One has also to keep in mind that neck rupture causes only rather small changes of the individual Hartree-Fock single-particle states. Their rate of change at scission is therefore not as large as the velocity of the macroscopic parameter ρ_{neck} might suggest.

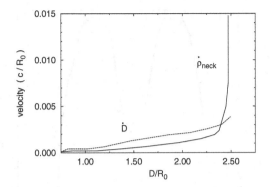

Fig. 4.6. Velocity of the center-of-mass distance $\dot{D}$ and of the neck radius $\dot{\rho}_{\text{neck}}$ versus D along a fission path of ^{215}Fr, starting with temperature $T = 2$ MeV and angular momentum 85 $\hbar$ (after Ref. [280]).

4.2.2 Mass and charge distributions of fission fragments

One may doubt the applicability of classical hydrodynamics for the rupture in nuclear fission. To also avoid the spurious discontinuities in standard calculations of potential surfaces around scission, mentioned above, the following model may be tried: All degrees of freedom of the fissioning system are in thermal equilibrium up to the geometrical scission point, where the proton and neutron numbers of the fragments and the motion in the center-of-mass direction decouple from the other shape and intrinsic degrees of freedom. The probability of populating a scission configuration with proton and neutron numbers of the two fragments Z_1, N_1 and $Z_2 = Z - Z_1, N_2 = N - N_1$, with two sets of multipole deformation parameters $\{\beta^{(1)}\}$ and $\{\beta^{(2)}\}$, and with translational kinetic energy $E_{\text{c.m.}}$ is given in terms of the level density of each fragment $\rho(Z_i, N_i, \beta^{(i)}, \epsilon_i)$, $i = 1, 2$, by

$$P(Z_1, N_1, Z_2, N_2, \beta^{(1)}, \beta^{(2)}, E_{\text{c.m.}}, E^*_{\text{sci}}) \propto$$
$$\rho(E_{\text{c.m.}}) \int_0^{E^*_{\text{sci}} - E_{\text{c.m.}}} \rho(Z_1, N_1, \beta^{(1)}, E_1)$$
$$\times \rho(Z - Z_1, N - N_1, \beta^{(2)}, E^*_{\text{sci}} - E_{\text{c.m.}} - E_1) dE_1,$$

where $\rho(E_{\text{c.m.}})$ is the density of translational states and E^*_{sci} the excitation energy at the scission point. The total kinetic energy in the exit channel (TKE) is deduced in this model from the Coulomb repulsion-energy of the two touching fragments with deformation parameters $\beta^{(1)}$ and $\beta^{(2)}$ and kinetic energy $E_{\text{c.m.}}$ at scission.

The model was first used by Fong [449, 450]. He considered only octupole deformations β_{30} for both fragments. Despite the somewhat crude modeling of the scission-point shape and a rather schematic treatment of shell effects in

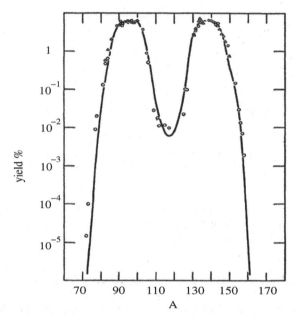

Fig. 4.7. Yield versus mass number A of fission products from thermal neutron-induced fission of ^{236}U. Solid line: calculated results, open circles: radiochemical data, and triangles: mass spectroscopic data (after Ref. [449]).

the level density, the success of the model, shown in Fig. 4.7, is remarkable and indicates its basic appropriateness, although neighboring nuclei are not so well described with the same parameters. When saddle and scission configurations lie very close to each other (for fissility $x < 0.7$) the scission-point model and the transition state theory coincide. Fong's results were later qualitatively confirmed by Ignatyuk [451] with more realistic input for the level density and for nuclear binding energies, including more sophisticated shell corrections. His scission configurations were assumed to be touching coaxial spheroids. However the width of the mass distribution comes out significantly smaller than experimentally observed.

Another version of a scission-point model was proposed by Wilkins, Steinberg, and Chasman and applied to low-energy fission of nuclei from polonium to fermium [452]. They assume, somewhat ad hoc, that the collective degrees of freedom decouple from the intrinsic ones somewhere between saddle and scission, but that the collective variables remain in statistical equilibrium among themselves until scission. Therefore there are two temperatures at scission in this model, one for the intrinsic system, T_{int} and one for the collective degrees of freedom, T_{coll}. The scission configuration is assumed to consist of two coaxial spheroids with a tip-to-tip distance s and deformation parameters $\beta_{20}^{(1)}$ and $\beta_{20}^{(2)}$.

The probability to find the proton and neutron numbers Z_1, N_1 and $Z_2 = Z - Z_1, N_2 = N - N_1$, and deformation parameters $\beta_{20}^{(1)}$ and $\beta_{20}^{(2)}$, was assumed to be given by a Boltzmann distribution

$$P(Z_1, N_1, Z_2, N_2, \beta_{20}^{(1)}, \beta_{20}^{(2)}, T_{\text{int}}) \propto$$
$$\exp[-V_{\text{sci}}(Z_1, N_1, Z_2, N_2, \beta_{20}^{(1)}, \beta_{20}^{(2)}, T_{\text{int}})/T_{\text{coll}}], \qquad (4.41)$$

where the exponent contains the potential energy of the scission configuration, including shell and pairing corrections for the two fragments at internal temperature T_{int}. The temperature-dependence of the shell corrections was calculated with a rather schematic formula, mentioned in Ref. [453], and that of the pairing corrections as described in Ref. [454]. Since the collective degrees of freedom are assumed to be in equilibrium among themselves at scission, no collective kinetic energy term appears in Eq. (4.41).

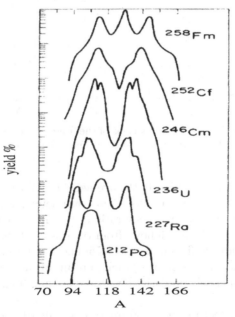

Fig. 4.8. Primary mass yield of fragments from various fissile nuclei, calculated with the scission-point model of Ref. [452].

There are three model parameters $T_{\text{int}}, T_{\text{coll}}$, and s. The washing-out of shell and pairing effects is controlled by T_{int} and therefore it determines in this model, besides other features, the peak-to-valley ratio of the double-humped yield curve for the fragment masses of actinides between uranium and californium. The overall width of the mass distribution is determined mostly by

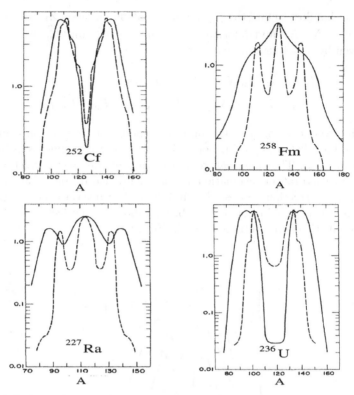

Fig. 4.9. Relative fission yields as functions of the fragment mass-number A for four compound nuclei, full lines experiment, dashed lines calculations (after Ref. [452]).

T_{coll} whereas s influences primarily the average TKE of the fission fragments. Only rough estimates were made for these parameters. $T_{\text{int}} = 0.75$ MeV is obtained by fitting the peak-to-valley ratio in the mass yield of spontaneous fission of ^{252}Cf, $T_{\text{coll}} = 1$ MeV follows from the observed charge dispersion of thermal neutron-induced fission of ^{236}U [455] and $s = 1.4$ fm corresponds approximately with the definition of scission in terms of the cut of the Fermi energy by the single-particle potential between emerging fragments, mentioned above.

There are many features in the primary mass distribution of fission fragments and their excitation energy which are qualitatively described with this single set of parameters: In Fig. 4.8 calculated mass distributions are plotted which show the observed trend of the mass-yield curves of low-energy fission from a single-humped distribution for elements below Rn over a triple humped curve around Ra, double-peaked distributions in the middle of the actinides, and single-peaked with broad shoulders for heavier actinides. (For data see Ref. [456] for the lighter nuclei and Ref. [457] for actinides.) Less well reproduced by the model is the fact that the position of the heavy-fragment peak is almost independent of the mass of the fissioning nucleus, whereas the position of the maximum for the light fragment increases with increasing mass of the parent nucleus. A more detailed comparison of model results with experimental data is presented in Fig. 4.9. The data for ^{252}Cf(sf) are from Ref. [457], for ^{257}Fm(n_{th},f) from Ref. [458], for ^{226}Ra(p,f) from Ref. [459], and for ^{236}U(n_{th},f) from Ref. [460].

Fig. 4.10 shows the dependence of the deformation of a fragment, calculated at scission, as function of its mass. One often identifies the intrinsic excitation energy with the deformation energy at scission and the mean deformation energy with the average number of emitted postscission neutrons $\bar{\nu}$. The scission-point model then yields an explanation for the sawtooth curve of $\bar{\nu}(A)$ as shown in Fig. 4.10. The position of the minimum in the sawtooth

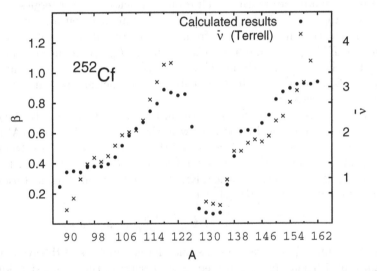

Fig. 4.10. The dots show averaged deformations of the fission fragments of ^{252}Cf, calculated with the scission-point model of Ref. [452]; crosses give the average number of neutrons $\bar{\nu}(A)$ emitted from fragments with mass number A. The scale in $\bar{\nu}$ on the right side of the figure is adjusted to fit the calculated quadrupole deformation β, data are taken from Ref. [461] (after Ref. [452]).

curve for $\bar{\nu}(A)$ is almost independent of the mass of the fissioning nucleus. That means for ^{260}Md the symmetric split corresponds to the minimum and for ^{236}U to the maximum. Also this trend can be reproduced by the model. The average TKE(A) is simply related to the deformation $\beta(A)$: large β imply large center-of-mass distance D and therefore smaller Coulomb repulsion at scission and smaller TKE(A). Wilkins, Steinberg and Chasman showed in detail how all these features can be traced back to the shell corrections for the neutron and proton subsystems at the scission configuration as functions of the nucleon numbers of the fragments and their deformation β_{20}.

Allowing the three fit parameters T_{int}, T_{coll}, and s to depend on mass number, charge and excitation energy of the fissile compound nucleus would certainly improve the agreement of the model with the data. In particular energy conservation requires that the sum of the energies $E(T_{int})$ and $E(T_{coll})$ connected with the temperatures T_{int} and T_{coll}, respectively, and the prescission kinetic energy E_{presci} in the degree of freedom of the centers-of-mass distance D should be equal to the sum of the excitation energy E^* and the gain in potential energy ΔE_{sci} between ground state and scission

$$E(T_{int}) + E(T_{coll}) + E_{presci} = E^* + \Delta E_{sci}. \tag{4.42}$$

Ruben, Märten and Seeliger [462] used a variant of the two-spheroid scission model to calculate the internal excitation and deformation energies of each of the two fragments which give rise to subsequent neutron evaporation. In this approach the energy $E(T_{int})$ from the energy balance (4.42) with the assumption $E(T_{coll}) = 0$ and a prescission energy E_{presci}, deduced from measured TKE(A), where the relation $E_{presci} = TKE-E_{Coul}-E_{prox}$ was employed, E_{Coul} and E_{prox} being the the Coulomb and the nuclear proximity energies at scission, respectively.

However, it seems that the variances in the mass yield and TKE (A) distributions are systematically too narrow in the two-spheroid model. Also the asymmetric peaks in the yield curves of the lighter actinides are too close to the peak at symmetry. There is further no plausible explanation for the empirical Viola systematics of the TKE, averaged over all fragmentations of a given fissioning system $\overline{\text{TKE}(A)}$ [463]

$$\overline{\text{TKE}(A)} = (0.1189\, Z_{comp}^2/A_{comp}^{1/3} + 7.3)\ \text{MeV},$$

where A_{comp} and Z_{comp} are the mass and charge numbers of the system and the constant 7.3 may be interpreted as representing the prescission kinetic energy.

Apart from the rather restricted shape class of β_{20} deformations for the fragments and the neglect of angular-momentum-bearing modes in this model, to be discussed in the next subsection, the implicit assumption is made that there is just one saddle point feeding the scission hypersurface. However, calculated potential-energy surfaces show several saddle points at different energies, corresponding to different shapes. They are expected to lead to different feeding patterns of various regions of the scission hypersurface, in particular as function of the excitation energy of the fissioning system.

Attempts have been made to bring the heavy-mass peak of the calculated mass distribution closer to the experimental value $A = 140$ for lighter actinides. In the framework of the shape class of the two-center shell-model, Mustafa, Mosel, and Schmitt observed that this can be achieved by moving the scission condition further inward where the neck radius is still about 1 fm [464]. Subsequent discussions of scission-point models have usually employed such a scission condition to define the scission hypersurface [465].

An example is the scission-point model of Prakash, Ramamurty, and Kapoor [466]. These authors treat the particle exchange between the nascent fragments as a diffusion process, governed by a master equation. Its stationary solution yields the mass and charge distributions of the fragments at scission. It is shown that this solution is mainly controlled by the difference between the chemical potentials for the heavy and the light fragment. Due to shell effects in the values of the chemical potentials, the most probable mass split yields $A_H = 140$ for the heavy fragment in thermal-neutron fission of ^{236}U. The width of the mass distribution is determined in this model by the temperature at scission. With the choice $T = 0.5$ MeV the observed width can be reproduced.

To bridge the gap between a prescission shape with a neck radius around 1 fm and a postscission configuration with a tip-to-tip distance around 3 fm Rubchenya and Yavshits use the following procedure [467]. They define moments

$$I_{lm} = 2\pi \int_{-\infty}^{\infty} dz\, z^l \int_{0}^{\infty} d\rho\, \rho^{m+1} \rho_{\text{nucl}}(\rho, z)$$

for an axially symmetric nucleon distribution $\rho_{\text{nucl}}(\rho, z)$. The latter is assumed to have a Woods-Saxon profile and a shape parametrization in terms of generalized Cassinian ovals for prescission shapes and in terms of coaxial spheroids after scission. For a given prescission shape the corresponding postscission shape is obtained by equating the first few moments I_{lm} (l, m = 0,0; 1,0; 2,0; 0,2; 0,4) of the prescission and postscission shapes since they characterize the bulk of the mass distributions. In order to determine the prescission shape, an energy surface is calculated in a temperature-dependent microscopic-macroscopic approach as function of the three Cassinian shape parameters α, α_1, and α_4. The temperature-dependent scission hypersurface is defined as the locus where the neck becomes unstable against pinch-off. After scission a diffusion process is assumed to start similar to the model of Ref. [466].

L. Meitner suggested in 1950 [468] that the mass yield in thermal-neutron-induced fission of ^{236}U compared with the yield of higher excited uranium and bismuth nuclei could be interpreted as superposition of two fission modes, a mass-symmetric mode, whose contribution to the total yield increases fast with increasing excitation energy of the compound nucleus, and an asymmetric mode, which is driven by the strong binding effect of one of the primary fragments with neutron number around 50. Meitner argued that shell effects are expected to vanish at high excitation energy so that eventually only the symmetric mode survives as predicted by the liquid-drop model. The same interpretation was given by Turkevich and Niday [469] to the peak-to-valley ratio of the fission fragments from the reaction ^{233}Th(n,f) in comparison with the yield from fission of uranium and plutonium isotopes.

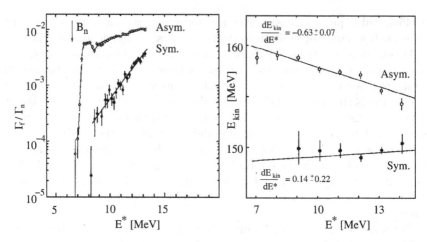

Fig. 4.11. Plotted against the excitation energy in the compound nucleus ^{227}Ac is the fission yield Γ_f/Γ_n in the left frame and the $\overline{\text{TKE}}$ from fission of ^{229}Th in the right frame, unfolded into contributions from the symmetric and from the asymmetric mode (after Ref. [470]).

Later experiments in the lighter actinides [470–472] showed that the mean total kinetic energy ($\overline{\text{TKE}}$), connected with the symmetric mode is smaller than that connected with the asymmetric mode, which points to a more compact scission configuration in this mode. As an example we show in Fig. 4.11 results obtained by Konecny, Specht, and Weber [470] for the reaction ^{226}Ra(^{3}He,df). In the left frame of the figure one sees the much faster increase of the fission yield Γ_f/Γ_n for the symmetric mode with increasing excitation energy than for the asymmetric mode. In the right frame the larger $\overline{\text{TKE}}$ of the latter mode compared to the $\overline{\text{TKE}}$ of the symmetric mode is seen to persist for all excitation energies. A bimodal structure of fission was later also identified in heavier actinides by Hulet et al. [473]. Particularly striking is the change of the mass distribution in the sequence of fermium isotopes between ^{256}Fm and ^{258}Fm. The lighter isotopes show an asymmetric mode with low $\overline{\text{TKE}}$ and a mass-symmetric mode with high $\overline{\text{TKE}}$. In ^{258}Fm the yield is only mass-symmetric, but can be unfolded into low $\overline{\text{TKE}}$ and high $\overline{\text{TKE}}$ components of approximately equal strengths [473]. In fair agreement with the data, bimodal fission in these nuclei was interpreted by Ćwiok et al. [474] in terms of two fission valleys in the potential-energy hypersurface. The latter was calculated in the macroscopic-microscopic frame, using the $\beta, \dots \beta_6$ shape class. The result was confirmed in a selfconsistent, constrained HFB scheme [264].

Brosa et al. [40, 475] generalized the bimodal concept to a multimodal picture of fission. The yields in each mode were assumed to be produced by separate bunches of trajectories branching off from the mainstream fission-path by bifurcation anywhere between the ground-state shape and the scission hypersurface. One may attempt to associate these modes with local energy minima in the cross section of the energy hypersurface with the scission hypersurface. The branching ratio of the flux into these fission modes can of course not be derived from equilibrium statistics. It remains a fit parameter in the scission model since it can of course only be calculated in the framework of a complete dynamical treatment of fission.

Pashkevich [476] made a calculation in the scission model within the shape class of generalized Cassinian ovals. His constraining condition is not the geometrical scission hypersurface, but a fixed value of a formal shape parameter whose definition singles out, somewhat arbitrarily, three points on the shape. The constraint leads to a shape subclass with a small, but finite neck radius. Using the macroscopic-microscopic approach to calculate the potential-energy hypersurface for several even Fm and Cf isotopes, three minima are found in this subclass with respect to variations of up to 20 of the Cassinian shape parameters. Each of these scission shapes was associated with a fission mode and the total mass and TKE distribution is constructed as a superposition of contributions from each mode. However, in this paper no connection is established between the phase space at scission and the individual contributions to the yield in each mode, as in Fong's approach.

The scission shapes I, II, III correspond only qualitatively to the scission shapes calculated by Brosa et al. [477] in the Lawrence shape-class for ^{252}Cf and called by the authors "super-short", "standard", and "super-long". They were obtained for the final points of three fission valleys at neck radii of about 1.7 fm. It appears that quantitative details of these prescission shapes depend on the choice of the underlying shape parametrization and the specific scission constraint used to determine the energy minimum.

In the end, Gaussian distributions in mass numbers and more complicated distributions with non-vanishing higher cumulants for the TKE distributions [475, 478] at scission were postulated and eventually fitted to the data, rather than really obtained from an underlying theory. Together with branching ratios and their dependence on the excitation energy and the mass and the charge numbers of the compound nucleus there are just too many phenomenological parameters in this model to provide more than a qualitative frame to understand the physics of fission yields.

As already noticed by Wilets [30], valleys and mountain ridges do not have a covariant meaning, in contrast to extremal points and the formal scission hypersurface, i.e. even within a given shape class with variables, say x_i, a transformation to new variables $x'_i = f_i(\mathbf{x})$ can transform the bottom of a valley into a ridge line and vice versa. In more physical terms, there is in general no "reasonable" choice of collective variables which automatically guaranties that the fissioning system would move along the bottom of a valley in these variables. In the hydrodynamical model the fission path is as much influenced by the inertial and friction tensors as by the energy surface, as shall be discussed in detail in the fifth chapter. These transport parameters are functionals of the total shape and not of some single points on the surface as for example radial minima and maxima of the shape in cylindrical coordinates or the outer tip-to-tip distance, used to define supposedly physically reasonable shape parameters [40, 476, 479]. Even the rather schematic graph of a potential landscape in Fig. 2.24 is to be interpreted with great caution if the valleys were understood as conduits of fission paths. This is only correct if the inertia connected with the coordinate D is much larger than the inertia connected with all other shape-degrees of freedom, lumped together in the variable h, and if, in addition, the actual motion is underdamped. There is just no simple substitute for a dynamical calculation, except in cases where one has reason to believe that observables are connected with properties of the thermal equilibrium with respect to some or all shape degrees of freedom.

4.2.3 Angular-momentum-bearing scission modes

So far fissioning nuclei were assumed to be axially symmetric, or at least rigidly aligned all along the fission path. If this were true, one could not understand why fission fragments carry a substantial amount of angular momentum, even from spontaneous fission of a 0^+ ground state as in the case of ^{252}Cf, where each of the fragments carries in the average $7\hbar$ angular momentum [480]. For a rigidly rotating nucleus a bending moment acts on the axis when it stretches and reduces its angular velocity at fixed angular momentum. The effect can be seen in Fig. 1.13 in the forth and fifth frame. A similar distortion of the axis occurs in the opposite direction in heavy-ion fusion until the rigidly rotating compound state is reached.

To account for modes bending the axis and other angular-momentum-bearing modes at least in the framework of the scission-point model, Moretto and Schmitt [481] considered a rather schematic model, which however allows to exhibit essential features of these modes. The model consists of two equal, rigid spheres with radius R and mass M in contact, neglecting sticking forces. Of the twelve degrees of freedom of two independent rigid spheres three account for the center-of-mass motion of the whole system, three for its rigid rotation, including the tilting mode. One degree of freedom is eliminated by the contact condition $D = 2R$. The remaining five degrees of freedom plus the tilting mode were first considered by Nix and Swiatecki [482]. They are sketched in Fig. 4.12.

In the twisting and bending modes the fragment spins have equal size and point into opposite directions. They are aligned with the symmetry axis in the twisting mode and perpendicular to the symmetry axis in the two degenerate bending modes. The spin of the two fragments is opposite in these modes. In the tilting and wriggling modes the spins are equal in size and direction. They are aligned with the total angular momentum and perpendicular to the symmetry axis in the two degenerate wriggling modes. In the tilting mode they are aligned with the symmetry axis and, in equilibrium, perpendicular to the total angular momentum. The direction of the spins of the fragments is the same in these modes. Thermal excitation of the tilting mode leads to $K > 0$ states, i.e. a tilting of the fission axis with respect to the total spin $\mathbf{J}$.

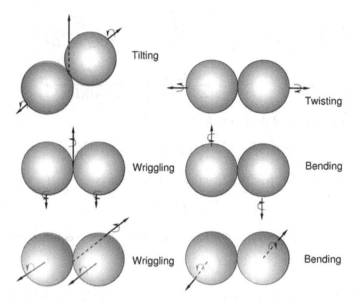

Fig. 4.12. The six angular-momentum-bearing scission modes (from Ref. [481]).

Such distribution of the axis is an alternative to the saddle-point expression (3.16), where it was assumed that K is fixed at the saddle point and does not change on the way from saddle to scission.

In terms of the total angular momentum $\mathbf{J}$, the spin of each fragment $\mathbf{S}$, the reduced mass $\mu = M/2$, and the moment of inertia of a rigid sphere $\mathcal{J} = (2/5)MR^2$ the total kinetic energy is in this model

$$E(\mathbf{S}) = \frac{(\mathbf{J} - 2\mathbf{S})^2}{2\mu D^2} + 2\frac{\mathbf{S}^2}{2\mathcal{J}}. \tag{4.43}$$

The fractionation of the total angular momentum $\mathbf{J}$ into spin and orbital moments is then obtained from a canonical partition sum $Z = \int \exp[-E(\mathbf{S})/T]d\mathbf{S}$. Making use of the relative orientation of the orbital and spin momenta in the six modes of Fig. 4.12 one has for the degenerate twisting and two bending modes

$$E = \mathbf{S}^2/\mathcal{J}, \qquad Z = \int e^{-E(\mathbf{S})/T}d^3\mathbf{S} = (\pi\mathcal{J}T)^{3/2},$$

$$\overline{\mathbf{S}} = 0, \qquad \overline{|\mathbf{S}|} = 2\sqrt{\mathcal{J}T/\pi}, \qquad \overline{\mathbf{S}^2} = (3/2)\mathcal{J}T. \tag{4.44}$$

Therefore the variance for each of the three modes is

$$\sigma^2 = (1/2)\mathcal{J}T. \tag{4.45}$$

For the two degenerate wriggling modes the spins lie in the plane perpendicular to the symmetry axis and are aligned with the total angular momentum J. One therefore has $E = (J - 2S)^2/(2\mu D^2) + \mathbf{S}^2/\mathcal{J}$ and

$$Z = \int_{-\infty}^{\infty} e^{-E(S)/T} dS = \sqrt{\frac{\pi \mu D^2 \mathcal{J} T}{2\mathcal{J} + \mu D^2}} \exp\left[-\frac{J^2}{2T(2\mathcal{J} + \mu D^2)}\right],$$

(4.46)

$$\overline{S} = \frac{\mathcal{J}}{2\mathcal{J} + \mu D^2} J, \qquad \overline{S^2} = \frac{\mu D^2 \mathcal{J} T}{2(2\mathcal{J} + \mu D^2)} + \frac{J^2}{(2\mathcal{J} + \mu D^2)^2} J^2.$$

Therefore the standard deviation for this mode is

$$\sigma^2 = \overline{S^2} - \overline{S}^2 = \mathcal{J}\mu D^2 T/(4\mathcal{J} + 2\mu D^2).$$

(4.47)

With $\mathcal{J} = (2/5)MR^2$, $\mu = M/2$, and $D = 2R$ one gets

$$\overline{S} = (1/7)J,$$

(4.48)

which is the limit of rigid rotation, as it should be for the equilibrium and

$$\sigma^2 = (5/14)\mathcal{J}T.$$

(4.49)

For the tilting mode one considers $E = (J^2 - K^2)/(2\mathcal{J}_\perp) + K^2/(2\mathcal{J}_\parallel)$ and obtains

$$Z = \int_{-J}^{J} e^{-E(K)/T} dK = \sqrt{2\pi \mathcal{J}_{\text{eff}} T} \exp\left(\frac{-J^2}{2\mathcal{J}_\perp T}\right) \text{erf}\left(\frac{J}{\sqrt{2\mathcal{J}_{\text{eff}} T}}\right),$$

$$\overline{K^2} = -\frac{1}{Z}\frac{\partial Z}{\partial(1/[2\mathcal{J}_{\text{eff}} T])}$$

$$= \mathcal{J}_{\text{eff}} T - J\sqrt{(2/\pi)\mathcal{J}_{\text{eff}} T} \exp\left(\frac{-J^2}{2\mathcal{J}_{\text{eff}} T}\right) \Big/ \text{erf}\left(\frac{J}{\sqrt{2\mathcal{J}_{\text{eff}} T}}\right) \quad (4.50)$$

with $\mathcal{J}_{\text{eff}} = (\mathcal{J}_\parallel^{-1} - \mathcal{J}_\perp^{-1})^{-1} = (14/5)\mathcal{J}$. For large J, $J^2 \gg 2\mathcal{J}_{\text{eff}} T$, the second term in Eq. (4.50) vanishes and $\overline{K^2} = (14/5)\mathcal{J}T$. In the tilting mode the rigid rotation of the binary object consists of a rotation around the common axis and a precession of the axis around the total angular momentum $\mathbf{J}$. Its component M, perpendicular to the axis, of size $M = \sqrt{J^2 - K^2}$ gives rise to a contribution to the spin of each fragment in the direction perpendicular to the axis and of size $M/7$ according to Eq. (4.48) and therefore of size $(2/7)\sqrt{J^2 - K^2}$ to both fragments. Together with the angular momentum K in the direction of the axis one obtains for the spin of both fragments

$$2S = (2J/7)\sqrt{1 + (45/4)(K^2/J^2)} =$$
$$(2J/7)\{1 + (45/8)(K^2/J^2) - 2(45/16)^2(K^4/J^4) + \mathcal{O}(K^6/J^6)\}.$$

In the limit of large J one has

$$\overline{S^2} = (1/196)(45\overline{K^2} + 4J^2) = J^2/49 + (9/14)\mathcal{J}T$$

and
$$\overline{S} = J/7 + (9/4)(\mathcal{J}T/J)$$

up to terms of linear order in $\mathcal{J}T/J$. Therefore, in this order, $\sigma^2 = \overline{S^2} - \overline{S}^2 = 0$.

In a body-fixed frame of reference with the z coordinate in the direction of the symmetry axis, one derives from Eqs. (4.45) and (4.49) the probability for the distribution of the three spin components S_x, S_y, S_z of one of the fragments

$$P(S_x, S_y, S_z) \propto e^{-[S_x^2/(2\sigma_x^2) + S_y^2/(2\sigma_y^2) + (S_z - \overline{S_z})^2/(2\sigma_z^2)]}$$

with $\sigma_x^2 = \sigma_y^2 = \sigma_{\text{bend}}^2 + \sigma_{\text{wrigg}}^2 = (6/7)\mathcal{J}T$, $\sigma_z^2 = \sigma_{\text{twist}}^2 + \sigma_{\text{tilt}}^2 = (1/2)\mathcal{J}T$, and $\overline{S_z} = (1/7)J$. If there were a sticking potential between the fragments, its contribution to the partition sum Z would cancel out in the averages $\overline{|S|}$ and $\overline{S^2}$. Its neglect does therefore not restrict the generality of the model.

The model has been extended to the case of unequal spheres by Schmitt and Pacheco [483]. In this case the kinetic energy is of course still a quadratic form of the spin and total angular momenta, the construction of the six normal modes is however not as trivial as in the symmetric case: The tilting and twisting modes remain eigenmodes, the bending and wriggling modes become mixed. There is an eigenmode with parallel fragment spins, which becomes the wriggling mode for equal spheres, and an eigenmode with antiparallel fragment spins, which is the bending mode for equal spheres. These two modes are both doubly degenerate.

The model of Moretto and Schmitt is meant to be used for rather large J as typically reached in heavy-ion fusion-fission reactions. In this case the classical treatment of angular momentum vectors and their coupling is a good approximation to the quantal coupling scheme. Also the internal excitation energy at scission is assumed to be rather high so that a canonical ensemble, characterized by the temperature T, makes sense. Likewise the treatment of the rotational energy of each fragment as that of a rigid rotor is only realistic in Thomas-Fermi approximation, i.e. at temperatures where pairing and shell effects are washed out.

Applications of the model to fully damped, deep-inelastic heavy-ion reactions are discussed in Ref. [484]. Only in some cases does the simple equilibrium model of Ref. [484] describe the data satisfactorily. The explanation given for the failure in other cases is a strong dependence of the model predictions on the mass asymmetry. In general there is a drift in this variable during the reaction time in heavy-ion collisions. The potential energy of two rigidly rotating spheres in contact shows that direction and size of the driving force of the drift does not only depend on the total mass and the mass asymmetry of the entrance channel, but also on the angular momentum, see corresponding graphs of the potential landscape in Ref. [289]. The slow drift does not always lead to an equilibrium situation, which is however one of the main conditions for the validity of a scission-point model.

In the transition state theory the angular distribution of the fission axis is assumed to be determined at the fission saddle-point according to the Halpern-Strutinsky formula (4.16). This equation was used, after insertion of Eq. (4.14), to determine B_{fiss} by fitting experimental angular distributions. From B_{fiss} empirical values for the parameter K_0 of Eq. (4.12) were deduced. They show that the resulting moments of inertia are compatible with saddle-point shapes only for low-energy fission. For larger energy and angular momenta the observed angular distributions indicate that the moments of inertia should rather correspond to a shape along the fission path between saddle and scission points [485, 486], but not really at scission (as defined by two spheres in contact, as in the Ericson model [325]).

If a fissioning nucleus would rotate rigidly until separation, the spins of the fragments should be parallel and perpendicular to the fission axis. Deviations from this alignment are an indication of an effect of the scission modes discussed above. Such misalignments of the fragment spins have indeed been observed [487]. The scission modes would also generally increase the spins of the fragments compared to the rigid rotor model.

Proposals have been made to describe the bending and wriggling modes at very low temperature under the assumption that there is a potential pocket at scission [488–490]. Hamiltonians for some scission modes may then be constructed with inertial parameters either from a rigid-rotation model or from a cranking calculation and with a potential term from the assumed potential pocket. The Hamiltonian for these normal modes at scission may be quantized in analogy to A. Bohr's treatment of collective rotational and vibrational states. Spin distributions for the fragments are obtained for the lowest quantal states of these modes. However, there is no evidence experimentally or theoretically that such potential pockets at scission exist in the first place, and that theoretical claims are not rather an artefact of an inappropriately restricted choice of fission variables. But even if they would exist, the system would have to live long enough that a collective ground state can establish itself.

It appears that no quantitatively reliable prediction of the distribution of angular momenta among fragment spins and orbital motion is possible for low-energy scission without dynamical calculations in all relevant angular variables plus the center-of-mass distance $D(t)$. One also has to keep in mind that excited states in the fragments with finite angular momentum must not necessarily be of a collective nature [491].

4.2.4 Cold fission

In all scission-point models discussed above it was assumed that the available phase space of collective and intrinsic excitations of the fragments at scission is large enough that phase-space factors dominate the dynamics. Considerable experimental efforts went into an investigation of fission events in which the phase space of internal excitations at scission is as small as possible [492, 493], starting with the work of Signarbieux and his collaborators [494]. These events have been called cold fission. There are two ways to select experimentally from all fission events the tiny fraction of cold events: one either looks for the fission events with the largest possible TKE, ideally equal to the fission Q-value for spontaneous fission and to the Q-value plus the excitation energy E^* for thermal neutron induced fission, so that no energy is left for internal excitation of the fission fragments. Or one puts a window on the events with the smallest TKE in the total distribution of the TKE values, which correspond to the largest possible distance D between the centers of mass of the fragments at scission. This implies the largest possible deformation of the fragments so that again no energy is left for internal excitation, except for the deformation energy of the fission fragments [495].

Cold fission events are observed in spontaneous fission of heavier actinides and thermal-neutron-induced fission of the lighter actinides. They are not seen for very asymmetric and for nearly symmetric mass splits. As "cold" fission events one takes those with less than 2-3 MeV excitation energy (with an experimental uncertainty of about 1 MeV in the measurement of the TKE [496]). They are only seen in the range $80 \leq A_L \leq 108$ for the mass of the light fragment and the optimal choice of Z_L.

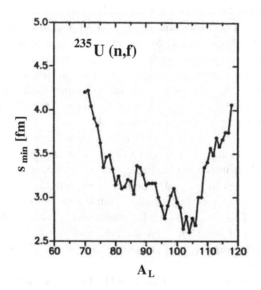

Fig. 4.13. Minimum tip distance as function of the light fragment mass-number A_L for the reaction ^{235}U(n_{therm},f) (from Ref. [497]).

The experimental results have been translated into scission-point geometry by the tip-to-tip distance (TTD) model of Gönnenwein and Börsig [497]. It is plausible that the fission path corresponding to cold fission events starts with a cold system at the fission saddle-point and does not lead to excitations between saddle and scission. The scission configuration is assumed in the TTD model to consist of the fragments in their ground states with some center-of-mass distance D. Fragments which are intrinsically deformed in their ground states are assumed to have their symmetry axes aligned with the fission axis. If their deformation-energy surface has an oblate and a prolate minimum of comparable depth, only the prolate minimum is assumed to be relevant. The scission configuration of the TTD model lies therefore on the bottom of the fusion valley and its energy is the sum of the ground-state energies of the fragments and the Coulomb and nuclear proximity interaction energies between the fragments, E_{Coul} and E_{prox}, respectively.

The crucial quantity of the TTD model is however the tip-to-tip distance s, rather than the center-of-mass distance D: It is assumed that only the most compact scission configurations, schematically characterized by small values of s, lead to cold fission. Because of energy conservation the observed TKE value is given by

$$\text{TKE} = E_{\text{Coul}}(s) + E_{\text{prox}}(s) + E_{\text{presci}}, \tag{4.51}$$

where E_{presci} is the prescission kinetic energy in the fission degree of freedom. The most compact configuration for given TKE occurs for $E_{\text{presci}} = 0$. Eq. (4.51) may then be solved for s for each mass and charge split once the TKE is measured. The functions $E_{\text{Coul}}(s)$ and $E_{\text{prox}}(s)$ are calculated for given ground-state shapes [215]. Minimizing $s(TKE)$ with respect to TKE for given mass and charge split, yields the s_{min} shown in Fig. 4.13. The smallest observed thermal excitation energies at scission are plotted in Fig. 4.14 as function of A_L. Choosing $s_{\text{crit}} \approx 3$ fm as upper limit for the tip-to-tip distance, the inequality $s \leq s_{\text{crit}}$ is seen to define roughly the mass range $80 \leq A_L \leq 108$ where cold fission is observed.

The mass and charge yields in cold fission show large odd-even effects: In the limit TKE$\rightarrow Q$ a split of an even system into two odd fragments occurs more frequently than into neighboring even-even fragments, just the opposite from low-energy fission in general. The trivial explanation is that the Q value of odd-odd splits is several MeV smaller than for even-even splits. To achieve TKE$\approx Q$ a larger TKE is necessary for even-even splits and therefore a more compact scission configuration than for odd-odd splits. Since splits become more rare the more compact the scision configuration is, the odd-odd splits are favored.

An alternative explanation is based on the fact that the limit TKE=Q can experimentally not be reached exactly. Therefore some, if not many excited states of the fragments are typically compatible with what is called "cold fission". This is in particular the case for odd systems with their large level density at small excitation energies. The split of an even system into two odd fragments is therefore favored compared to the split into neighboring even fragments because of the larger phase-space factor. Only if the limit of true cold fission could be reached, pairing-correlation dynamics is no longer hidden

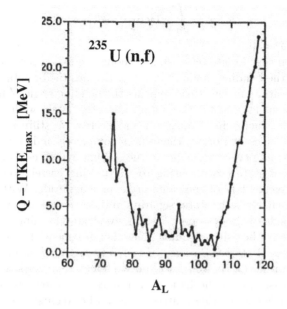

Fig. 4.14. Difference between the Q-value and the observed maximum total kinetic energy-release TKE_{max} versus the light fragment mass-number A_L, for the reaction $^{135}U(n_{therm},f)$ (from Ref. [497]).

behind dominant phase-space factors. In this case cold fission would provide a unique tool to study the pairing phase transition which occurs around scission.

Since only very schematic calculations of the pairing mode at scission exist [445], it shall be sufficient to sketch roughly the conceptual frame of the theoretical treatment of this phase transition. Pairing phase transitions are somewhat difficult to visualize. It is therefore helpful to make use of the close analogy with the phase transition from spherical to deformed nuclei, see e.g. Ref. [108] Section 11.2. The violation of angular momentum conservation in a deformed, but still axially-symmetric, intrinsic state corresponds to the particle-number violation of the BCS wave-function. The orientation angle ϕ of the symmetry axis corresponds to the gauge angle ϕ of the BCS state and a measure of the degree of symmetry breaking is given by the deformation parameter δ in the case of a deformed single-particle potential and by the pairing gap parameter Δ in the BCS case. States with proper symmetry in the lab system are obtained by projection on states with well-defined angular momentum M and with well-defined particle number N. This can be achieved by an ansatz for the wave function Ψ as superposition of intrinsic states $\psi(\phi)$ for all angles ϕ, $0 \leq \phi < 2\pi$, and a weight $f(\phi)$ which should minimize the total energy,

$$\Psi = \int_0^{2\pi} f(\phi)\psi(\phi)d\phi \tag{4.52}$$

and $\delta\langle\Psi|H|\Psi\rangle/\delta f = 0$.

We have seen that for stretched shapes along the fission path a decreasing stiffness allows the bending mode to develop. Besides the orientation angle ϕ of the rigidly rotating nucleus there appears the bending angle ϕ' between the axes of the nascent fragments. The amplitudes of ϕ' begin with small values and increase with increasing elongation and decreasing stiffness until, after scission, the two axes can orient themselves completely independently. This phenomenon has its perfect analogue in the pairing dynamics. For a compact shape all pairs of states contributing to the pairing correlations lead to a pairing matrix element G of the same order of magnitude. With increasing elongation of the nucleus the states separate in those which are predominantly located on one side of the fissioning system, eventually forming one fragment and those on the other side, forming the other fragment. Let us call them class A and class B states. Because of the very short range of the pairing potential the pairing matrix element G between a state of class A and a state of class B decreases along the fission path since the states are localized on opposite sides, whereas pairing matrix elements between states of the same class increase because of their increased compactness in one side of the volume of the fissioning nucleus.

As a consequence the correlation between the gauge angles of the paired subsystems A and B decreases along the fission path until they become uncorrelated after scission. Before scission however, the relative gauge angle is a quantity of limited fluctuation with the consequence that one has to deal with a superposition of pairing fields between saddle and scission of a fissionable nucleus and not with just one intrinsic BCS state. The ansatz (4.52) has then to be generalized

$$\Psi = \int_0^{2\pi} d\phi \int_0^{2\pi} d\phi' f(\phi, \phi')\psi(\phi, \phi').$$

Minimization of the energy expectation value $\langle\Psi|H|\Psi\rangle$ with respect to the weight function $f(\phi, \phi')$ leads to the factorization

$$f(\phi, \phi') = e^{i\phi(N/2)}g(\phi')$$

because the total number of protons or neutrons N is a conserved quantity during fission [108]. The factor $g(\phi')$ is the solution of a Hill-Wheeler equation and becomes also a mere phase factor $\exp(i\phi'\Delta N/2)$ beyond scission, where ΔN is the difference of the proton or neutron numbers of the fragments and N is the sum of these numbers.

4.2.5 Ternary fission

In thermal neutron-induced fission a light third particle is observed besides the fission fragments in about $2 \cdot 10^{-3}$ of all fission events. Roughly 90% are alpha particles, either primary fission products or decay products of primary,

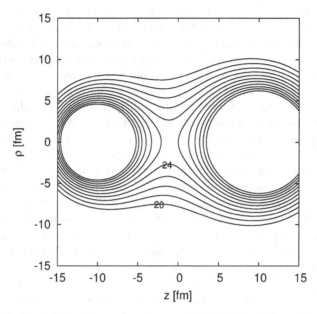

Fig. 4.15. Coulomb field of two point charges with $Z_1=36$ and $Z_2 = 54$ at a distance of D=20.7 fm, corresponding to the most probable scission configuration, considered in Ref. [499]. Distance between contour lines is 1 MeV.

particle-unstable products, like ^{5}He or ^{8}Be (pseudo-quaternary events). The primary particles are emitted approximately perpendicular to the fission axis with a slight angular shift towards the direction of the light fragment. Their energy distribution is approximately Gaussian with a mean value significantly larger than that of alpha particles from the usual α-decay of fissile nuclei or from its fission products. One has argued that they originate from the neck of the scission configuration because it is easy to see that the electrostatic potential in the neck region is significantly larger than on the surface of the compound nucleus in its ground state. For a rough estimate one may compare the Coulomb potential produced by a point charge eZ in the center of a sphere with radius $R_0 = r_0 A^{1/3}$ on the surface, $E_{\mathrm{spher}} = eZ/R_0$, with the potential at the center between two point charges, $eZ/2$ each, at a distance $R_{\mathrm{frag}} = r_0(A/2)^{1/3}$ from the center, $E_{\mathrm{sci}} = 2\,e(Z/2)/R_{\mathrm{frag}} = (eZ/R_0)2^{1/3}$ so that $(E_{\mathrm{sci}}-E_{\mathrm{spher}})/E_{\mathrm{spher}} = 2^{1/3}-1 = 0.26$ in this crude model of a symmetric scission configuration. From the potential energy of two point charges, shown in Fig. 4.15, it is also clear that light charged particles originating in the neck region are focused in a direction roughly perpendicular to the fission axis with a slight bias towards the direction of the light fission fragment.

Somewhat more quantitative calculations for the energy and angular distribution of the light fragments have been made by Boneh, Fraenkel, and Nebenzahl [498] and by Guet, Nifenecker, Signarbieux, and Asghar [499]. These authors consider classical trajectories of three point particles with the masses of the two heavy fragments and of an alpha particle in the Coulomb field produced by their charges. Trajectory calculations require 18 initial conditions for the coordinates and velocities of the three particles of which 6 are fixed by the conservation of the total linear and angular momenta, which are zero in the chosen frame of reference. In addition, realistic distributions of the mass and charge splits of the fissioning nucleus have to be assumed.

As a typical example we choose the trajectory calculations for thermal neutron-induced ternary fission of ^{236}U, reported in Ref. [499]. In this paper probability distributions for the initial conditions are assumed and Monte-Carlo calculations are performed to generate final energy and angular distributions to be compared with experimental data. The distribution of the mass number of the light fragment A_L is taken to be Gaussian with mean value $\overline{A}_L = 92$ amu and variance $\sigma_{A_L} = (\overline{A_L^2} - \overline{A}_L^2)^{1/2} = 6$ amu representing the experimental finding. In terms of A_L the mass number of the heavy fragment is $A_H = 236 - A_L - 4$ amu and the charges of the heavy fragments follow from $Z_L/A_L \approx Z_H/A_H \approx 92/236$. The initial values of the center-of-mass distances D were assumed to have also a Gaussian distribution at scission. Its mean value $\overline{D}$ and variance σ_D were eventually fitted to observed data. From this a distribution of the tip-to-tip distance $s = D - r_0(A_L^{1/3} + A_H^{1/3})$ is derived, where $r_0 = 1.16$ fm is assumed. No allowance is made for the deformation of the heavy fragments at scission. The initial kinetic energy of the heavy fragments, the prescission energy E_{presci}, is also taken to be Gaussian-distributed with mean value $\overline{E}_{\text{presci}}$ and variance $\sigma_{E_{\text{presci}}}$. Because of the relation TKE$=E_{\text{presci}} + e^2 Z_L Z_H/D$ the distributions for TKE and D imply the distribution of E_{presci} for given Z_L and Z_H.

The initial distribution of the momentum of the alpha particles is assumed to be isotropic and Gaussian with mean value zero and with variance σ_{p_α} for each of its Cartesian components. Therefore the mean energy is $\overline{E}_{\alpha_0} = 3\sigma_{p_\alpha}^2/(2M_\alpha)$. The initial positions of the alpha particle may also have a Gaussian distribution in the ρ direction in cylindrical coordinates with mean value zero and variance related to that of the conjugate momentum p_ρ via the uncertainty principle. Actually however, all trajectories are assumed to start with $\rho = 0$ so that the motion of the three bodies remains in a plane. Finally, a rectangular distribution of the initial positions along the fission axis is assumed starting 2.2 fm away from the tip of the left fragment (taken as spherical) and ending 2.2 fm before the tip of the right fragment (also taken to be spherical). Nuclear proximity forces are not considered, but the surface of the heavy fragments is assumed to be black for alpha particles.

From energy conservation follows – eventwise – the relation

$$\text{TKE} + E_{\alpha_\infty} = E_{\text{presci}} + E_{\alpha_0} + V_{\text{Coul}}(z_0), \tag{4.53}$$

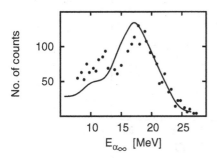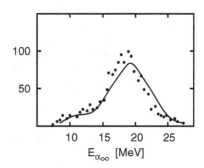

Fig. 4.16. Distribution of measured (points) and calculated (full lines) final kinetic energies of the alpha particle $E_{\alpha\infty}$ for emission under an angle of 68° with respect to the direction of the light fragment (left frame) and under 98° (right frame) (after Ref. [499])

where $V_{\mathrm{Coul}}(z_0)$ is the Coulomb energy of the scission configuration and $E_{\alpha\infty}$ is the final kinetic energy of the alpha particle. In terms of the distance z_0 of the birth place of the alpha particle on the fission axis, counted from the midpoint between the tips of the heavy fragments

$$V_{\mathrm{Coul}}(z_0) = Z_L Z_H \frac{e^2}{D} + \frac{2Z_L e^2}{R_L + (s/2) - z_0} + \frac{2Z_H e^2}{R_H + (s/2) + z_0}.$$

Equation (4.53) implies a constraining condition between the observed distributions of TKE and $E_{\alpha\infty}$ on the one hand and of the initial distributions of E_{presci}, E_{α_0}, and D at scission on the other hand.

The five parameters $\overline{D}$, σ_D, $\overline{E}_{\mathrm{presci}}$, $\sigma_{E_{\mathrm{presci}}}$, and $\overline{E}_{\alpha_0}$, have been varied to fit, for a given mass and charge split, the TKE distribution of the fission fragments, the angular and final energy distributions of the alpha particles, and the cross correlation of these distributions with the TKE distribution. The fit gave $\overline{D} = 20.7$ fm, $\sigma_D = 0.9$ fm, $\overline{E}_{\mathrm{presci}} = 8$ MeV, $\sigma_{E_{\mathrm{presci}}} = 0.5$ MeV, and $\overline{E}_{\alpha_0} = 1.54$ MeV. Results of the fit are shown in Figs. 4.16 and 4.17. The agreement of the calculations with the data, in particular in Fig. 4.16, is only qualitative. Later work by Pik-Pichak [500] showed that the deformation of the fission fragments at scission, neglected in the model by Guet et al., does have a non-negligible effect on the final distribution of the alpha particles. Roshchin, Rubchenya, and Yavshits therefore included vibrational and rotational degrees of freedom for the heavy fragments in their trajectory calculations [501]. Unfortunately, this still increased the required number of initial conditions. They also added a nuclear proximity force to the Coulomb force and introduced some kind of proximity friction in the equations of motion to enforce an orbiting motion of some alpha particles. The latter were

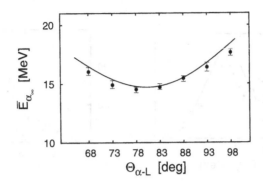

Fig. 4.17. Average final energy of the alpha particles $\overline{E}_{\alpha\infty}$ as function of the emission angle $\theta_{\alpha-L}$ with respect to the light fragment, dots are measured data, full line gives the calculated result (from Ref. [499]).

supposed to produce the observed small maximum in the angular distribution of the alpha particles in the direction of the fission axis, the "polar" alphas. The same region of the final phase space of the alpha particles is however also populated by evaporation particles from the accelerated, excited fragments.

It had been the original claim of the work with trajectory calculations that the distribution of all geometrical parameters needed to describe the scission configuration could be inferred from the observed TKE distribution and the observed angular and energy distributions of the alpha particles. In view of the many ad hoc assumptions concerning the initial distributions this claim is not justified. In fact, trajectory calculations with somewhat different initial conditions [502–504] fitted the data with comparable quality. In all trajectory calculations only planar solutions of the three-body problem were considered, without convincing experimental or theoretical justification. Often nuclear proximity forces were disregarded and the reabsorption of the alpha particle by one of the fragments was only simulated in a crude way. Unfortunately, correlations between the initial coordinates and momenta, imposed by the uncertainty principle, were never taken into account.

For the actinides the ratio of ternary fission yields to binary yields Y_t/Y_b depends only little on the mass and charge numbers of the fissioning nucleus. Halpern found empirically a rather linear dependence of the ratio on the variable $4Z - A$ [505]. This is shown in Fig. 4.18. The ratio does also not depend strongly on the excitation energy E^* in the range $0 \leq E^* < 6$ MeV, i.e. between spontaneous and thermal neutron-induced fission as also seen in Fig. 4.18. There is no convincing theoretical model for these empirical findings.

Besides alpha particles, the next most frequently observed charged third particle in fission is ^{5}He, which decays immediately into ^{4}He+n. The next frequent ternary particles are the hydrogen isotopes. The probability for the appearance of charged particles, heavier than ^{4}He and sufficiently long-lived

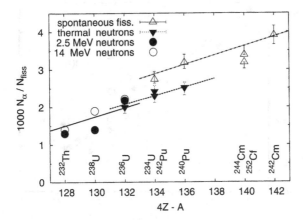

Fig. 4.18. Number of ternary events per 10^3 binary events as function of $4Z - A$ (after Ref. [505]).

to be detected, is rapidly decreasing when their mass increases. To describe this trend Halpern proposed the following scission-point model [505]. In view of the fact that the potential energy, liberated within a few 10^{-22} seconds at neck rupture, cannot be shared by all internal degrees of freedom of the system in such a short time, it is assumed that it is only distributed among the various binary and ternary fragmentations, accessible at the moment of scission. Halpern therefore postulates that the ratio of fission yields for any ternary particle to that of ^{4}He is given by the ratio of their respective Boltzmann factors $\exp(-E_c/T)$, where T is an effective temperature and E_c is the removal energy needed to separate a fragment with mass and charge numbers A_t and Z_t, respectively, from one of the heavy fragments with mass number A_1 and charge number Z_1 and to place it between the tips of the heavy fragments with a center-of-mass distance x to fragment 2 and x' to the residue of fragment 1 (with charge $Z_1 - Z_t$)

$$E_c = B_i + \Delta E_{\text{def}} + K + E_{\text{Coul}}(Z_1, Z_2, Z_t, x, x') - e^2 Z_1 Z_2/(2x). \qquad (4.54)$$

In this equation B_i is the separation energy of the third particle from the nucleus A_1, Z_1, which can be taken from mass tables. The energy ΔE_{def} is the change in the deformation energies of the heavy fragments at scission for binary fission as compared to ternary fission. This term is usually neglected. The term K is the initial kinetic energy of the third particle, which is also neglected in the following. The energy

$$E_{\text{Coul}}(Z_1, Z_2, Z_t, x) = e^2 [(Z_1 - Z_t)Z_2/(x + x') + Z_t Z_2/x + (Z_1 - Z_t)Z_t/x']$$

is the Coulomb energy of three point charges Z_2, Z_t and $(Z_1 - Z_t)$ in a row with distances x and x' between them. The last term in Eq. (4.54) is

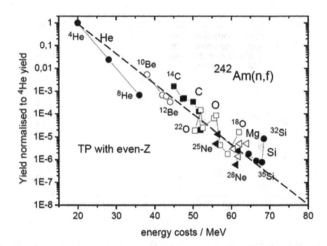

Fig. 4.19. Relative yields of ternary fission events with ternary particles having even charge numbers Z_t as function of the energy cost E_c for the reaction ^{242}Am(n_{therm},f). The mass numbers of the helium isotopes are: A = 4,6,8; of the beryllium isotopes: A = 10,9,11,12; of the carbon isotopes: A = 14,15,16,17,18; of the oxygen isotopes: A = 18,19,20,21,22,23,24; of the neon isotopes: A = 23,24,25,28; of the magnesium isotopes: A = 28,30,31,32; and of the silicon isotopes: A = 32,33,34,35. The dashed line is drawn to guide the eye. Data from Ref. [506].

the Coulomb energy of the scission configuration of the corresponding binary fission to which ternary yields are traditionally normalized.

Halpern assumed that fragment 2 is "born" with a rather compact shape leaving most of the neck material with fragment 1, from which the third particle is formed. Therefore the center of mass of the residue fragment and the third particle is positioned at a distance $2x$ from the center of fragment 1. From this construction follows $x' = x[1 + A_t/(A_1 - A_t)]$. The parameters x and T remain as free parameters in Halpern's model, associated with a given initial compound state and kept fixed for all its ternary decays.

The logarithm of the yield ratio for ternary fragments heavier than ^{4}He to the yield for alpha particles is plotted in Figs. 4.19 and 4.20 as function of the energy cost E_c for the reaction ^{242}Am(n_{therm}, f) [507]. In the calculation of E_c fragment 2 was identified with the double magic nucleus ^{132}Sn and for the parameter x the rather large value $x = 11$ fm was used to simulate the fact that binary events with rather long necks are those to be compared with ternary fission when constructing E_c. The general trend, represented by the dashed line, is fairly convincingly reproduced by the relation yield(Z_t, A_t)/yield(alpha)=exp($E_c(Z_t, A_t)/T$) with an effective temperature $T = 3.3$ MeV, although details in the construction of E_c appear to be somewhat arbitrary.

Several modifications of Halpern's model have been proposed, in particular by introducing elements of nuclear fluid dynamics [467]. Since the latter is

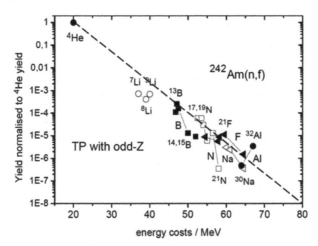

Fig. 4.20. Relative yields of ternary fission events with ternary particles having odd charge numbers Z_t as function of the energy cost E_c for the reaction ^{242}Am(n_{therm},f). Mass numbers of the lithium isotopes: $A = 7,8,9$; of the boron isotopes: $A = 11,12,13,14,15$; of the nitrogen isotopes: $A = 16,17,18,19,20,21$; of the fluorine isotopes: $A = 20,21,22,23,24$; of the sodium isotopes: $A = 27,28,30$; and of the aluminum isotopes: $A = 32,33$. Data from Ref. [506].

intended to describe adiabatic collective motion, its application to the very rapid scission process, however, appears to be not quite justified.

5

Fission Dynamics

We have seen in the preceding chapter that many observables connected with fission can only be calculated in a dynamical description of the fission process. This holds for the calculation of spontaneous fission lifetimes and of branching ratios between competing fission modes. A dynamical theory is also needed to obtain mass, charge, and angular momentum distributions of the fission fragments if one does not want to rely on doubtful assumptions about statistical equilibria at scission or the misleading assumption that fission paths follow the bottom of fission valleys in somebody's shape parametrization.

In two limiting cases equations of motion for the fission process have been proposed: There are microscopic equations of motion, obtained in the framework of the time-dependent Hill-Wheeler (TDHW) theory [87,508,509] which are suitable to describe cluster emission and to calculate spontaneous fission lifetimes. In the other limit, fission of excited nuclei has been modeled by classical, thermodynamical transport equations in the shape variables. Since they rely on canonical, rather than microcanonical averages, they can only be approximately valid when the temperature T of the system satisfies the condition (4.1) $T \gg 10/A$ MeV on the whole fission path. Below we shall give a motivation for this approach on a purely phenomenological basis. Unfortunately, no general transport theory in the microcanonical frame has been developed so far.

The program for a dynamical theory of fission, as described above, implies a continuous evolution of the shape of the fissioning nucleus to be described by a set of "collective" variables, for which equations of motion are assumed to exist. This is not self-evident. The spectroscopy of low-lying collective modes does not suggest that multipole vibrations have a continuation to large amplitudes, i.e. multi-phonon states, since three-phonon states are usually already so broad and fragmented that they are no longer collective. And in fact, in the standard theory for the decay of a compound state of, say, 6 MeV excitation energy one would ask for the probability to find a saddle-point configuration among the many states of which the compound state is composed. The assumption that this probability is determined by the ratio of

H. J. Krappe and K. Pomorski, *Theory of Nuclear Fission*,
Lecture Notes in Physics 838, DOI: 10.1007/978-3-642-23515-3_5,
© Springer-Verlag Berlin Heidelberg 2012

phase-space volumes of the A-particle system lead to the transition-state theory and eventually to the fission-rate formula of Bohr and Wheeler as we saw in the preceding chapter. In the present chapter we assume that dynamical, collective shape variables exist. This leads to a different fission-rate formula, which was first derived by Kramers [510].

A similar ambiguity exists in the understanding of cluster and α decay. One may either consider the ground state of the parent nucleus as containing many complicated cluster amplitudes – the preformation amplitudes – of which one consists of the appropriate clustering and connects through the Coulomb barrier with the observed decay channel. This is the picture of the standard α decay model [511–513] and was later extended to cluster decay by Blendowske, Fliessbach, and Walliser [514]. In contrast, cluster decay was described by Greiner et al. [515] assuming a continuous evolution of shape variables in time. In this chapter we will focus on this latter approach.

5.1 Adiabatic motion

5.1.1 The cranking approach

Inertial tensor

To describe spontaneous fission one needs obviously a Schrödinger equation in the collective (shape) variables to obtain tunneling. An often used approach to its construction starts from a constrained HF wave function $|k; q\rangle$ which is a solution of the static, constrained HF equation

$$\hat{H}_q|k; q\rangle = E_k(q)|k; q\rangle$$

with the Routhian $\hat{H}_q = \hat{H} - \lambda\hat{q}$. For the sake of simplicity we consider only one collective variable in the following. The time evolution of adiabatic motion is assumed to proceed through the time dependence of the constraint $q(t)$ only. In a small neighborhood of $q(t)$ any time-dependent Slater determinant $|t\rangle$ can be expanded in the form

$$|t\rangle \equiv |q(t)\rangle = \sum_k a_k(t)e^{i\phi_k(t)}|k; q(t)\rangle \qquad (5.1)$$

due to Thouless's theorem (cf. Ref. [108], Appendix E3), where we use the time-dependent phase $\phi_k(t) = -(1/\hbar)\int_{t_0}^{t} E_k(q(t'))dt'$. Inserting this expansion into the time-dependent Schrödinger equation

$$\hat{H}_q|t\rangle = i\hbar\partial_t|t\rangle, \qquad (5.2)$$

we obtain

$$\sum_k a_k E_k e^{i\phi_k}|k; q(t)\rangle = i\hbar \sum_k (\dot{a}_k + ia_k\dot{\phi}_k + a_k\dot{q}\partial_q)e^{i\phi_k}|k; q(t)\rangle$$

and with $\dot{\phi}_k = -E_k(q)/\hbar$

$$\sum_k (\dot{a}_k + a_k \dot{q} \partial_q) e^{i\phi_k(t)} |k; q\rangle = 0 .$$

Multiplying with the state $\langle l; q | e^{-i\phi_l}$, yields a set of linear differential equations for the expansion coefficients $a_l(t)$

$$\dot{a}_l = -\dot{q} \sum_k \langle l; q | \partial_q | k; q \rangle e^{i(\phi_k - \phi_l)} a_k . \tag{5.3}$$

The HF ground state shall have $k = 0$. The system (5.3) is to be solved with the initial condition $a_k(q(t)) = \delta_{k0}$. After some time the coefficient a_0 will still be close to 1 (if no pseudo level-crossing appears during that time) and the other a_k will be small of the order $\dot{q}$. To linear order in $\dot{q}$ the system (5.3) simplifies to

$$\dot{a}_l = -\dot{q} \langle l; q | \partial_q | 0; q \rangle e^{i(\phi_0 - \phi_l)}, \qquad l \neq 0 . \tag{5.4}$$

Assuming that the variation of the product $\dot{q} \langle l | \partial_q | 0 \rangle$ is much slower than that of the exponential in Eq. (5.4) the equation can be integrated and yields

$$a_l(t) = -\dot{q} \langle l; q | \partial_q | 0; q \rangle \int_{t_0}^{t} e^{i(\phi_0 - \phi_l)} dt' = \frac{i\hbar \dot{q}}{E_l - E_0} \langle l; q | \partial_q | 0; q \rangle e^{i(\phi_0 - \phi_l)} , \tag{5.5}$$

where $dt' = \hbar/(E_0 - E_l) \, d(\phi_0 - \phi_l)$ was used and $E_l - E_0$ is the excitation energy of the particle-hole state l.

The energy expectation-value in the state $|q(t)\rangle$ becomes in this approximation

$$\langle q(t) | \hat{H}_q | q(t) \rangle = E_0 + \sum_{l \neq 0} (E_l - E_0) |a_l|^2 ,$$

which may be written as a classical Hamiltonian

$$\langle q | \hat{H}_q | q \rangle = \frac{1}{2} m^{(\text{coll})}(q) \dot{q}^2 + V(q) \tag{5.6}$$

with the potential energy $V(q) = E_0(q)$ and a mass

$$m^{(\text{coll})}(q) = 2\hbar^2 \sum_{l \neq 0} \frac{|\langle l; q | \partial_q | 0; q \rangle|^2}{E_l - E_0} . \tag{5.7}$$

In the following the superscript (coll) on the collective inertia will be dropped. One uses the fact that $|l; q\rangle$ is an eigenstate of $\hat{H}_q$ and therefore obtains

$$\langle l; q | \partial_q \hat{H}_q | 0; \rangle = \langle l; q | [\hat{H}_q, \partial_q] | 0; q \rangle = (E_l - E_0) \langle l; q | \partial_q | 0; q \rangle . \tag{5.8}$$

Eq. (5.7) can therefore be brought into the more convenient form

$$m(q) = 2\hbar^2 \sum_{l \neq 0} \frac{|\langle l; q|\partial_q \hat{H}_q|0; q\rangle|^2}{(E_l - E_0)^3} \,. \tag{5.9}$$

A collective Hamilton operator is derived from Eq. (5.6) by quantizing the classical Hamiltonian according to the Pauli prescription [516–518]

$$\hat{\mathcal{H}}^{(\mathrm{coll})} = -\frac{1}{2}\hbar^2 \frac{1}{\sqrt{m(q)}} \frac{d}{dq} \frac{1}{\sqrt{m(q)}} \frac{d}{dq} + V(q)\,. \tag{5.10}$$

The kinetic energy follows from the requirement that the classical Hamiltonian (5.6) should be obtained in the classical limit and that the kinetic energy shall be a scalar under coordinate transformations [516]. Note that $q(t)$ is given from outside in this approach, rather than determined selfconsistently. This scheme was first applied to collective rotation, where it amounts to probing the inertial reaction forces to a cranking of a deformed nucleus (Inglis [318]). The model is therefore called cranking model and the term was also used when it was later applied to collective vibrations by Bès and Kerman [519, 520] and to fission by Nilsson and his collaborators in Lund and Warsaw [103, 521].

The cranking formalism is easily generalized to multidimensional, collective motion. The classical, collective kinetic energy becomes in this case

$$E_{\mathrm{kin}}^{(\mathrm{coll})} = \frac{1}{2} \sum_{ij} m_{ij}(\mathbf{q}) \dot{q}_i \dot{q}_j$$

with a mass tensor

$$m_{ij}(\mathbf{q}) = 2\hbar^2 \sum_{k \neq 0} \frac{\langle 0; \mathbf{q}|\partial_{q_i}|k; \mathbf{q}\rangle \langle k; \mathbf{q}|\partial_{q_j}|0; \mathbf{q}\rangle}{E_k - E_0} \,. \tag{5.11}$$

After Pauli quantization one obtains the collective Hamiltonian

$$\hat{\mathcal{H}}^{(\mathrm{coll})} = -\frac{1}{2}\hbar^2 \sum_{ij} \frac{1}{D} \frac{\partial}{\partial q_i} D(m^{-1})_{ij} \frac{\partial}{\partial q_j} + V(\mathbf{q})\,, \tag{5.12}$$

where $D = \sqrt{\det m_{ij}(\mathbf{q})}$. The Hamiltonian $\hat{\mathcal{H}}^{(\mathrm{coll})}$ is Hermitian with the volume element $d\tau = Dd\mathbf{q}$.

It can be seen directly from Eqs. (5.7) or (5.11) that the cranking model fails when for one particle-hole state l $E_0 = E_l$, i.e. when an unoccupied level crosses the Fermi surface along the fission path. This problem is less serious if pairing correlations create a gap. To take pairing into account one defines the states $|\mathbf{q}\rangle$ as HF+BCS states (2.40)

$$|\mathbf{q}\rangle = \prod_{\nu > 0} (u_\nu + v_\nu a_\nu^\dagger a_{-\nu}^\dagger)|0; \mathbf{q}\rangle\,. \tag{5.13}$$

The derivatives of the BCS wave function (2.40) with respect to the deformation parameters can be written as

$$\frac{\partial}{\partial q_i}|\mathbf{q}\rangle \equiv P_i|\mathbf{q}\rangle = \sum_{\mu\nu} (P_i)_{\nu\mu}\, \alpha_\mu^\dagger \alpha_{-\nu}^\dagger |\mathbf{q}\rangle\,, \tag{5.14}$$

where $\alpha_\mu^\dagger$ and $\alpha_{-\nu}^\dagger$ are the quasiparticle creation operators (2.38) and

$$(P_i)_{\nu\mu} =$$
$$-\frac{\langle\mu|\partial_{q_i}\hat{H}_\mathbf{q}|\nu\rangle}{\mathcal{E}_\mu + \mathcal{E}_\nu}(u_\mu v_\nu + u_\nu v_\mu) + \frac{\delta_{\mu\nu}}{2}\left(\frac{\Delta}{\mathcal{E}_\mu^2}\frac{\partial\lambda}{\partial q_i} + \frac{\epsilon_\mu - \lambda}{\mathcal{E}_\mu^2}\frac{\partial\Delta}{\partial q_i}\right). \tag{5.15}$$

The last term in the above equations comes from the dependence of the pairing gap Δ and the Fermi energy λ on the deformation parameters q_i. Both quantities are obtained from the set of BCS equations (2.47) and (2.48).

The collective mass tensor becomes in the BCS framework

$$m_{ij} = 2\hbar^2 \sum_{\nu,\mu} \frac{(P_i)_{\mu\nu}^* (P_j)_{\mu\nu}}{\mathcal{E}_\nu + \mathcal{E}_\mu}\,. \tag{5.16}$$

For a derivation of the last four equations see Ref. [103], Sec. IX.3. For odd systems the blocking effect has to be taken into account. The appropriate explicit formulae for the mass tensor in this case can be found e.g. in Ref. [522].

In this treatment of pairing it was assumed that for each value of the collective variables $\mathbf{q}$ the pairing parameters Δ and λ assume their equilibrium values. However, in the BCS framework the pairing-vibration degree of freedom can be coupled to the dynamics in the shape degrees of freedom as proposed by Góźdź et al. [158] and specifically to fission dynamics, discussed by Moretto and Babinet [523] and by Staszczak et al. [524, 525]. The collective degrees of freedom consist in this case of the shape variables $\mathbf{q}$ and the gap parameters Δ_p and Δ_n for protons and neutrons, respectively. The shift operator P_i, introduced in Eq. (5.14), gets the additional components

$$\frac{\partial}{\partial\Delta}|\mathbf{q},\Delta\rangle \equiv P_\Delta|\mathbf{q},\Delta\rangle = \sum_{\mu\nu}(P_\Delta)_{\mu\nu}\alpha_\mu^\dagger\alpha_{-\nu}^\dagger|\mathbf{q},\Delta\rangle \tag{5.17}$$

with

$$(P_\Delta)_{\mu\nu} = \delta_{\mu\nu}\frac{(\epsilon_\nu - \lambda) + \Delta(\partial\lambda/\partial\Delta)}{2\mathcal{E}_\nu} \tag{5.18}$$

for the proton and the neutron gap Δ. Simultaneously the last term in Eq. (5.15) proportional to $\partial\Delta/\partial q_i$ has to be dropped.

In the cranking approach extra degrees of freedom, connected with the collective variables, are introduced, in addition to the $3A$ degrees of freedom in the HF or HFB states $|\mathbf{q}\rangle$. This has the odd consequence that one has to "quantize" an equation which results already from a fully quantal description of the A-body system in terms of the Schrödinger equation (5.2). It has the further consequence that additional zero-point energies appear in connection with the collective variables. As already pointed out in the discussion of Eq. (2.159), they are spurious.

5.1.2 The Hill-Wheeler-Griffin theory

The success of the static mean-field theory with appropriate effective two-body interactions led to the expectation that the time-dependent Hartree-Fock (TDHF) or Hartree-Bogolyubov (TDHB) equation may be the appropriate dynamical equation, which does not involve supernumerary degrees of freedom. In this approach the time-dependent wave function $\Phi(t)$ of the A-body system is obtained from the variational equation

$$\frac{\delta \int_{t_1}^{t_2} \langle \Phi(t) | \hat{H} - i\hbar \partial_t | \Phi(t) \rangle \, dt}{\delta \Phi} = 0 \,, \tag{5.19}$$

restricting the variation to the set of HF or HFB states. In general these states do not span a linear space. Thouless's theorem [108] guarantees only in a small neighborhood of such state that any state can be represented again by one HF or HFB state. The TDHF equation is therefore useful to describe small oscillations around the ground state, and in fact, it is equivalent in this limit to the very successful RPA approximation. On the other hand, the violation of the quantum-mechanical superposition principle for large-scale collective motion has, among other deficiencies, the consequence that TDHF cannot account for the split of a wave packet into two waves, moving in opposite directions. This happens, for instance, when a wave hits a potential barrier and is split into a reflected and a transmitted wave. The TDHF (or TDHB) scheme is therefore too narrow to describe tunneling.

A minimal extension of the mean-field ansatz in Eq. (5.19) is therefore to take for $\Phi(t)$ all linear superpositions of HF or HFB states, needed to describe at least the quantal motion in the collective variables correctly. This leads to the ansatz

$$|\Phi(t)\rangle = \int f(\mathbf{q}, t) |\mathbf{q}\rangle \, d\mathbf{q} \tag{5.20}$$

in terms of a time-dependent weight function $f(\mathbf{q}, t)$ and $|\mathbf{q}\rangle$ is a constrained HF or HFB state, depending on a set of generator coordinates $\mathbf{q} = (q_1, \ldots q_N)$, which are in our case the shape constraints, needed to represent the fission path. The weight function $f(\mathbf{q}, t)$ is varied in Eq. (5.19) for fixed states $|\mathbf{q}\rangle$ to determine an adiabatic path. In this approach one does not introduce redundant variables and has a great deal of flexibility in choosing the collective parameters. On the other hand this has the disadvantage that the method by itself does not indicate what the relevant collective variables are.

Excursion: the stationary version of the Hill-Wheeler-Griffin theory

It is useful to interrupt here the discussion of the time-dependent Hill-Wheeler theory in order to introduce some approximations and notations to be borrowed from the generator-coordinate formalism in the time-independent case.

The latter was formulated by Griffin and Wheeler [526], then further developed by Brink and Weiguny [527] and others.

The energy surface of many nuclei shows several minima of comparable depth as function of a constraining parameter, for instance the components of the quadrupole moment $\mathbf{q}$. In this case it is natural to make the ansatz

$$\Phi = \int d\mathbf{q}\, f(\mathbf{q})|\mathbf{q}\rangle \tag{5.21}$$

in terms of constrained HF or HFB states $|\mathbf{q}\rangle$ and a weight function $f(\mathbf{q})$ [528]. Formally the ansatz (5.21) is similar to the ansatz (2.91) or (2.94) for a projection of a state with broken symmetry onto states with proper symmetry. However, in that case the weight function $f(\mathbf{q})$ is known to be just a representation of the broken symmetry group, whereas in Eq. (5.21) the weight function has to be determined from the variational equation

$$\frac{\delta}{\delta f}\langle\Phi|\hat{H}|\Phi\rangle/\langle\Phi|\Phi\rangle = 0\,,$$

which yields the Hill-Wheeler equation

$$\int d\mathbf{q}'\left(\langle\mathbf{q}|\hat{H}|\mathbf{q}'\rangle - E\langle\mathbf{q}|\mathbf{q}'\rangle\right)f(\mathbf{q}') = 0\,. \tag{5.22}$$

It is convenient to introduce the notation

$$\mathcal{H}(\mathbf{q},\mathbf{q}') = \langle\mathbf{q}|\hat{H}|\mathbf{q}'\rangle\,, \qquad \mathcal{N}(\mathbf{q},\mathbf{q}') = \langle\mathbf{q}|\mathbf{q}'\rangle \tag{5.23}$$

for the overlap matrix-elements, where $\mathcal{N}(\mathbf{q},\mathbf{q}) = 1$. The matrix $\mathcal{N}$ is Hermitian with nonnegative eigenvalues n_k

$$\int \mathcal{N}(\mathbf{q},\mathbf{q}')u_k(\mathbf{q}')d\mathbf{q}' = n_k u_k(\mathbf{q})\,.$$

In terms of the eigenstates u_k one can define the matrix

$$\mathcal{N}^{1/2}(\mathbf{q},\mathbf{q}') = \sum_k u_k(\mathbf{q})\sqrt{n_k}\, u_k^*(\mathbf{q}')\,. \tag{5.24}$$

When the basis states $|\mathbf{q}\rangle$ are linear dependent (or for numerical purposes almost so), some eigenvalues n_k are zero (or become very small). In order to form the inverse of $\mathcal{N}^{1/2}$, one has therefore either to restrict the sum in Eq. (5.24) to those states whose eigenvalues are larger than some small numerical limit ϵ, or alternatively, regularization techniques [529] may be used if they are more convenient. Keeping this in mind, one can introduce instead of the weight function $f(\mathbf{q})$ a function $g(\mathbf{q})$ with

$$f = \mathcal{N}^{-1/2}g\,, \tag{5.25}$$

in terms of which the Hill-Wheeler equation (5.22) becomes an Hermitian eigenvalue equation

$$\mathcal{N}^{1/2} \mathcal{H} \mathcal{N}^{1/2} g = E g . \tag{5.26}$$

In the basis of the so-called natural states

$$|k\rangle = \frac{1}{\sqrt{n_k}} \int d\mathbf{q} \, u_k(\mathbf{q}) \, |\mathbf{q}\rangle$$

this equation takes the form

$$\sideset{}{'}\sum_{k'} \langle k|\hat{H}|k'\rangle g_{k'} = E g_k , \tag{5.27}$$

where the sum is restricted to states with $n_{k'} > \epsilon$.

By discretization one can convert the integral equation (5.26) into a system of linear equations or solve the finite set of eigenvalue equations (5.27). Alternatively, an approximate solution of the Hill-Wheeler equation (5.22) is sought, using the fact that $\mathcal{N}(\mathbf{q}, \mathbf{q}')$ as well as $\mathcal{H}(\mathbf{q}, \mathbf{q}')$ are sharply peaked functions for $\mathbf{q} = \mathbf{q}'$. It is therefore natural to introduce new variables $\bar{\mathbf{q}} = (\mathbf{q} + \mathbf{q}')/2$ and $\mathbf{s} = \mathbf{q} - \mathbf{q}'$. With the shift operator $P_i = \partial_{\bar{q}_i}$ one can formally write

$$|\bar{\mathbf{q}} - \mathbf{s}/2\,\rangle = e^{-(1/2) \quad _i s_i P_i} |\bar{\mathbf{q}}\rangle$$

and therefore

$$\mathcal{N}(\mathbf{q}, \mathbf{q}') = \mathcal{N}(\bar{\mathbf{q}} + \mathbf{s}/2 \,,\, \bar{\mathbf{q}} - \mathbf{s}/2) = \langle \bar{\mathbf{q}}| \exp\left(-\frac{1}{4} \sum_{i,j} s_i \, \overleftarrow{P_i} \, s_j \, \overrightarrow{P_j} \,\right)|\bar{\mathbf{q}}\rangle . \tag{5.28}$$

The Gaussian overlap assumption (GOA) consists in the ansatz

$$\mathcal{N}(\bar{\mathbf{q}}, \mathbf{s}) = \langle \bar{\mathbf{q}}| \exp\left(-\frac{1}{2} \sum_{ij} \gamma_{ij}(\bar{\mathbf{q}}) s_i s_j \right) |\bar{\mathbf{q}}\rangle . \tag{5.29}$$

Expanding Eqs. (5.28) and (5.29) in powers of s_i and comparing the coefficients of the quadratic terms, one finds

$$\gamma_{ij}(\bar{\mathbf{q}}) = \langle \bar{\mathbf{q}}| \overleftarrow{P_i} \overrightarrow{P_j} |\bar{\mathbf{q}}\rangle . \tag{5.30}$$

Since $\mathcal{N}(\mathbf{q}, \mathbf{q}')$ has a maximum for $\mathbf{q} = \mathbf{q}'$, it follows that $\langle \bar{\mathbf{q}}|P_i|\bar{\mathbf{q}}\rangle = 0$.

It is useful to consider the one-dimensional case first, since it does not require the use of methods from tensor calculus as does the multi-dimensional GOA. The tensor $\gamma_{ij}(\bar{\mathbf{q}})$ has now only one component $\gamma(\bar{q})$. One introduces a new collective coordinate α by the relation [158, 530]

$$\alpha(q) = \int^q \sqrt{\gamma(q')} dq' , \tag{5.31}$$

in terms of which the overlap matrix $\mathcal{N}$ becomes

$$\mathcal{N}(\alpha, \alpha') = \exp\left(-\frac{1}{2}(\alpha - \alpha')^2\right) \tag{5.32}$$

and its square root is

$$\mathcal{N}^{1/2}(\alpha, \xi) = \left(\frac{2}{\pi}\right)^{1/4} e^{-(\alpha-\xi)^2} \tag{5.33}$$

since one obtains by elementary integration

$$\mathcal{N}(\alpha, \alpha') = \int\limits_{-\infty}^{\infty} \mathcal{N}^{1/2}(\alpha, \xi)\mathcal{N}^{1/2}(\xi, \alpha')d\xi. \tag{5.34}$$

It has been found that the ratio $h(\alpha, \alpha') = \mathcal{H}(\alpha, \alpha')/\mathcal{N}(\alpha, \alpha')$ is a smooth function of its arguments and may be expanded around a point $\alpha = \alpha' = \xi$ in a Taylor series in powers of $\Delta = \alpha - \xi$ and $\Delta' = \alpha' - \xi$ up to the second order

$$h(\alpha, \alpha') = h(\xi, \xi) + h_\alpha \Delta + h_{\alpha'} \Delta' + h_{\alpha\alpha'} \Delta\Delta' + \frac{1}{2}(h_{\alpha\alpha}\Delta^2 + h_{\alpha'\alpha'}\Delta'^2), \tag{5.35}$$

where we use the notation

$$h_\alpha = \frac{\partial h(\alpha, \xi)}{\partial \alpha}\bigg|_{\alpha=\xi}, \qquad h_{\alpha\alpha} = \frac{\partial^2 h(\alpha, \xi)}{\partial \alpha^2}\bigg|_{\alpha=\xi}$$

and similar expressions for derivatives with respect to α'. With Eqs. (5.21), (5.23), (5.34), and the definition of $h(\alpha, \alpha')$, the energy expectation value can be written

$$\langle \Phi|\hat{H}|\Phi\rangle = \int d\xi \int d\alpha \int d\alpha' f^*(\alpha)\mathcal{N}^{1/2}(\alpha, \xi)h(\alpha, \alpha')\mathcal{N}^{1/2}(\xi, \alpha')f(\alpha'). \tag{5.36}$$

We abbreviate the integrations with respect to α and α' by

$$g(\xi) = \int f(\alpha)\mathcal{N}^{1/2}(\alpha, \xi)d\alpha.$$

Inserting the expansion (5.35) into Eq. (5.36) and using the identities

$$(\alpha - \xi)\mathcal{N}^{1/2}(\alpha, \xi) = \frac{1}{2}\partial_\xi \mathcal{N}^{1/2}(\alpha, \xi),$$

$$(\alpha - \xi)^2\mathcal{N}^{1/2}(\alpha, \xi) = \frac{1}{4}(2 + \partial_\xi^2)\mathcal{N}^{1/2}(\alpha, \xi)$$

and the symmetry property $h(\alpha, \alpha') = h(\alpha', \alpha)$, (which implies $h_\alpha = h_{\alpha'}$ and $h_{\alpha\alpha} = h_{\alpha'\alpha'}$) one obtains

$$\langle \Phi | \hat{H} | \Phi \rangle = \int d\xi \, g^*(\xi)$$

$$\left[h + \frac{\overleftarrow{d}}{2d\xi} h_\alpha + h_\alpha \frac{\overrightarrow{d}}{2d\xi} + \frac{1}{2} h_{\alpha\alpha} + \frac{1}{8} \left(\frac{\overleftarrow{d^2}}{d\xi^2} + \frac{\overrightarrow{d^2}}{d\xi^2} \right) h_{\alpha\alpha} + \frac{\overleftarrow{d}}{2d\xi} h_{\alpha\alpha'} \frac{\overrightarrow{d}}{2d\xi} \right] g(\xi).$$

This expression can be simplified by partial integration of the second and the fifth term in the bracket, making use of the relation $\partial_\xi h_\alpha = h_{\alpha\alpha} + h_{\alpha\alpha'}$ and neglecting third derivatives of h

$$\langle \Phi | \hat{H} | \Phi \rangle = \int d\xi \, g^*(\xi) \left[h - \frac{1}{2} h_{\alpha\alpha'} + \frac{1}{4} \frac{\overrightarrow{d}}{d\xi} (h_{\alpha\alpha} - h_{\alpha\alpha'}) \frac{\overrightarrow{d}}{d\xi} \right] g(\xi). \qquad (5.37)$$

We have finally to transform back to the original variables q and q' using the relation $d\alpha = \sqrt{\gamma(q)} \, dq$ and obtain

$$d\xi = \sqrt{\gamma(q)} dq, \qquad \frac{\overrightarrow{d}}{d\xi} = \frac{1}{\sqrt{\gamma(q)}} \frac{\overrightarrow{d}}{dq}, \qquad h_\alpha = \frac{1}{\sqrt{\gamma}} \frac{\partial h}{\partial q},$$

$$h_{\alpha\alpha} = \frac{1}{\sqrt{\gamma(q)}} \frac{\partial}{\partial q} \left(\frac{1}{\sqrt{\gamma(q)}} \frac{\partial h}{\partial q} \right), \qquad h_{\alpha\alpha'} = \frac{1}{\sqrt{\gamma(q)}} \frac{\partial}{\partial q} \left(\frac{1}{\sqrt{\gamma(q')}} \frac{\partial h}{\partial q'} \right).$$

Inserting these relations into Eq. (5.37) results in an equation of the form

$$\langle \Phi | \hat{H} | \Phi \rangle = \int g^*(q) \, \hat{\mathcal{H}}^{(\text{coll})}(q, \partial_q) \, g(q) \, \sqrt{\gamma(q)} dq \qquad (5.38)$$

with

$$\hat{\mathcal{H}}^{(\text{coll})} = -\frac{\hbar^2}{2} \frac{d}{\sqrt{\gamma(q)} \, dq} \sqrt{\gamma(q)} m^{-1}(q) \frac{d}{dq} + V(q), \qquad (5.39)$$

where the inverse inertial function is

$$m^{-1}(q) = \frac{1}{2\hbar^2 \gamma^2(q)} \left(\frac{\partial^2 h(q, q')}{\partial q \partial q'} - \frac{\partial^2 h(q, q')}{\partial q^2} + \frac{\gamma'(q)}{2\gamma(q)} \frac{\partial h(q, q')}{\partial q} \right)_{q=q'} \qquad (5.40)$$

and the effective potential becomes

$$V(q) = \langle q | \hat{H} | q \rangle - \frac{1}{2\gamma(q)} \frac{\partial^2 h(q, q')}{\partial q \partial q'} \bigg|_{q=q'}. \qquad (5.41)$$

We will now discuss the multi-dimensional case and assume as in Ref. [531] that the metric in $\overline{q}$-space, given by $\gamma_{ij}(\overline{q})$, describes a flat space, i.e. we assume that the Riemannian tensor derived from γ_{ij} vanishes identically. Under this condition a transformation to new coordinates $\alpha_i(q)$, $i = 1, \ldots N$ exists, in which the metric becomes Euclidean, $\gamma_{ij}^{\text{transf}} = \delta_{ij}$. One can then proceed in a way very much analogous to the steps between Eqs. (5.31) and (5.37). Defining

$$\mathcal{N}^{1/2}(\boldsymbol{\alpha}, \boldsymbol{\xi}) = \left(\frac{2}{\pi}\right)^{N/4} e^{-(\boldsymbol{\alpha}-\boldsymbol{\xi})^2}, \qquad (5.42)$$

one verifies by elementary integrations that

$$\mathcal{N}(\boldsymbol{\alpha}, \boldsymbol{\alpha}') = \int_{-\infty}^{\infty} \mathcal{N}^{1/2}(\boldsymbol{\alpha}, \boldsymbol{\xi}) \mathcal{N}^{1/2}(\boldsymbol{\xi}, \boldsymbol{\alpha}') d^N \boldsymbol{\xi}. \qquad (5.43)$$

The energy overlap $\mathcal{H}(\boldsymbol{\alpha}, \boldsymbol{\alpha}')$ is also sharply peaked as in the one-dimensional case and the reduced energy overlap

$$h(\boldsymbol{\alpha}, \boldsymbol{\alpha}') = \frac{\mathcal{H}(\boldsymbol{\alpha}, \boldsymbol{\alpha}')}{\mathcal{N}(\boldsymbol{\alpha}, \boldsymbol{\alpha}')} \qquad (5.44)$$

is typically a smooth function of its arguments and may be expanded around a point $\boldsymbol{\xi}$ in a Taylor series in both arguments up to second order terms

$$h(\boldsymbol{\alpha}, \boldsymbol{\alpha}') = h + \sum_i h_{\alpha_i}(\alpha_i - \xi_i) + \sum_i h_{\alpha_i'}(\alpha_i' - \xi_i)$$

$$+ \frac{1}{2} \sum_{ij} h_{\alpha_i \alpha_j}(\alpha_i - \xi_i)(\alpha_j - \xi_j) + \sum_{ij} h_{\alpha_i \alpha_j'}(\alpha_i - \xi_i)(\alpha_j' - \xi_j)$$

$$+ \frac{1}{2} \sum_{ij} h_{\alpha_i' \alpha_j'}(\alpha_i' - \xi_i)(\alpha_j' - \xi_j) + \dots \qquad (5.45)$$

with abbreviations

$$h = h(\boldsymbol{\xi}, \boldsymbol{\xi}) , \quad h_{\alpha_i} = \partial_{\alpha_i} h(\boldsymbol{\alpha}, \boldsymbol{\xi})|_{\boldsymbol{\alpha}=\boldsymbol{\xi}}$$

$$h_{\alpha_i \alpha_j} = \partial^2_{\alpha_i \alpha_j} h(\boldsymbol{\alpha}, \boldsymbol{\xi})|_{\boldsymbol{\alpha}=\boldsymbol{\xi}} , \quad h_{\alpha_i \alpha_j'} = \partial^2_{\alpha_i \alpha_j'} h(\boldsymbol{\alpha}, \boldsymbol{\alpha}')|_{\boldsymbol{\alpha}=\boldsymbol{\alpha}'=\boldsymbol{\xi}}$$

and analogous expressions where α_i is replaced by α_i'.

Using Eqs. (5.21), (5.23), (5.43), and (5.44) the energy expectation value becomes

$$\langle \Phi | \hat{H} | \Phi \rangle = \int d^N \xi \int d^N \alpha \int d^N \alpha' f^*(\boldsymbol{\alpha}) \mathcal{N}^{1/2}(\boldsymbol{\alpha}, \boldsymbol{\xi}) h(\boldsymbol{\alpha}, \boldsymbol{\alpha}') \mathcal{N}^{1/2}(\boldsymbol{\xi}, \boldsymbol{\alpha}') f(\boldsymbol{\alpha}'), \qquad (5.46)$$

which we want to rewrite in the form

$$\langle \Phi | \hat{H} | \Phi \rangle = \int d^N \xi g^*(\boldsymbol{\xi}) \hat{\mathcal{H}}^{(\text{coll})}(\boldsymbol{\xi}, \partial_{\boldsymbol{\xi}}) g(\boldsymbol{\xi}) \qquad (5.47)$$

in terms of a collective Hamiltonian $\hat{\mathcal{H}}^{(\text{coll})}$ depending on the collective variables and their partial derivatives up to second order and a wave function

$$g(\boldsymbol{\xi}) = \int f(\boldsymbol{\alpha}) \mathcal{N}^{1/2}(\boldsymbol{\alpha}, \boldsymbol{\xi}) d^N \alpha.$$

We therefore insert the expansion (5.45) into Eq. (5.46) making use of the identities

$$(\alpha_i - \xi_i)\mathcal{N}^{1/2}(\boldsymbol{\alpha}, \boldsymbol{\xi}) = \frac{1}{2}\partial_{\xi_i}\mathcal{N}^{1/2}(\boldsymbol{\alpha}, \boldsymbol{\xi}), \tag{5.48}$$

$$(\alpha_i - \xi_i)(\alpha_j - \xi_j)\mathcal{N}^{1/2}(\boldsymbol{\alpha}, \boldsymbol{\xi}) = \frac{1}{4}(2\delta_{ij} + \partial^2_{\xi_i\xi_j})\mathcal{N}^{1/2}(\boldsymbol{\alpha}, \boldsymbol{\xi}) \tag{5.49}$$

and the fact that $h(\boldsymbol{\alpha}, \boldsymbol{\alpha}')$ is symmetric and therefore $h_{\alpha_i} = h_{\alpha'_i}$ and $h_{\alpha_i\alpha_j} = h_{\alpha'_i\alpha'_j}$ and obtain

$$\langle\Phi|\hat{H}|\Phi\rangle = \int d^N\xi \left[g^*hg + \frac{1}{2}\sum_i \frac{\partial g^*}{\partial\xi_i}h_{\alpha_i}g + \frac{1}{2}\sum_i g^*h_{\alpha_i}\frac{\partial g}{\partial\xi_i}\frac{1}{2}\sum_i g^*h_{\alpha_i\alpha_i}g \right.$$

$$\left. + \frac{1}{8}\sum_{ij}\left(\frac{\partial^2 g^*}{\partial\xi_i\partial\xi_j}h_{\alpha_i\alpha_j}g + 2\frac{\partial g^*}{\partial\xi_i}h_{\alpha_i\alpha'_j}\frac{\partial g}{\partial\xi_j} + g^*h_{\alpha_i\alpha_j}\frac{\partial^2 g}{\partial\xi_i\partial\xi_j} \right) \right].$$

In the next step all derivatives with respect to ξ are moved to the right by partial integration. Use is made of the relation $\partial_{\xi_i}h_{\alpha_j} = h_{\alpha_i\alpha_j} + h_{\alpha_i\alpha'_j}$ and all third and higher derivatives of h are neglected. One obtains

$$\langle\Phi|\hat{H}|\Phi\rangle = \int d\xi\, g^* \left[\left(h - \frac{1}{2}\sum_i h_{\alpha_i\alpha'_i} \right) g \right.$$

$$\left. + \frac{1}{4}\sum_{ij}(h_{\alpha_i\alpha_j} - h_{\alpha_i\alpha'_j})\partial^2_{\xi_i\xi_j}g \right]. \tag{5.50}$$

In order to return to the original variables $\mathbf{q}$ and $\mathbf{q}'$ it is convenient to arrange the ξ-derivatives more symmetrically in the second term

$$\sum_{ij}(h_{\alpha_i\alpha_j} - h_{\alpha_i\alpha'_j})\partial^2_{\xi_i\xi_j}g \approx \sum_{ij}\partial_{\xi_i}(h_{\alpha_i\alpha_j} - h_{\alpha_i\alpha'_j})\partial_{\xi_j}g, \tag{5.51}$$

where again third derivatives of h have been neglected. Then the second derivatives of h, $h_{\alpha_i\alpha_j}$ and $h_{\alpha_i\alpha'_j}$ must be replaced by covariant derivatives (defined for instance in Ref. [532] or Ref. [201], Sec. 16.10). The term involving the mixed derivatives of h in Eq. (5.50) becomes

$$\sum_i h_{\alpha_i\alpha'_i} \rightarrow \sum_{ij} \frac{\partial h(\mathbf{q}, \mathbf{q}')}{\partial q^i \partial q'^j}\gamma^{ij}, \tag{5.52}$$

where γ^{ij} is defined as the inverse of γ_{ij}

$$\sum_i \gamma_{ik}\gamma^{ij} = \delta^j_k.$$

In covariant tensor products we employ in the following the usual Einstein convention that summing over pairs of upper and lower indices is implied. The term $h_{\alpha_i\alpha_j}$ becomes a covariant tensor

$$h_{\alpha_i \alpha_j} \rightarrow \left.\frac{\partial^2 h(\mathbf{q}, \mathbf{q}')}{\partial q^i \partial q^j}\right|_{\mathbf{q}=\mathbf{q}'} - \left\{{n \atop ij}\right\} \left.\frac{\partial h(\mathbf{q}, \mathbf{q}')}{\partial q^n}\right|_{\mathbf{q}=\mathbf{q}'}, \qquad (5.53)$$

where the expression in braces is the Christoffel symbol of the second kind. Note that the two derivatives with respect to q^i and q'^j in Eq. (5.52) refer to different functional spaces. Only after differentiation were they related to each other by putting $\mathbf{q} = \mathbf{q}' = \bar{\mathbf{q}}$. Therefore there appears no term with a Christoffel symbol in Eq. (5.52) in contrast to Eq. (5.53), where both derivatives refer to the same functional space.

The derivative with respect to ξ_j in Eq. (5.51) yields the covariant vector

$$\partial_{\xi_j} g \rightarrow \frac{\partial g}{\partial q^j}, \qquad (5.54)$$

the derivative with respect to ξ_i becomes the divergence of a contravariant vector

$$\partial_{\xi_i}(\cdots) \rightarrow \frac{1}{\sqrt{G}}\frac{\partial \sqrt{G}(\cdots)}{\partial q^i}, \qquad (5.55)$$

with $G(\mathbf{q}) = \det(\gamma_{ij})$ and the volume element $d^N\xi$ in Eq. (5.21) becomes

$$d^N\xi \rightarrow \sqrt{G(\mathbf{q})}d^N q.$$

Collecting the results (5.50)-(5.55) we obtain a collective Hamiltonian

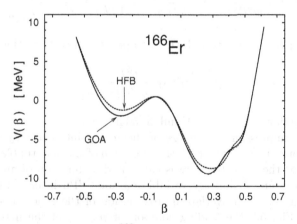

Fig. 5.1. Collective potential of ^{166}Er obtained in Ref. [533] as function of the quadrupole deformation β with the generator-coordinate method in Gaussian overlap plus cranking approximation (5.61) (solid line) and with the constrained HFB approximation (dashed line).

$$\hat{\mathcal{H}}^{(\text{coll})} = -\frac{1}{2}\hbar^2 \frac{\partial}{\sqrt{G}\,\partial q^i}\sqrt{G}(m^{-1})^{ij}\frac{\partial}{\partial q^j} + V(\mathbf{q}) \qquad (5.56)$$

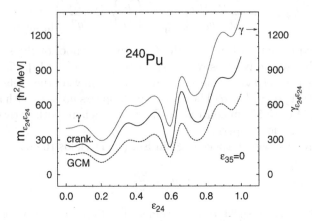

Fig. 5.2. The component $m_{\epsilon_{24}\epsilon_{24}}$ of the mass tensor in the cranking approximation (5.16) and in the generator-coordinate method (GCM) with Gaussian overlap plus cranking approximation (5.61) as function of the elongation variable ϵ_{24} for ^{240}Pu. Also shown is the component $\gamma_{\epsilon_{24}\epsilon_{24}}$ of the width tensor (after Ref. [158]).

with the inverse of the inertial tensor

$$(m^{-1})^{ij} =$$
$$\frac{1}{2\hbar^2}\gamma^{ik}(\mathbf{q})\left(\frac{\partial^2 h(\mathbf{q},\mathbf{q}')}{\partial q^k \partial q'^l} - \frac{\partial^2 h(\mathbf{q},\mathbf{q}')}{\partial q^k \partial q^l} + \left\{\begin{matrix} n \\ kl \end{matrix}\right\}\frac{\partial h(\mathbf{q},\mathbf{q}')}{\partial q^n}\right)\gamma^{lj}(\mathbf{q}) \quad (5.57)$$
$$\hspace{10cm}_{\mathbf{q}=\mathbf{q}'}$$

and $V(\mathbf{q}) = \langle \mathbf{q}|\hat{H}|\mathbf{q}\rangle - \epsilon_0(\mathbf{q})$ with the zero-point energy-correction

$$\epsilon_0(\mathbf{q}) = \frac{1}{2}\gamma^{ij}(\mathbf{q})\left.\frac{\partial^2 h}{\partial q^i \partial q'^j}\right|_{\mathbf{q}=\mathbf{q}'}. \quad (5.58)$$

Eq. (5.57) was first reported by Kamlah [534].

To give an impression of the size of the zero-point energy ϵ_0 we show in Fig. 5.1 the potential energy for ^{166}Er as function of the quadrupole parameter β. In Ref. [158] the mass tensor was calculated as function of an elongation variable ϵ_{24} and an asymmetry variable ϵ_{35} for ^{240}Pu in cranking approximation and in Gaussian overlap approximation of the generator-coordinate approach. In Fig. 5.2 the leading component $m_{\epsilon_{24}\epsilon_{24}}$ of the mass tensor is shown as function of ϵ_{24} for $\epsilon_{35} = 0$ in the cranking (5.19) and in the generator coordinate approach (GCM). Also plotted is the matrix element $\gamma_{\epsilon_{24}\epsilon_{24}}$ of the width tensor in the Gaussian norm overlap.

Partial derivatives of the reduced energy overlap $h(\mathbf{q},\mathbf{q}')$ with respect to its arguments can be expressed in terms of the matrix elements of the A-body Hamiltonian $\hat{H}$ and the shift operators $P_i = \partial/\partial \bar{q}^i$. Taking into account that $\mathcal{N}(\bar{\mathbf{q}},\bar{\mathbf{q}}) = 1$ and $\langle\bar{\mathbf{q}}|P_i|\bar{\mathbf{q}}\rangle = 0$, and that

$$\frac{\partial \mathcal{H}(\mathbf{q}, \overline{\mathbf{q}})}{\partial q^k}\bigg|_{\mathbf{q}=\overline{\mathbf{q}}} = \langle\overline{\mathbf{q}}| \overleftarrow{P_k} \hat{H}|\overline{\mathbf{q}}\rangle \quad \text{and} \quad \frac{\partial^2 \mathcal{H}(\mathbf{q}, \overline{\mathbf{q}})}{\partial q^k \partial q^l}\bigg|_{\mathbf{q}=\overline{\mathbf{q}}} = \langle\overline{\mathbf{q}}| \overleftarrow{P_k}\overleftarrow{P_l} \hat{H}|\overline{\mathbf{q}}\rangle$$

one obtains from the definition (5.44)

$$\frac{\partial^2 h(\mathbf{q}, \overline{\mathbf{q}})}{\partial q^k \partial q^l}\bigg|_{\mathbf{q}=\overline{\mathbf{q}}} = \langle\overline{\mathbf{q}}| \overleftarrow{P_k}\overleftarrow{P_l} \hat{H}|\overline{\mathbf{q}}\rangle - \langle\overline{\mathbf{q}}| \overleftarrow{P_k}\overleftarrow{P_l} |\overline{\mathbf{q}}\rangle\langle\overline{\mathbf{q}}|\hat{H}|\overline{\mathbf{q}}\rangle . \tag{5.59}$$

The expression on the right-hand side of this equation is sometimes called linked matrix element and abbreviated by $-\langle\overline{\mathbf{q}}| \overleftarrow{P_k}\overleftarrow{P_l} \hat{H}|\overline{\mathbf{q}}\rangle_L$ [108]. Similarly

$$\frac{\partial^2 h(\mathbf{q}, \mathbf{q}')}{\partial q^k \partial q'^l}\bigg|_{\mathbf{q}=\mathbf{q}'=\overline{\mathbf{q}}} = -\langle\overline{\mathbf{q}}| \overleftarrow{P_k} \hat{H} \overrightarrow{P_l} |\overline{\mathbf{q}}\rangle_L . \tag{5.60}$$

Often the A-body Hamiltonian $\hat{H}$ on the r.h.s. of Eqs. (5.59) and (5.60) is replaced for convenience by a single-particle Hamiltonian $\hat{H}_{\mathbf{q}}$ of Hartree-Fock or Hartree-Fock-plus-BCS type with a shape-dependent mean field [535]. This implies a serious limitation of the fluctuations in the mean field in contrast to the original intention of the Hill-Wheeler ansatz (5.20). Not surprisingly, in this scheme one can express the matrix elements in Eqs. (5.57) and (5.58) in terms of the matrix elements (5.15) of the shift operator $(P_i)_{\mu\nu}$ in the cranking approach

$$\gamma_{ij} = \sum_{\mu\nu}(P_i)^*_{\mu\nu}(P_j)_{\mu\nu}$$

$$\tag{5.61}$$

$$-\langle\mathbf{q}| \overleftarrow{P_i} \hat{H}_{\mathbf{q}}^{(BCS)} \overrightarrow{P_j} |\mathbf{q}\rangle_L = \sum_{\mu\nu}(P_i)^*_{\mu\nu}(\mathcal{E}_\mu + \mathcal{E}_\nu)(P_j)_{\mu\nu} .$$

The other matrix elements conveniently vanish.

One should remember that the metric tensor γ_{ij} was assumed to describe a flat space. There is however no general reason why this should be the case. Numerical calculations can only show that it is approximately true [536].

For translational motion the inertia from Eqs. (5.40) or (5.57) does not yield the total mass of the system as one might have expected. In view of the fact that the quantal evolution proceeds in phase space, rather than in ordinary space, Peierls and Thouless [537] proposed to use as constraining parameters not only the collective coordinates, but in addition the corresponding collective momenta.

For an implementation of the generator coordinate method with general collective coordinates $\mathbf{q}$ and momenta $i\hbar\partial/\partial\mathbf{q}$ we refer to the work of Goeke and Reinhard [538]. The frequently used result for the mass tensor $m_{ij}(\mathbf{q})$ in the cranking approximation of this theory is somewhat misleadingly called (cranking) adiabatic time-dependent Hartree-Bogoliubov (ATDHB) approximation. It is given by the expression

$$m_{ij} = 2\hbar^2 \sum_{kl} (M_{(1)}^{-1})_{ik} (M_{(3)})^{kl} (M_{(1)}^{-1})_{lj}$$

with

$$(M_{(\alpha)})^{kl} = \sum_{\mu\nu} \frac{\langle \mathbf{q} | \hat{q}^i | \mathbf{q}, \mu\nu \rangle \langle \mu\nu, \mathbf{q} | \hat{q}^j | \mathbf{q} \rangle}{(\mathcal{E}_\mu + \mathcal{E}_\nu)^\alpha} \qquad \alpha = 1, 3,$$

where $\hat{q}^i$ are the constraining operators of the collective variables q^i, $|\mathbf{q}\rangle$ is the HFB ground state with the constraint $\mathbf{q}$, and $|\mu\nu, \mathbf{q}\rangle = \alpha_\mu^+ \alpha_\nu^+ |\mathbf{q}\rangle$ is a two-quasiparticle state. Numerical comparisons of some microscopic expressions for the mass tensor, corresponding to various collective modes, have been reported in Refs. [508, 539, 540].

5.1.3 Decay rates

One-dimensional dynamics

Adiabatic, collective Hamiltonians are used to evaluate lifetimes of fissile nuclei against spontaneous fission and cluster decay. The Schrödinger equation is usually solved in this case in WKBJ approximation. For an introduction into the WKBJ scheme see e.g. Landau and Lifshitz, Nonrelativistic Quantum Mechanics, Ref. [413], Chap. VII. The WKBJ approximation is justified if the relative change of the de Broglie wave length $\lambda(q)$ is very small over one wave length

$$\frac{1}{2\pi} \left| \frac{d\lambda}{dq} \right| \ll 1. \tag{5.62}$$

In terms of the modulus of the local momentum

$$|p(q)| = \frac{2\pi\hbar}{\lambda} = [\, 2m(q) | (E_0 - V(q)) | \,]^{1/2} \tag{5.63}$$

the condition (5.62) becomes

$$\frac{\hbar}{p^2(q)} \left| \frac{dp}{dq} \right| \ll 1. \tag{5.64}$$

In Eq. (5.63) $m(q)$ is the collective mass, E_0 is the energy of the initial state, in our case the ground-state energy of the decaying nucleus, and $V(q)$ is the collective potential.

Special care is needed for the correct interpretation of the potential $V(\mathbf{q})$ in Eq. (5.62) (and later in Eq. (5.71)). Since the motion along the fission path is an "external", rather artificial degree of freedom in the cranking approach, its zero-point energy ϵ_0 has to be subtracted from the constrained HF (or HFB) expectation value of the effective A-body Hamiltonian $\langle \mathrm{HF}(\boldsymbol{\lambda}) | \hat{H} | \mathrm{HF}(\boldsymbol{\lambda}) \rangle$ so that the fission mode starts with the correct energy. (The HF wave functions depend on Lagrange multipliers $\boldsymbol{\lambda}$, which in turn are functions of the constraining parameters $\bar{\mathbf{q}}$ as explained in subsection 2.4.1.)

The situation is more transparent in the Hill-Wheeler approach which avoids introducing supernumerary degrees of freedom. Eqs. (5.41) and (5.58) show that the effective potential in the collective Hamiltonian contains already automatically a zero-point correction $\epsilon_0(\mathbf{q})$. Since the function $\epsilon_0(\mathbf{q})$ depends on dynamical properties of the system, it is not appropriate to include the effect of ϵ_0 already in the mass formula by a constant E_{zp} term, Eq. (2.159).

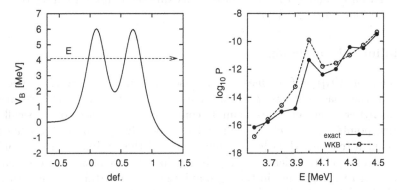

Fig. 5.3. Exact and WKBJ decay rates are shown in the right frame as functions of the excitation energy E for the double humped barrier, plotted in the left frame. The inertia parameter m was taken to be constant and equal to $300/\hbar^2$ [MeV^{-1}] (after Ref. [541]).

The condition (5.64) is fulfilled when the energy is sufficiently far below the potential barrier, which is typically the case in spontaneous fission and cluster decay, except for the vicinity of the classical turning points $q = a$ and $q = b$, the entry and exit point, respectively, of the "tunnel", where $p(a) = p(b) = 0$ and the l.h.s. of the inequality (5.64) diverges. In §50 of Ref. [413], the decay rate of the quasi-bound ground state

$$r_{\mathrm{WKB}} = \frac{\omega}{2\pi} \exp\left[-\frac{2}{\hbar} \int_a^b p(q)dq \right] \qquad (5.65)$$

is derived in semiclassical approximation, where ω is the frequency of the harmonic oscillator potential osculating the ground-state minimum of the collective potential $V(q)$. If the potential is flat in the inside region, as it is assumed e.g. in Gamov's theory of α decay [542], the preexponential factor is v/R with the velocity v and the radius $R = b$.

For a detailed derivation of this result we recommend Fermi's Lectures on Nuclear Physics [543], Chap. III. Instead of the decay rate r_{WKB} often the half-life time

$$T_{1/2} = \frac{\ln 2}{r_{\mathrm{WKB}}} \qquad (5.66)$$

is used to describe the decay property of a quasi-bound state.

For double-humped potential barriers four classical turning points may occur when the initial energy is higher than the second minimum, as shown in the left frame of Fig. 5.3. A general discussion of the semiclassical approximation in this case was presented by Fröman and Dammert [544]. Dudek [541] checked the accuracy of the WKBJ expression (5.65) by comparison with the numerical solution of the Schrödinger equation. The result is shown in Fig. 5.3. The deviation is smaller than the uncertainty due the limited accuracy of the input functions $m(q)$ and $V(q)$.

Multi-dimensional dynamics

The condition $V(\mathbf{q}) = E$ defines in this case for a given energy E a hypersurface in the space of collective coordinates. For simplicity we will assume that this hypersurface encloses a simply-connected, convex area of classically forbidden points $V(\mathbf{q}) > E$. We will first consider the case of a well-defined entry point $\mathbf{a}$ (connected with the ground state of the decaying nucleus) and a given exit point $\mathbf{b}$ (which corresponds e.g. in cluster emission to the configuration of two fragments in their ground states at a certain distance with zero relative velocity). We will further assume that the system is integrable and energy-conserving in the collective degrees of freedom as is appropriate for adiabatic motion.

A trajectory is a function $\mathbf{q}(t)$ whose components satisfy Hamilton's equations

$$\dot{q}^i = \frac{\partial H}{\partial p_i} \quad \text{and} \quad \dot{p}_i = -\frac{\partial H}{\partial q^i} \tag{5.67}$$

with

$$p_i = m_{ij}\dot{q}^j \,.$$

It is convenient to use the mass tensor m_{ij} as metrical tensor in $\mathbf{q}$-space. The line element ds along a trajectory is then given by

$$ds = \sqrt{dq^i m_{ij} dq^j} \,.$$

Using the terminology of Brack and Bhaduri, Ref. [114], we define the action corresponding to a trajectory between two points $\mathbf{a}$ and $\mathbf{b}$ in $\mathbf{q}$-space with an energy E by

$$\mathcal{S}(\mathbf{a}, \mathbf{b}, E) = \sum_i \int_{a^i}^{b^i} p_i(\mathbf{q}) dq^i = \int_{t_a}^{t_b} p_i \dot{q}^i \, dt \,, \tag{5.68}$$

where $\mathbf{q}(t_a) = \mathbf{a}$ and $\mathbf{q}(t_b) = \mathbf{b}$ (Ref. [114], Chap. 2). Since

$$p_i \dot{q}^i = 2E_{\text{kin}} = 2[E - V(\mathbf{q})] \,, \tag{5.69}$$

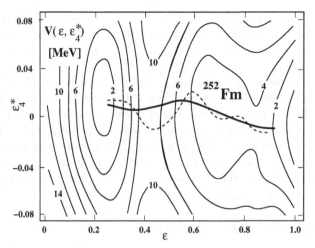

Fig. 5.4. A contour plot of the energy surface for ^{252}Fm is shown in the two-dimensional (ϵ, ϵ_4) parametrization of the shape class (2.212). The dashed line is the path of steepest descent from the saddle points to the ground state and to the scission line, the thick line is the minimal-action path. Note that the latter does not even go exactly through the saddle points. For graphical convenience the ordinate $\epsilon_4^* = \epsilon_4 - 0.2\epsilon + 0.06$ was used. The potential was calculated in the macroscopic-microscopic approach and the inertial tensor was evaluated in the cranking model with a Nilsson potential (after Ref. [545]).

we find for the action

$$\mathcal{S}(\mathbf{a}, \mathbf{b}, E) = 2 \int_{t_a}^{t_b} E_{\mathrm{kin}} dt = 2 \int_{t_a}^{t_b} [E - V(\mathbf{q}(t))] \, dt . \tag{5.70}$$

In order to emphasize the similarity with the one-dimensional case, as e.g. in Baran, Ref. [545], the line element ds can be introduced instead of dt. With

$$p_i \dot{q}^i = \dot{q}^i m_{ij} \dot{q}^j = \frac{dq^i}{ds} m_{ij} \frac{dq^j}{ds} \dot{s}^2$$

and the abbreviation $B_s = (dq^i/ds) m_{ij} (dq^j/ds)$ one gets from Eqs. (5.68) – (5.70)

$$\mathcal{S}(\mathbf{a}, \mathbf{b}, E) = \int_0^s \sqrt{2B_s(\mathbf{q}(s'))[E - V(\mathbf{q}(s'))]} \, ds' , \tag{5.71}$$

where the trajectory length counts from zero at $\mathbf{a}$ to s at $\mathbf{b}$. In the classically forbidden region $E - V(\mathbf{q})$ in Eqs. (5.70) and (5.71) has to be replaced by $|E - V(\mathbf{q})|$. In close analogy to Eq. (5.65) the decay rate is given by

$$r_{\mathrm{WKB}} = \frac{\omega}{2\pi} \exp\left[-\frac{2}{\hbar} \mathcal{S}\right] . \tag{5.72}$$

Instead of using Eqs. (5.67) one may employ the minimal action principle to construct the path $\mathbf{q}(t)$ by minimizing the action $\mathcal{S}(\mathbf{a}, \mathbf{b}, E)$ with respect to the path. Representing the paths by a discrete set of expansion parameters x_i one has to find the minimum of $\mathcal{S}(x_i)$ [103, 545–548].

In spontaneous fission of actinides the exit point $\mathbf{b}$ out of the tunnel lies between saddle and scission points. Since the lifetime against fission is only determined by the fission path in the classically forbidden region, no matter what states are later populated at the scission hypersurface, one has in this case to determine not only the trajectory, but also its terminal point $\mathbf{b}$ so that the action becomes minimal. Note that such minimal-action trajectories do in general not follow the steepest-descend path from the saddle point. An example is shown in Fig. 5.4.

5.1.4 Spontaneous fission half-lives

Measured spontaneous fission half-lives of actinides are plotted in the left frame of Fig. 5.5 as functions of Z^2/A which is proportional to the fissil-ity x. Swiatecki [549] observed that the quantity $\log T_{1/2}[\text{years}] + k\delta m$ de-pends almost linearly on x for even-even nuclei as shown in the right frame of Fig. 5.5. In Swiatecki's ansatz δm is the difference between the empirical ground-state binding energy and the binding energy in Green's liquid-drop model [550]. The empirical fit parameter k is about 5000 (Mass Units)$^{-1}$. The neutron-rich fermium and transfermium isotopes do however no longer follow the general trend. At the same time a symmetric mass distribution of fission fragments appears with substantially larger TKE than typical for asymmetric fission events. This empirical situation is schematically illustrated in Fig. 5.6. The change from asymmetric to symmetric mass splits can be attributed to a change in the fission path leading at scission from elongated fragments with asymmetric splits to rather compact, symmetric scission configurations in the heavy actinides and transactinides. It may be tempting to associate the two types of fission modes with two different fission valleys in a two-dimensional energy surface with some choice of collective fission coordinates [552]. How-ever, as we showed in Fig. 2.22 such identification can be misleading.

Half-lives of odd nuclei are systematically larger than those of neighboring even nuclides. This increase of $T_{1/2}$ is often expressed in terms of a hindrance factor χ. A definition of χ proposed by Hoffman [554] is

$$\chi = \begin{cases} \dfrac{T_{1/2}(^A X_Z)}{[T_{1/2}(^{A-1} X_{Z-1}) \times T_{1/2}(^{A+1} X_{Z+1})]^{1/2}} & \text{for o-e nuclei} \\[4mm] \dfrac{T_{1/2}(^A X_Z)}{[T_{1/2}(^{A-1} X_Z) \times T_{1/2}(^{A+1} X_Z)]^{1/2}} & \text{for e-o nuclei.} \end{cases} \quad (5.73)$$

The definition makes of course only sense when the fission paths of the three nuclides, the half-lives of which enter the expression (5.73), are of the same

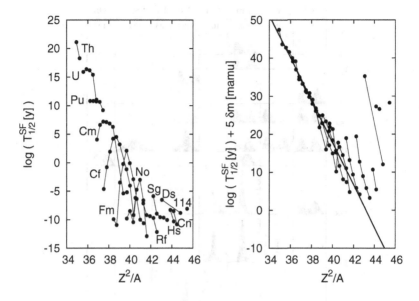

Fig. 5.5. Logarithm of fission half-lives (l.h.s. plot) and logarithms of half-lives corrected for the special ground-state binding energy $k\delta m$ (r.h.s. plot). Recent ground-state masses and fission lifetimes from Ref. [551] have been used in these plots (after Ref. [549]).

type. The logarithms of hindrance factors χ for measured spontaneous fission half-lives of some o-e and e-o actinides are shown in Fig. 5.7 as functions of odd-proton or odd-neutron numbers. In general, the measured hindrance factors are about 10^5 for both, odd protons and odd neutrons.

Early approaches to calculate fission lifetimes started with the determination of the energies of the ground state, of the saddle points and of secondary minima in some shape parametrization. A fission path was then determined either by a smooth polynomial connection of these extrema [96,555,556] or by "minimizing out" all deformation parameters, except for one fission variable, e.g. the quadrupole moment q_{20} [93]. In both cases the path depends on the chosen shape parametrization and is somewhat arbitrary. To calculate fission rates the one-dimensional WKBJ expression (5.65) was then used along the chosen fission path with a mass parameter from the cranking formula (5.9). This seemingly parameter-free approach was not successful in describing measured fission lifetimes of actinides.

Of the many possible sources of uncertainty in this approach it was the inertial parameter which was blamed for the problem and Fiset and Nix [557] therefore proposed the following phenomenological ansatz

$$m^{(\text{coll})}(q) = m_{\text{red}} + k(m_{\text{irrot}}(q) - m_{\text{red}}),\qquad(5.74)$$

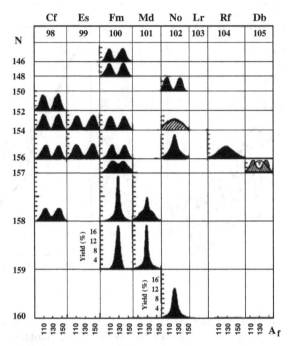

Fig. 5.6. Schematic representation of the mass split in spontaneous fission of californium and transcalifornium nuclids (after Ref. [553]).

where m_{red} is the reduced mass in the exit channel, m_{irrot} is the irrotational mass of the hydrodynamical model, taken in Werner-Wheeler approximation (cf. Sec. 5.2.3), q is a length parameter along the chosen fission path, and $k > 0$ is a fudge factor to fit observed fission lifetimes. For ^{240}Pu $k = 8.9$ was obtained in Ref. [557]. The idea behind the ansatz (5.74) is that beyond scission $m_{\mathrm{irrot}} \to m_{\mathrm{red}}$ and for compact shapes m_{irrot} is seen as a lower limit for the collective inertia. Formula (5.74) just interpolates between these limits.

Randrup et al. [556] replaced the function $m_{\mathrm{irrot}}(q)$ in Eq. (5.74) by a phenomenological ansatz (with an additional fit parameter) and calculated fission life-times for various even-even isotopes of elements with $92 \le Z \le 106$. For $k = 11.5$ the authors got agreement with empirical life-times within one order of magnitude; for uranium and plutonium, however, disagreements up to several orders of magnitude were obtained for the decay from the ground state as well as from the fission-isomeric state.

The ansatz (5.74) was first fitted for actinides, decaying asymmetrically. For the symmetric fission path, dominant for the heavier, neutron-rich actinides, another phenomenological inertial function was proposed by Möller, Nix, and Sierk [552]

$$m^{(\mathrm{coll})}(D) = m_{\mathrm{red}} + f(D)k(m_{\mathrm{irrot}}(D) - m_{\mathrm{red}}) \tag{5.75}$$

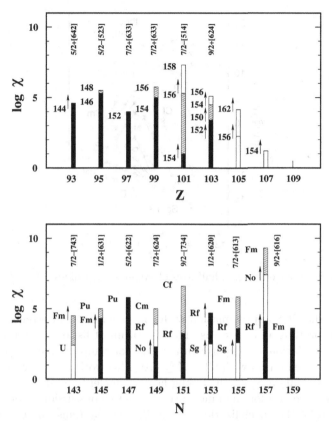

Fig. 5.7. The white, black, and hatched bars represent the logarithms of the hindrance factor χ for various Nilsson single-particle states, occupied by the odd neutron (upper frame) and the odd proton (lower frame) of odd-A actinides. Lower limit values are indicated by arrows. For the white bars only one even-even neighbor was available to calculate χ (after Ref. [553]).

in terms of the center-of-mass distance between nascent fragments D and a function

$$f(D) = \begin{cases} [(D_{\text{sci}} - D)/(D_{\text{sci}} - D_0)]^2 & D < D_{\text{sci}} \\ 0 & D \geq D_{\text{sci}} \end{cases} \quad (5.76)$$

with $D_0 = 0.75$ fm and the scission distance D_{sci} as additional fit parameters.

More satisfactory are calculations based on minimal-action paths in the multidimensional deformation-parameter space as proposed in Ref. [103]. Baran, Pomorski, Łukasiak, and Sobiczewski [547] calculated minimal-action paths for actinides in the Nilsson shape class (2.211) - (2.213) including the deformation parameters $\epsilon, \epsilon_3, \epsilon_4$, and ϵ_5. For the inertial tensor the cranking formula (5.16) was used and the potential was obtained in the macroscopic-microscopic approach. The droplet model with parameters from Ref. [212] and shell corrections from the Nilsson model were employed. Pairing effects were

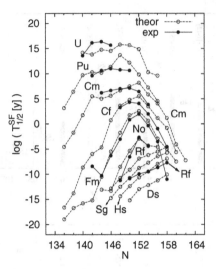

Fig. 5.8. Spontaneous fission half-lives of even-even actinides (after Ref. [547]).

treated in the BCS approximation with shape-independent strength parameters G for protons and neutrons. The logarithm of fission lifetimes resulting from this calculation and experimental values for comparison are shown in Fig. 5.8 for even-even nuclei with $92 \leq Z \leq 110$. This work was later extended by Smolańczuk et al. [558, 559] including heavier nuclei beyond the actinides.

The main reason for the hindrance of spontaneous fission in odd nuclei is assumed to be the specialization energy in these systems, discussed in Subsection 2.4.4. In contrast to the calculation of shell corrections, the evaluation of the specialization energy depends very sensitively on details of the single-particle potential used in the calculation. Already early attempts to determine the specialization energy by Randrup et al. [555] showed that it is not trivial to obtain the observed ground-state spin and parity of odd actinides. Even more uncertainties are connected with the correct sequence and energy differences of single-particle states above the Fermi level at saddle-point deformations. Empirically the hindrance factor becomes smaller with increasing charge number Z. This effect is related to the width of the fission barrier which on average becomes smaller with increasing Z.

A systematic investigation of the effect of an odd-particle on fission lifetimes of odd nuclei (o-e and e-o) was presented by Pomorski in Ref. [560] for nuclei with $Z \geq 96$ and for odd and odd-odd nuclei by Łojewski and Baran in Ref. [522]. The latter authors fitted a Nilsson single-particle potential specifically to reproduce (most of) the ground-state spins of transactinides. In Fig. 5.9 calculated half-lives, obtained in Ref. [560] for Z-odd (left frame) and for N-odd (right frame) nuclei, are compared with theoretical values for neighboring even-even isotopes. Results from Ref. [560] for Z=100-107 isotopes are compared in Fig. 5.10 with experimental data.

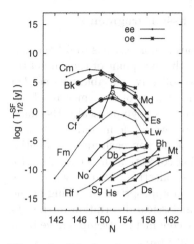 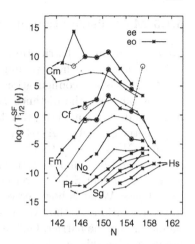

Fig. 5.9. Calculated estimates of the spontaneous fission half-lives for Z-odd (left frame) and N-odd (right frame) nuclei compared with the estimates for the even-even systems (points) [560]. The crosses correspond to estimates obtained with the experimentally known spin and parity of the odd-particle, while the circles refer to results obtained with theoretically predicted quantum numbers of the valence orbitals.

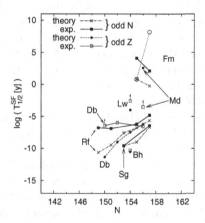

Fig. 5.10. Comparison of experimental data with theoretical estimates for the logarithm of the half-lives of some odd-A actinides, obtained in Ref. [560].

Möller et al. [217] calculated specialization energies of odd nuclei from Pu to Ds for symmetric fission. They used a folded-Yukawa single-particle potential and experienced similar problems as with the Nilsson potential in predicting the parity and angular momentum projection Ω of the odd nucleon. These calculations showed that the specialization energy can increase fission half-lives between one and seven orders of magnitude, depending on the barrier width.

5.1.5 Mass distribution of spontaneous fission fragments

To determine mass distributions in spontaneous fission one has to solve equations of motion for the fissioning system from the exit point of the tunnel towards the scission configuration. At least two collective variables are needed for this purpose, the distance D between the centers of the emerging fragments and an asymmetry variable, which becomes asymptotically the mass asymmetry $\xi = (A_1 - A_2)/(A_1 + A_2)$. Desirable would be in addition a neck variable. With a mass tensor, say, from a cranking approach and a potential from a liquid-drop model with shell and pairing corrections it is easy to write a time-dependent Schrödinger equation in the collective variables with these ingredients. What is however very difficult is an appropriate matching of the initial conditions at the exit point to the semiclassical description of the dynamics under the barrier. Only a few, not fully convincing attempts have been made to solve this problem in some approximate way.

As an example we will discuss the work of Maruhn and Greiner [561]. They assume that the dynamics in the D degree of freedom can be described by a classical Newtonian equation of motion, including a (constant) friction term, with the initial condition that $D(t = 0)$ is the D coordinate of the exit point and $\dot{D}(t = 0) = 0$. In the asymmetry variable ξ a time dependent Schrödinger equation is postulated with the initial condition of a local ground-state wavefunction. Coupling terms between the two degrees of freedom through the mass tensor are neglected, only couplings in the potential are retained. The Schrödinger equation for the wave function $\psi(\xi, t)$ in the ξ degree of freedom therefore contains a potential $V(\xi; D(t))$, driven by the motion in D which produces excitations in the ξ subsystem. The potential for the motion in D is $\langle \psi(t)|V(D; \xi)|\psi(t) \rangle$ and therefore depends in turn on $\psi(t)$. Of interest is the function $P(\xi; D) = \psi^*(\xi; D)\psi(\xi; D)$, describing the distribution in the asymmetry at each point D along the fission path, in particular the "quantum spread" at scission for $D = D_{\text{sci}}$.

It is clear that this approach solves the problem of coupling the quantal dynamics under the barrier with the one beyond the barrier only in a rather vague sense. This is even more so in the model of Samaddar et al. [562] who postulate that the dynamics behind the barrier is described by classical diffusion equations in the collective variables, with initial distributions in all variables and in their velocities to be delta functions. The model yields at scission what is sometimes called the "thermal spread".

For a fissioning nucleus with an excitation energy above the barrier, one does not need to describe subbarrier motion and there are no continuation problems between classically allowed and forbidden regions. For this situation the time-dependent generator-coordinate method was applied to the decay of the excited nucleus ^{238}U by Goutte, Berger, Casoli, and Gogny [536]. The assumption of any adiabatic theory that quasiparticle excitations are rare and therefore negligible becomes increasingly problematic the larger the excitation energy is. This is one limitation of dynamical models as in Ref. [536]. A second uncertainty comes from the treatment of the initial state. The initial compound state is composed of many complicated states among which there may be states in which all excitation energy is stored in the collective degrees of freedom. In Ref. [536] the following ansatz is proposed for these states

$$|\Phi_{P,K=0,J,M}\rangle = \frac{1}{\sqrt{2\pi}} Y_{JM}(\Omega) \int d\mathbf{q} \, f_n^\pi(\mathbf{q}, t = 0)|\mathbf{q}\rangle \,,$$

where Y_{JM} is a spherical harmonic, Ω are angles characterizing the position of the fission axis in space, P is the parity, related to the intrinsic parity π by $P = (-1)^J \pi$, J is the total spin of the compound state, and $\mathbf{q} = (q_{20}, q_{30})$ is the vector of the two collective variables, which are considered in this paper. The amplitudes f_n^π are related to wave functions $g_n^\pi(\mathbf{q})$ by Eq. (5.25). The g_n^π are solutions of the stationary eigenvalue problem

$$\hat{\mathcal{H}}_{\text{modif}}^{(\text{coll})} \, g_n^\pi(\mathbf{q}) = E_n g_n^\pi(\mathbf{q}) \,,$$

where $\hat{\mathcal{H}}_{\text{modif}}^{(\text{coll})}$ is obtained from $\hat{\mathcal{H}}^{(\text{coll})}$ of Eq. (5.56) by closing the fission barrier artificially. There is a considerable ambiguity in this construction of the initial state. It is shown that in particular the choice of the parity of the initial state has a drastic effect on the mass distribution after scission.

5.1.6 Cluster radioactivity and α-decay

A very mass asymmetric spontaneous fission process where the mass number of the light fragment is smaller than about 34 is called cluster radioactivity. It was predicted in 1980 by Sandulescu, Poenaru and Greiner [563] and four years later experimentally confirmed by Rose and Jones [564]. They observed the spontaneous emission of ^{14}C from ^{223}Ra. The cluster emission is a very rare process. Its relative branching ratio to the α-decay is of the order 10^{-10} to 10^{-17}. Nevertheless, in the last two decades one has observed clusters from ^{14}C to ^{34}Si emitted by nuclei from ^{221}Fr to ^{242}Cm [565]. In all cases the residual nucleus was close to the double magic ^{208}Pb and the light cluster was always an even-even nucleus. Cluster radioactivity was also predicted in the mass region $A \approx 110$ by Furman, Kadmensky and Tchuvilsky [570]. Here one expects that the residual nucleus will be close to ^{100}Sn. Only an upper experimental limit for the decay of ^{114}Ba $\rightarrow^{12}$ C $+^{102}$ Sn was determined in Ref. [571]. Measured half-lives for cluster emission are compared in Fig. 5.11 with theoretical predictions from Refs. [566–568].

For α-decay a relation between the half-life $T_{1/2}$ and the energy E of the emitted particle was empirically determined, described by the Geiger-Nuttall rule

$$\log_{10} T_{1/2} = aZE^{-1/2} + \text{const} \,, \tag{5.77}$$

where Z is the charge number of the daughter nucleus. The rule follows from Gamov's theory of α-decay [542]. In the following we show how the theory in

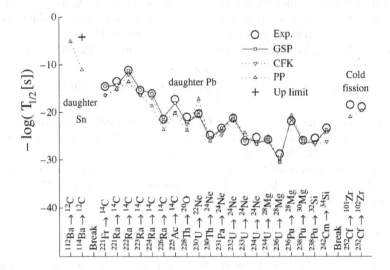

Fig. 5.11. Cluster decay half-lives. The experimental data (circles) are compared with the theoretical predictions by Poenaru et al. [566] (GSP), Kadmensky et al. [567] (CFK), and Pik-Pichak [568] (PP), (after Ref. [569]).

its simplest form can be generalized to cluster emission [572] if the assumptions of the Gamov theory and its later refinements [513, 573, 574] are satisfied as claimed by several authors [514, 575–577]. The ground state of the emitter is supposed to consist with probability S of the daughter nucleus and the cluster. The decay dynamics is controlled by a potential $V(D)$, which is flat for small D and becomes a repulsive Coulomb potential for large D. Schematically one often assumes [543]

$$V(D) = \begin{cases} -V_0 & 0 \leq D \leq R \\ (Z_1 Z_2 e^2)/D & D > R, \end{cases} \tag{5.78}$$

where V_0 is roughly the zero-point energy in the D degree of freedom, R is close to the radius of the emitter, but it is in practice a fit parameter, accounting also for the neglect of the nuclear proximity force and deformation effects of the daughter nucleus and the cluster. In this schematic model R is the entrance point of the "tunnel". The charge numbers of the daughter nucleus and the cluster are Z_1 and Z_2, respectively. When the asymptotic energy is E, the exit point from the tunnel is at $D = b = Z_1 Z_2 e^2 / E$. For simplicity we consider only zero angular momentum in the D degree of freedom.

The decay rate r_{clust} is taken as the product of the preformation factor S and – as in Eq. (5.65) – a frequency unit $\nu = v_{\text{int}}/R$, and a transmission coefficient $\mathcal{T} = \exp(-G)$

$$r_{\text{clust}} = \nu S \mathcal{T} \tag{5.79}$$

with the velocity $v_{\text{int}} = [2(E + V_0)/\mu]^{1/2}$ in the internal region $D < R$, the reduced mass $\mu = M_{\text{nucl}} A_1 A_2/(A_1 + A_2) \approx A_2 M_{\text{nucl}}$ (since $A_2 \ll A_1$), and the WKBJ expression for the action integral

$$G = \frac{2}{\hbar} \int_R^b \sqrt{2\mu(V(D) - E)} \, dD = \frac{2}{\hbar} \sqrt{2\mu Z_1 Z_2 e^2} \int_R^b \sqrt{\frac{1}{D} - \frac{1}{b}} \, dD$$

$$= \frac{2}{\hbar} \sqrt{2\mu Z_1 Z_2 e^2 b} \left(\arccos \sqrt{\frac{R}{b}} - \sqrt{\frac{R}{b} - \frac{R^2}{b^2}} \right). \tag{5.80}$$

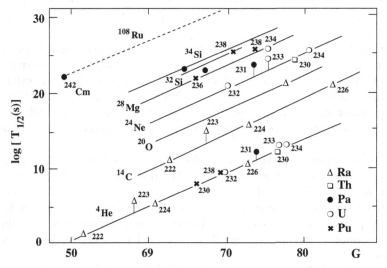

Fig. 5.12. Geiger-Nuttall rule for cluster decays. The logarithm of the empirical half-life is plotted versus the calculated exponent G of Eq. (5.80). The dashed line would correspond to spontaneous cold fission of ^{242}Cm$\rightarrow$^{108}Ru+^{134}Te if it should fit into the systematics of cluster decays (after Ref. [578]).

Typically $b \gg R$ and one can therefore write

$$\arccos \sqrt{\frac{R}{b}} \approx \frac{\pi}{2} - \sqrt{\frac{R}{b}} \quad \text{and} \quad \sqrt{\frac{R}{b} - \frac{R^2}{b^2}} \approx \sqrt{\frac{R}{b}}.$$

In this approximation

$$G = \frac{2}{\hbar} \sqrt{2\mu Z_1 Z_2 e^2 b} \left(\frac{\pi}{2} - 2\sqrt{\frac{R}{b}} \right) = \frac{v_0}{v_{\text{asympt}}} - \frac{4}{\hbar} \sqrt{2 M_{\text{nucl}} R Z_1 Z_2 e^2 A_2}$$

with $v_0 = 2\pi Z_1 Z_2 e^2/\hbar$ and the asymptotic velocity $v_{\text{asympt}} = \sqrt{2E/\mu}$. Using Eq. (5.66) one obtains

$$\log_{10} T_{1/2} = -0.159 - \log_{10} \nu - \log_{10} S_{A_2}$$
$$+0.434 \left(\frac{v_0}{v_{\text{asympt}}} - \frac{4}{\hbar} \sqrt{2 M_{\text{nucl}} R Z_1 Z_2 e^2 A_2} \right) \qquad (5.81)$$

with $\log_{10} \ln 2 = -0.159$ and $\log_{10} e = 0.434$. For a fixed cluster $^{A_2}Z_2$ the energy dependence comes from the term v_0/v_{asympt} in the bracket. Since the dependence of the remaining terms on A_1 and Z_1 of the daughter nucleus is not very strong over the limited range of cluster emitters, one may take them as constant for each cluster and one obtains Eq. (5.77) with $E^{1/2} \propto v_{\text{asympt}}$. To compare life times for different clusters one can use a semiempirical rule [576] for the relation between the preformation factor S_{A_2} for a cluster with nucleon number A_2 and the corresponding preformation factor S_α for the α particle

$$S_{A_2} = S_\alpha^{(A_2 - 1)/3}. \qquad (5.82)$$

Inserting this relation into Eq. (5.81) one expects parallel straight lines for the plot of $\log T_{1/2}$ vs v_0/v_{asympt}. This result is in fact confirmed as shown in Fig. 5.12, where observed $\log T_{1/2}$ values are plotted versus calculated values for the exponent G of Eq. (5.80). Decays involving even-mass emitters of the same cluster follow straight lines, neighboring odd-mass decays lie somewhat above these lines, presumably because of the same reason as in ordinary spontaneous fission.

The systematics reaches up to the decay ^{242}Cm $\rightarrow ^{34}$ Si$+^{208}$Pb, which does not seem to follow this trend any more. The fragments resulting from cluster decay are almost exclusively formed in their ground states. It was therefore suggested to compare cluster emission with cold fission. However only one case of spontaneous, cold fission has been observed and the energy resolution was too poor to guarantee that there was no low-energy excitation in the fragments.

In Ref. [576] the schematic potential (5.78) was replaced by a more realistic expression allowing for a phenomenological nuclear proximity potential and a numerical solution of the Schrödinger equation was used instead of the WKBJ approximation to determine the transmission coefficient $\mathcal{T}$. Using semiphenomenological scaling rules for ν_{A_2} and S_{A_2} half-lives were predicted deviating from later measured values by less than a factor of 3 as shown in table 5.1. A longer list of calculated decay times was published in Ref. [514].

Nevertheless conceptual as well as numerical uncertainties remain with the evaluation of the spectroscopic factors S_{A_2} and the frequencies ν_{A_2} [419, 513]. It appears therefore natural to describe cluster decay as an extremely asymmetric spontaneous fission event by methods described in the previous subsection. This was in fact done in the first treatment of cluster decay [563] and a number of subsequent papers [568, 581–588]. In these approaches a one-dimensional fission model was used with the WKBJ fission rate (5.65) and $p(q) = \sqrt{2\mu[V(D) - Q]}$, where Q is the reaction Q-value and $D \equiv q$.

Shi and Swiatecki [581] use the reduced mass for the inertial parameter μ. For distances D larger than the sum of the half-density radii of cluster

Table 5.1. Calculated and measured half-lives of some cluster decays (from Ref. [514]).

$^{A}Z \rightarrow {}^{A_1}Z_1 + {}^{A_2}Z_2$	E/MeV	$\log \tau_{\text{th}}$	$\log \tau_{\text{exp}}$	Ref.
$^{236}\text{Pu} \rightarrow {}^{28}\text{Mg} + {}^{208}\text{Pb}$	79.90	21.3	21.7	[579]
$^{238}\text{Pu} \rightarrow {}^{32}\text{Si} + {}^{206}\text{Hg}$	91.47	25.8	25.3	[580]
$^{238}\text{Pu} \rightarrow {}^{28}\text{Mg} + {}^{210}\text{Pb}$	76.16	25.6	25.7	[580]
$^{238}\text{Pu} \rightarrow {}^{30}\text{Mg} + {}^{208}\text{Pb}$	77.26	25.8	25.7	[580]

and daughter nucleus the potential $V(D)$ is assumed to be the sum of the Coulomb and the proximity potential from Ref. [58]. For smaller distances a smooth power-law interpolation to the ground-state energy of the parent nucleus was used. Only rough estimates for the frequency factor ν were made. The transmission probability T depends of course sensitively on the choice of radii and on the deformation of the parent nucleus.

Greiner et al. [515] use a shape parametrization in terms of two spheres, intersecting at short distances, where radii are scaled to keep the volume fixed. Various expressions for the inertial parameter μ were used [582], including the hydrodynamical Werner-Wheeler approximation, to be discussed in Sec. 5.2.3. Also a variety of potentials $V(D)$ were tried, among them the Yukawa-plus-exponential potential (2.132)-(2.134) with Strutinsky shell and pairing corrections and subtraction of a zero-point energy ϵ_0 [588]. Earlier also the potentials from Ref. [589] were used and the liquid drop model of Ref. [224] before scission and a pure Coulomb potential beyond scission [584]. A simplified variant of it, given in Ref. [583], was called by the authors "analytic, superasymmetric fission model". The frequency factor $\nu = \omega/(2\pi)$ was derived from the zero-point energy ϵ_0 and was essentially used to fit the data, including even-odd effects [585]. It was not related to the chosen potential well. Tables of half-lives were published in Ref. [566].

Further approaches based on the rate formula (5.65) were listed by Price [580] and Poenaru and Greiner [587]. The latter authors pointed out in Ref. [586] that the validity of the Geiger-Nuttall rule for cluster decay is essentially determined by the transmission coefficient T. It is therefore valid for both descriptions of cluster decay, by the rate formula (5.79) as well as by Eq. (5.65).

5.2 Fission dynamics of excited nuclei

We will assume in this section that the excitation energy is large enough all along the fission path that quantal effects are washed out. This implies in particular that pairing effects have disappeared. We will further assume that the intrinsic degrees of freedom of the fissioning nucleus are in thermodynamic equilibrium at each point of the fission path. More specifically, it will be assumed that the intrinsic system can be approximated by a canonical ensemble characterized by a temperature. This is a far-reaching assumption: In a canonical ensemble the total energy can deviate from the average energy by more than the average energy. This is of course due to the coupling of the canonical ensemble to a heat bath, which is assumed to transfer any required amount of energy instantaneously in order to keep the temperature constant. Since a fissioning nucleus is not coupled to a heat bath, the canonical description can only be an approximation to the real, microcanonical situation. For its validity one has to make sure that at least the variance of the excitation energy is always much smaller than its average. This puts a lower limit on admitted temperatures at each point of the fission path. A rough estimate for the limit was given in Eq. (4.1). This means that the dynamical theory developed below is not the continuation of the adiabatic theory to finite excitation energies. For example thermal-neutron induced fission of ^{236}U should not be described by the following macroscopic approach.

5.2.1 The Langevin process

We will consider a shape class with deformation parameters q^i, $i = 1, \ldots N$ of which we believe a priori that it contains all actual fission paths. In order to make sure that the dynamical equations of motion we are going to construct are at least covariant under coordinate transformations within the chosen shape class, we will use tensorial notation, in which the q^i represent a contravariant vector. We will use in the following the Einstein convention where summation over an equal pair of upper and lower indices is implied. We start with the construction of a scalar Lagrangian for a classical dynamical problem with dynamical variables q^i. Connected with large-scale changes of the shape of a nucleus there is a flow of matter and therefore a kinetic energy, which is to be a quadratic form of the velocity vector $\dot{q}^i$

$$E_{\text{kin}} = \frac{1}{2} m_{ij} \dot{q}^i \dot{q}^j \tag{5.83}$$

with the positive-definite, symmetric inertial tensor $m_{ij}(\mathbf{q})$. It depends in general on the coordinates.

The conservative forces can be obtained from the derivatives of a potential $V(\mathbf{q})$, for which the liquid-drop energy is frequently taken. We can then write the Lagrangian for the classical system $\mathcal{L} = E_{\text{kin}} - V$ and obtain the momenta

$$p_i = \frac{\partial \mathcal{L}}{\partial \dot{q}^i} = m_{ij} \dot{q}^j . \tag{5.84}$$

One does not expect that the shape degrees of freedom decouple completely from the intrinsic degrees of freedom. The simplest way to model the effect of their coupling to the collective motion is the introduction of a friction force proportional to the velocity

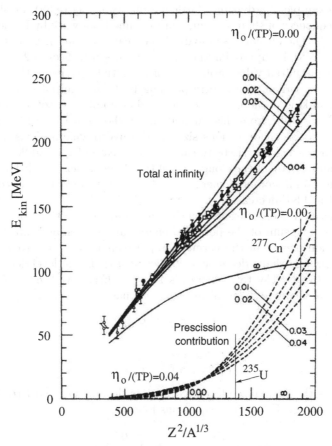

Fig. 5.13. Total kinetic energies, calculated as function of the quantity $Z^2/A^{1/3} \propto E_{\text{Coul}}$ for viscosity constants η_0 between 0 and ∞. Also shown, by the dashed lines is the prescission kinetic energy. The symbols with error bars refer to average values of measured data from medium-energy fission (after Ref. [82]).

$$K_i^{(\text{fric})} = \eta_{ij}\dot{q}^j$$

in terms of a covariant, symmetric, positive-semidefinite friction tensor $\eta_{ij}(\mathbf{q})$. The equation of motion becomes in this case

$$\frac{d}{dt}\frac{\partial\mathcal{L}}{\partial\dot{q}^i} - \frac{\partial\mathcal{L}}{\partial q^i} = \dot{p}_i - \frac{1}{2}\frac{\partial m_{kl}}{\partial q^i}\dot{q}^k\dot{q}^l + \frac{\partial V}{\partial q^i} = -\eta_{ij}\dot{q}^j . \qquad (5.85)$$

Note that for a coordinate-dependent inertial tensor $m_{ij}(\mathbf{q})$

$$\dot{p}_i = m_{ij}\ddot{q}^j + \dot{q}^k\frac{\partial m_{ij}}{\partial q^k}\dot{q}^j \quad \text{and} \quad \dot{q}^k\frac{\partial m_{kl}}{\partial q^i}\dot{q}^l = -p_k\frac{\partial(m^{-1})^{kl}}{\partial q^i}p_l .$$

As an example of a dynamical calculation, using Eq. (5.85), we showed in Fig. 4.4. sequences of shapes as functions of time from work by Davies, Sierk, and Nix [82]. The authors used hydrodynamical mass and friction tensors in the Werner-Wheeler approximation, to be discussed in Secs. 5.2.3 and 5.2.4, and solved the initial value problem with an initial velocity in the fission direction at the saddle point, corresponding to 1 MeV initial kinetic energy. Translational kinetic energies at scission and at infinity i.e. total kinetic energies (TKE) resulting from this calculation are shown in Fig. 5.13 as function of $Z^2/A^{1/3} \propto E_{\text{Coul}}$ for various sizes of the hydrodynamical viscosity constant η_0. Increasing the viscosity causes an increase of the saddle-to-scission time and an increase of the elongation of the shape at scission. It follows that more of the gain in potential energy is dissipated with increasing viscosity and therefore the TKE decreases.

To describe fission of excited nuclei one characterizes them by a temperature $T(\mathbf{q})$. The sum of the potential energy at zero temperature $V(\mathbf{q})$ and the excitation energy in the Fermi-gas approximation $E^* = a(\mathbf{q})T^2$ will be called $U(\mathbf{q})$. The shape-dependent coefficient $a(\mathbf{q})$ is the level-density parameter, Eq. (3.28). The total energy is $E_{\text{total}} = E_{\text{kin}} + V + E^* = E_{\text{kin}} + U$. From general thermodynamical relations one has

$$dU = TdS - K_i dq^i \tag{5.86}$$

and

$$dF = -SdT - K_i dq^i \,, \tag{5.87}$$

where K_i is the generalized force vector, S is the entropy, and $F(\mathbf{q}; T)$ is the Gibbs free energy. From these equations follows

$$-K_i = \left(\frac{\partial U}{\partial q^i}\right)_{S=\text{const}} = \left(\frac{\partial F}{\partial q^i}\right)_{T=\text{const}} . \tag{5.88}$$

Conservative forces K_i should be obtained from the derivatives of U at fixed entropy i.e. they are identical with the generalized forces in Eqs. (5.86) and (5.87). In these equations finite-size effects are neglected, see the discussion in Sec. 3.1. We assume that the temperature T can be defined at each point of the fission trajectory in our approximate canonical ensemble. This implies an instantaneous thermalization of the excitation energy E^* on the time scale of the motion along the trajectory.

In a small system fluctuations connected with friction are not negligible as they are in macroscopic systems with virtually infinitely many degrees of freedom. Assuming that the intrinsic motion, essentially the single-particle motion, is much faster than the collective motion, their coupling is assumed to give rise to a stochastic force, the Langevin force,

$$K_i^{(\text{Langevin})}(\mathbf{q}; t) = \sum_j g_{ij}(\mathbf{q})\Gamma_j(t) \,, \tag{5.89}$$

to be added to the right-hand-side of Eq. (5.85). The Langevin force is written in Eq. (5.89) as product of a strength factor $g_{ij}(\mathbf{q})$ depending only on the collective variables $\mathbf{q}$ and a stochastic factor $\Gamma_j(t)$, which is a function of time only. The Γ_j are Gaussian-distributed random numbers with vanishing mean value $\langle \Gamma_j(t) \rangle = 0$ and delta-correlated variance

$$\langle \Gamma_i(t)\Gamma_j(t') \rangle = 2\delta_{ij}\delta(t - t')\,. \tag{5.90}$$

Under coordinate transformations the expression $\sum_j g_{ij}\Gamma_j$ is a covariant vector, however the Γ_j are not covariant. Instead, Eq. (5.90) is required to be valid in any coordinate system. The delta function in time in Eq. (5.90) idealizes the assumption that the intrinsic dynamics is random and changes very fast on the time scale of the collective motion. This is also described as the coupling of "white" noise to the slow motion in the collective variables because it has frequency-independent Fourier amplitudes (Markov process). In order to allow for a finite "memory time" τ, the delta function would have to be replaced by a Gaussian of width τ (non-Markovian process).

Friction and Langevin forces have the same origin: The stochastic coupling of the collective motion to the intrinsic dynamics. One may therefore expect a relation between the transport parameters η_{ij} and g_{ij}. From the fluctuation-dissipation theorem [590] follows the relation

$$\sum_k g_{ik}g_{jk} = T\eta_{ij} \overset{\text{def}}{=} D_{ij}\,. \tag{5.91}$$

It will be shown in Sec. 5.2.2, Eq. (5.115), that D_{ij} is the diffusion coefficient of the Fokker-Planck equation. We postpone the question of how to calculate inertial and friction tensors to Secs. 5.2.3 and 5.2.4. In view of the assumed finite temperature of the system we do not consider quantum corrections to Eq. (5.91). By adding the Langevin force (5.89) to the deterministic equations of motion (5.85) the latter become stochastic equations. Instead of giving rise to one trajectory for each initial condition, they generate a bundle of trajectories. The latter create a probability distribution $w(\mathbf{q}, \mathbf{p}, t)$ to find the system after time t at a given point $\mathbf{q}, \mathbf{p}$ in its phase space.

Even though we have been using statistical mechanics of the canonical ensemble, it would be unrealistic to keep the temperature constant during the whole Langevin process. Instead, one requires that the energy transferred per unit time into the intrinsic system by the friction and Langevin forces shall be equal to the rate of increase of the internal energy

$$\frac{dE^*}{dt} = \eta_{ij}\dot{q}^i\dot{q}^j - \dot{q}^i \sum_j g_{ij}\Gamma_j \tag{5.92}$$

with the local excitation energy $E^* = U(\mathbf{q}, T) - V(\mathbf{q}) = aT^2$. This equation may be seen as the equation of motion for the temperature. We consider first the case of a shape-independent level-density parameter a. Then Eq. (5.92) guarantees the conservation of the total energy E_{total}. To see this we use $\partial_{q^i} F|_{T=\text{const}} = \partial_{q^i} V$ and multiply the balance of all forces

$$m_{ij}\ddot{q}^j + \frac{1}{2}\partial_{q^i} m_{jk}\dot{q}^j\dot{q}^k + \partial_{q^i} V + \eta_{ij}\dot{q}^j - \sum_j g_{ij}\Gamma_j = 0 \qquad (5.93)$$

by $\dot{q}^i$ and obtain

$$\frac{d}{dt}\left(\frac{1}{2}m_{jk}\dot{q}^j\dot{q}^k + V + E^*\right) = \dot{E}_{\text{total}} = 0.$$

If the q-dependence of a is taken into account, one obtains an additional term $-(\partial a/\partial q^i)T^2$ from $\partial_{q^i} F$. The term is compensated by a corresponding term from $\dot{E}^*$ if Eq. (5.92) is replaced by

$$a(\mathbf{q})\frac{dT^2}{dt} = \eta_{ij}\dot{q}^i\dot{q}^j - \dot{q}^i\sum_j g_{ij}\Gamma_j . \qquad (5.94)$$

This means that the excitation energy is changed by dissipative effects only by a change of the temperature, not through the shape-dependence of the level-density parameter a.

The fluctuations Γ_j can be positive or negative. It may therefore happen that a large negative fluctuation would pump more energy out of the internal system than it actually contains. To avoid such unphysical events the tail of the Gaussian distribution $P(\mathbf{\Gamma})$, from which the Γ_j are taken, has to be cut off

$$P(\mathbf{\Gamma}) \propto e^{-\tau} {}_i \Gamma_i \Gamma_i \Theta(E^*).$$

This awkward recipe is the price one has to pay for using the convenient canonical formalism instead of the microcanonical one, required in principle by the boundary condition of a fixed total energy. The second of the inequalities (4.1) ensures that such truncation of the Gaussian distribution happens only rarely.

The set of equations

$$\dot{q}^i = (m^{-1})^{ij}p_j, \qquad (5.95)$$

$$\dot{p}_i = -\partial_{q^i} F|_{T=\text{const}} + \frac{\partial(m^{-1})^{jk}}{2\partial q^i}p_j p_k - \eta_{ij}(m^{-1})^{jk}p_k$$

$$+ \sum_j g_{ij}\Gamma_j(t), \qquad (5.96)$$

$$2a(\mathbf{q})T\dot{T} = \eta_{ij}\dot{q}^i\dot{q}^j - \dot{q}^i\sum_j g_{ij}\Gamma_j \qquad (5.97)$$

shall be referred to as Langevin equations. One can prove the identity

$$\partial_{q^i} F|_{T=\text{const}} = -T\partial_{q^i} S|_{U=\text{const}}$$

with the temperature $T(\mathbf{q})$, the free energy $F = V - a(\mathbf{q})T^2$, and the entropy $S = 2a(\mathbf{q})T$ in terms of the level-density parameter $a(\mathbf{q})$. Some authors [591–595] therefore write $T\partial_{q^i} S$ instead of $-\partial_{q^i} F$ in Eq. (5.96).

Summarizing the arguments needed to obtain these equations, we assumed that (a) the system should exhibit dynamics on two different time scales (b) the slow motion should be connected with the collective degrees of freedom and should follow classical mechanics (c) the friction force should be linear in the velocities, and (d) the intrinsic system should obey the second law of thermodynamics. No other specific assumption about the nature of the intrinsic system was made. This will only be necessary when expressions for the mass and friction tensors as functions of the collective variables are derived.

The numerical solution of the Langevin equations requires a discretization of the time in steps of size τ. It was found [281] that it is advantageous to use the first order Euler procedure to discretize Eqs. (5.96) and the second order Heun procedure [596] for Eqs. (5.95)

$$q^i(t_{n+1}) - q^i(t_n) = \frac{1}{2}\{m^{-1}(\mathbf{q}(t_n))\}^{ij}\left[p_j(t_n) + p_j(t_{n+1})\right]\tau \tag{5.98}$$

$$p_i(t_{n+1}) - p_i(t_n) = -\left[\eta_{ij}(\mathbf{q}(t_n))\{m^{-1}(\mathbf{q}(t_n))\}^{jk}\,p_k(t_n) + \partial_{q^i}F(\mathbf{q}(t_n))\right.$$
$$\left. -\frac{1}{2}\frac{\partial\{m^{-1}(\mathbf{q}(t_n))\}^{jk}}{\partial q^i}p_j(t_n)p_k(t_n)\right]\tau + I_i\,, \tag{5.99}$$

where the integral

$$I_i = \int_{t_n}^{t_{n+1}} \sum_j g_{ij}(\mathbf{q}(t'))\Gamma_j(t')dt'$$

on the right-hand side of Eq. (5.99) cannot be represented by a sum in the limit $\tau \to 0$ since $\Gamma_j(t')$ is a distribution rather than an integrable function as it has to satisfy Eq. (5.90). We therefore introduce the auxiliary quantity $W_j(t) = \int_0^t \Gamma_j(t')dt'$, in terms of which I_i can be rewritten as a Stieltjes integral

$$I_i = \int \sum_j g_{ij}(\mathbf{q}(t'))dW_j(t')\,,$$

which is approximated by

$$I_i \approx \sum_j g_{ij}(\mathbf{q}(t_n))[W_j(t_{n+1}) - W_j(t_n)]\,. \tag{5.100}$$

This expression is not unique. It is called the Ito version for the discretization of the stochastic equation (5.96). Alternative discretization of the form

$$I_i \approx \sum_j g_{ij}(\mathbf{q}(t_n + \lambda[t_{n+1} - t_n]))[W_j(t_{n+1}) - W_j(t_n)]$$

with various values for λ have been proposed. We refer for further discussions of this point to the general literature [590,597] and to a paper by Gettinger and Gontchar [595]. Since the $W(t)$ are defined as sums of Gaussian-distributed random numbers, they are themselves random numbers with Gaussian distribution. We want to express them in terms of normalized, Gaussian-distributed random numbers $w_j(n)$ with $\langle w_j \rangle = 0$, $\langle (w_j)^2 \rangle = 2$. In the ansatz

$$W_j = a w_j$$

the scaling factor a is determined by equating the variances of $a w_j$ and of $\int \Gamma_j(t') dt'$

$$a^2 \langle w_i w_j \rangle = \int_t^{t+\tau} dt_1 \int_t^{t+\tau} dt_2 \, \langle \Gamma_i(t_1) \Gamma_j(t_2) \rangle$$

$$2 a^2 \delta_{ij} = 2 \delta_{ij} \int_t^{t+\tau} dt_1 \int_t^{t+\tau} dt_2 \, \delta(t_1 - t_2) = 2\tau \delta_{ij}\,.$$

Therefore $a = \sqrt{\tau}$ and one obtains finally

$$I_i = \sum_j g_{ij}(\mathbf{q}(t_n))[w_j(n+1) - w_j(n)]\sqrt{\tau}\,.$$

For a general discussion of errors in the numerical solution of stochastic differential equations we refer to Ref. [598].

The third of the Langevin equations is satisfied by readjusting the temperature T after each time step so that

$$T^2(\mathbf{q}) = E^*/a = [E_{\text{total}} - E_{\text{kin}}(\mathbf{q}) - V(\mathbf{q})]/a(\mathbf{q})\,.$$

Abe, Grégoire, and Delagrange, were the first to suggest the use of Langevin equations to describe fission [599]. They used the elongation as the only shape variable. As an example for a solution of the Langevin equations in two dimensions we show in Fig. 5.14 a bundle of trajectories together with the contour lines of the free energy at $T = 2$ MeV for the system ^{215}Fr with an angular momentum of rigid rotation $I = 85\hbar$. This system has no fission barrier because of its high angular momentum. The shape parametrization is based on mirror-symmetric Cassinian ovals [278]; the resulting shape class is shown in Fig. 5.15. All trajectories start at a spherical shape with initial velocities corresponding to a thermal distribution $\propto \exp\{-m_{ij}\dot{q}^i\dot{q}^j/(2T)\}$. They follow a broad, down-sloping valley towards the scission line, located at the upper edge of Fig. 5.14, giving rise to a distribution of scission shapes with different distances between the centers of charge of the nascent fragments. This translates into a distribution of the TKE. When a reflection-asymmetric shape variable is included as in Refs. [600–605], one similarly generates a fragment-mass distribution at the scission surface. Note that the neutron to proton ratio is fixed in this model to its initial value since no dynamical variables have been introduced to change it.

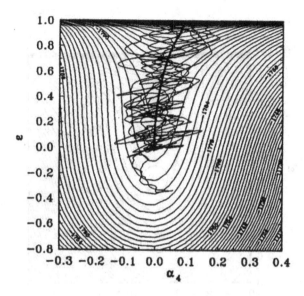

Fig. 5.14. Ten trajectories calculated with the Langevin equations, starting at the spherical shape $\epsilon = \alpha_4 = 0$. The scission line is $\epsilon = 1$. Also shown are contour lines for the free energy of ^{215}Fr and $I = 85\hbar$. The thick line is obtained from Eq. (5.85) (from Ref. [280]).

Along each Langevin trajectory one has to run simultaneously an evaporation cascade for light-particle emission, controlled by the rates (4.4) or (4.28) and their generalization to deformed, rotating emitters [426–429, 606]. After each emission event the fission dynamics has to be restarted with the charge, mass, excitation energy and angular momentum that are left in the residual system. The implicit assumption is made here that after each emission event the residual system thermalizes instantaneously on the time scale of the fission motion. The first of the inequalities (4.1) is expected to ensure that emission rates remain sufficiently low that between successive particle-emission events thermal equilibrium is reestablished.

There is a strong coupling between evaporation and fission dynamics in this model: The energy and angular momentum carried away by particles before the trajectory has passed the fission barrier leads to an increase of the barrier height and therefore to an increase of the fission life-time. But also behind the barrier the scission configuration depends on the angular momentum and the temperature. On the other hand the emission rates depend on the shape through the shape-dependence of the level density, of the light particle binding-energies [607], and of the transmission coefficients for charged particles, see e.g. Ref. [427].

A discontinuous evaporation cascade along each trajectory was so far used mostly in connection with one-dimensional trajectory calculations [593, 608].

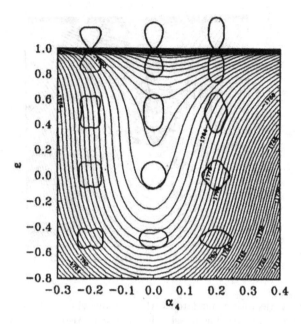

Fig. 5.15. Shown are the shapes, corresponding to the ϵ, α_4 parametrization in the shape class of Cassinian ovals (from Ref. [280]).

Only Schmitt et al. [604] combined it with two-dimensional Langevin dynamics. In the – less realistic – continuous evaporation model, proposed by Delagrange, Grégoire, Scheuter, and Abe [609], a considerably smaller set of random trajectories needs to be calculated to achieve statistically significant results. In this approach one introduces average numbers $\overline{n}_{\alpha\beta}^{\nu}(t)$ of particles of type ν (=n, p, d, t, α, γ) emitted until time t in the energy bin α, typically of size 0.5 MeV, with angular momentum in the bin β of size $1\hbar$. The $\overline{n}_{\alpha\beta}^{\nu}(t)$ are continuous functions of the time t and satisfy the equation of motion

$$\frac{d\overline{n}_{\alpha\beta}^{\nu}(t)}{dt} = r_{\alpha\beta}^{\nu}\big(\mathbf{q}(t), E^*(t), I_{\mathrm{comp}}(t)\big) \qquad (5.101)$$

with the emission rate $r_{\alpha\beta}^{\nu}$ of particles of type ν from a compound nucleus with deformation $\mathbf{q}$, excitation energy E^*, and angular momentum I_{comp}. The updating of the potential F and the tensors m_{ij} and η_{ij} after each time step involves in general non-integer values of ΔN, ΔZ, and $\Delta I_{\mathrm{comp}}/\hbar$. Semiclassical expressions for F, m_{ij}, and η_{ij}, typically used in these calculations, depend analytically on N, Z, and I_{comp}. One can update them therefore also by non-integer parameter values.

Of particular interest is the multiplicity of particles of type ν, emitted before scission. For each trajectory the quantity

$$m^\nu\{\mathbf{q}(t)\} = \sum_{\alpha\beta} \overline{n}^\nu_{\alpha\beta}(t_{\text{sci}})$$

is determined, where t_{sci} is the scission time for that trajectory. Averaging m^ν over all N_{sci} trajectories reaching the scission hypersurface yields the prescission multiplicity

$$M^\nu\{\mathbf{q}(t)\} = \frac{1}{N_{\text{sci}}} \sum_{\alpha\beta} \overline{n}^\nu{}_{\alpha\beta}(t_{\text{sci}}) \,. \tag{5.102}$$

Also more exclusive observables can be calculated, for instance the multiplicity of particles ν in fission events from a given TKE bin. In this case the averaging has to be restricted to those trajectories which lead to the TKE values of the given bin. Of course, the requirement on the statistics, i.e. the number of calculated, statistically independent trajectories, increases, the more exclusive the desired observable is.

Numerical solutions of the Langevin equations require the use of a generator for pseudorandom numbers. To avoid spurious results it is recommended to check whether the quality of the generator used is sufficient for the size of the task one intends to deal with. Criteria can for instance be found in Ref. [610] and in the pertinent mathematical literature [611].

Applications of the Langevin process to fusion-fission reactions

In heavy-ion fusion-fission reactions measured angular distributions of light particles allow an unfolding into precompound particles and thermally distributed particles from three distinct sources: those emitted in the center-of-mass frame of the fused system and those emitted by the two accelerated fission fragments [612–617]. Attempts to account for the multiplicity of the thermal neutrons, emitted in the center-of-mass frame, the "prescission neutrons", by a multiple-chance fission, cascade calculation gave considerably lower multiplicities than observed.

Two reasons have been considered to explain multiplicities larger than predicted by the cascade model with Bohr-Wheeler fission rates (4.5). When one describes fission as a diffusive motion over the fission barrier, it might happen that a system which has already passed the barrier - and would be counted as fission event in the Bohr-Wheeler transition state formalism - is scattered back by Langevin forces and returns to a presaddle shape. As first pointed out by Kramers [510] this diminishes the fission rate. It therefore increases the light-particle multiplicity. A further source of additionally evaporated light particles are emissions on the way from saddle to scission if the fission motion is sufficiently slow. The multiplicity has therefore been considered as a clock for the fission motion of a hot nucleus [617]. In addition, particle evaporation and precompound emission must also be expected from the fusion step of the reaction.

It is this physical picture of the fusion-fission reaction, which has often been modeled by a Langevin process, augmented by light-particle emission.

There are several review articles on this subject [594,618–621] and Chap. 12.4 in the textbook by Fröbrich and Lipperheide [421]. We will therefore only comment on a few aspects of these calculations. Apart from the task of getting reliable values for the transport tensors m_{ij} and η_{ij}, the main problem with this approach is the choice of appropriate initial conditions for the Langevin equations. It would be natural to formulate initial conditions for the entrance channel of the reaction. Unfortunately our present understanding of essential details of the fusion mechanism are rather incomplete. From an analysis of the differential cross section of deep-inelastic heavy-ion reactions in terms of classical trajectories one knows that most of the available kinetic energy in the entrance channel is converted into internal excitation on a time scale of 10^{-22} s [622,623]. On this time scale the particle-hole states acting as doorway states for the energy dissipation [624] cannot be thermalized, nor is there a still sufficiently faster internal motion to produce Langevin noise. Also Eq. (5.91) is only valid in a rather stationary situation [625].

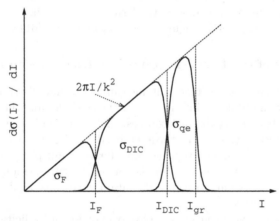

Fig. 5.16. Schematic plot of the differential cross section $d\sigma(I)/dI$ as function of the orbital angular momentum I for fusion (σ_{fus}), deep-inelastic collisions (σ_{DIC}), and quasielastic reactions (σ_{qe}) (after Ref. [594]).

After hard contact of the reaction partners there are substantial changes of the shape, leading to a compact object. At the same time the doorway states have to approach thermal equilibrium and the angular momentum of the orbital motion has to be converted into rigid rotation. It is not known on which time scale these processes occur, what the corresponding equations of motion are, and whether a compound state in the local ground-state potential-minimum is eventually reached for all partial waves contributing to the subsequent fission reaction. In particular for the largest angular momenta, considered in the Langevin process, the potential minimum is very shallow, or has even disappeared. It appears to be not very plausible that the static ground-state shape is ever reached in these cases.

Despite these concerns many Langevin calculations use an initial distribution for the dynamical parameters of the form (e.g. [282])

$$w(\mathbf{q}, \mathbf{p}, I; t = 0) \propto \exp\left\{-\frac{(m^{-1}(\mathbf{q}_0)^{ij} p_i p_j}{2T_0}\right\} \delta(\mathbf{q} - \mathbf{q}_0)\frac{d\sigma_{\text{fus}}(I)}{dI}, \qquad (5.103)$$

where $\mathbf{q}_0$ is the ground-state minimum of the potential-energy surface for the partial wave I and T_0 is the initial temperature, obtained from the initial excitation energy $E^*(t = 0) = a(\mathbf{q}_0)T_0^2$. The fusion cross section for the partial wave I is $\sigma_{\text{fus}}(I)$ and follows from a fusion theory for the entrance channel. A schematic plot of the differential cross section $d\sigma(I)/dI$ is shown in Fig. 5.16. The start parameters for the Langevin trajectories are drawn at random from the distribution (5.103).

The rather schematic assumption of a sharp initial distribution of the shape variable $\mathbf{q}$ has the consequence that only after some time a stationary flow over the barrier is established. In Fig. 5.17 the histogram shows the decay rate as function of time. Grangé, Weidenmüller, and Li Jun-Qing called

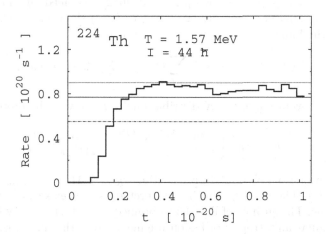

Fig. 5.17. Calculated decay rates for ^{224}Th as functions of time. The histogram is the result of a two-dimensional Langevin calculation, the full and dot-dashed lines are the two-dimensional and one-dimensional Kramers limits, respectively. The dotted line includes quantum corrections to the two-dimensional Kramers limit. The calculation was made for an initial temperature of 1.57 MeV and a rotational angular momentum of $I = 44\hbar$ (after Ref. [626]).

this time "transient time" [627, 628]. In one-dimensional dynamical calculations with the initial distribution (5.103) the transient time is of the order of

10^{-20} sec in the overdamped case [629] and an order of magnitude smaller in two-dimensional calculations with hydrodynamical viscosity [280]. In contrast to the rate of stationary flow, the transient time depends sensitively on the assumed initial distribution and therefore contains considerable arbitrariness.

The effect of the transient time on the prescission multiplicities would be large if the inverse fission rate were of the same order of magnitude as the transient time. This is however only the case for a very shallow potential pocket. However, the system is less likely to be caught by a shallow pocket on the way from the fusion entrance channel to the fission saddle-point [630]. Because of its rather ill-defined nature the transient time is not very meaningful as a free parameter to fit observed light-particle multiplicities or other fission data.

Another uncertainty in Langevin calculations is connected with the probability $P_J(K)$ of the K quantum number for a system with total angular momentum $\mathbf{J}$, or in terms of macroscopic dynamics, the tilting angle between the fission axis and the vector $\mathbf{J}$. For the saddle point the two distributions (4.14) and (4.15) have been used. As pointed out in Ref. [382] these distributions lead to different fission times and consequently to different prescission multiplicities. Also the angular distribution of the fission fragments according to Eq. (4.16) depends on $P_J(K)$. More recently Drozdov et al. suggested to allow the K quantum number to fluctuate in the local canonical ensemble along the fission path [631]. Karpov et al. [632] combined this idea with a three-dimensional Langevin process in the (c, h, α') shape class. The authors added to the Langevin equations the master equation

$$\dot{P}_J(K;t) = -P_J(K;t)\sum_{K'} w_{KK'} + \sum_{K'} P_J(K';t)w_{K'K}$$

for the time evolution of the K distribution $P_J(K;t)$ with the probability

$$w_{KK'} = \tau_K^{-1}\begin{cases} \exp(-\Delta E_{KK'}/T) & \Delta E_{KK'} > 0 \\ 1 & \Delta E_{KK'} \leq 0 \end{cases}$$

for a transition from K to K' during the time τ_K and $\Delta E_{KK'} = E_{\text{rot}}(\mathbf{q}, J, K') - E_{\text{rot}}(\mathbf{q}, J, K)$. The time τ_K can be interpreted as the relaxation time of the tilting mode. The authors of Ref. [632] obtained $\tau_K = 4 \times 10^{-21}$s by fitting observed angular anisotropies of fission fragments from the three fusion-fission reactions ^{16}O+^{232}Th, ^{16}O+^{238}U, and ^{16}O+^{248}Cf. This time is comparable with the saddle-to-scission time in dynamical calculations.

A trajectory which starts in a deep potential pocket may take a long time until it eventually finds its way over the saddle to continue towards scission. Since each time step τ of the numerical integration of the Langevin equations is connected with a numerical error, the ratio between the total time over which the trajectory is calculated and the time step τ should not become too large. Otherwise the statistics at scission is dominated by rounding errors rather than the Langevin force. It has therefore been proposed by Mavlitov, Fröbrich, and Gontchar [633] to use the Langevin equations only until a stationary flux is

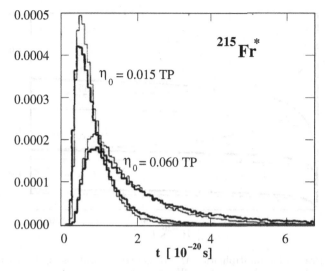

Fig. 5.18. Distribution of ground-state-to-scission times for ^{215}Fr* calculated with the two viscosity constants $\eta_0 = 0.015$ TP and $\eta_0 = 0.06$ TP. The thin histograms were obtained without evaporation, the thick histograms include evaporation, (after Ref. [282]).

established and switch thereafter to a cascade calculation with an appropriate version of Kramers' formula for the fission rate. The rate derived from the multidimensional Fokker-Planck equation with corrections for the change of the temperature between the compound state and the saddle point is

$$r_{\text{Kramers}} = \frac{\omega_K}{2\pi} \left(\frac{\det[\partial_{\mathbf{q}_i}\partial_{\mathbf{q}_j}F(\mathbf{q}_0)|_{T=T_0}]}{\det m_{ij}(\mathbf{q}_0)} \frac{\det m_{ij}(\mathbf{q}_{\text{sd}})}{|\det[\partial_{\mathbf{q}_i}\partial_{\mathbf{q}_j}F(\mathbf{q}_{\text{sd}})|_{T=T_{\text{sd}}}]|} \right)^{1/2}$$

$$\times \left(\frac{T_{\text{sd}}}{T_0} \right)^n \exp\left\{ -\left[\frac{F(\mathbf{q}_{\text{sd}};T_{\text{sd}})}{T_{\text{sd}}} - \frac{F(\mathbf{q}_0;T_0)}{T_0} \right] \right\}, \qquad (5.104)$$

which describes approximately the diffusion rate of the system in the stationary regime [634]. (In this reference the temperature is however assumed to be constant along the fission path). In Eq. (5.104) $\mathbf{q}_{\text{sd}}$, T_{sd} and $\mathbf{q}_0$, T_0 are the fission saddle-point and ground state coordinates and temperatures, respectively and the Kramers frequency ω_K is the solution of the equation

$$\det[\omega_K^2 m_{ij}(\mathbf{q}_{\text{sd}}) + \omega_K \eta_{ij}(\mathbf{q}_{\text{sd}}) + \partial_{\mathbf{q}_i}\partial_{\mathbf{q}_j}F(\mathbf{q}_{\text{sd}})|_{T=T_{\text{sd}}}] = 0. \qquad (5.105)$$

A derivation and detailed discussion of these equations will be given in Sec. 5.2.2. The switching procedure is particularly important to obtain evaporation-residue yields. When the cascade calculation yields a fission event, its further history is again followed by a Langevin process between saddle and scission to obtain the prescission particles emitted on this part of the fission path and not accounted for by the cascade calculation.

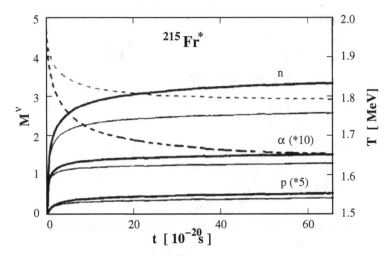

Fig. 5.19. Prescission multiplicities M^ν for neutrons, protons, and α particles from the decay of the compound nucleus ^{215}Fr* as functions of time. The dashed lines give the average temperature at the scission line of those fission fragments which reached the scission line up to this time; thin lines are obtained for $\eta_0 = 0.015$ TP, thick lines for $\eta_0 = 0.06$ TP. The α multiplicities are multiplied by 10, the proton multiplicities by 5 (after Ref. [282]).

As a typical example for results obtained from the solution of the Langevin equations with the initial distribution (5.103) we show in Fig. 5.18 the distribution of passage times from ground state to scission. The calculation was made for the compound nucleus ^{215}Fr* produced in the fusion reaction ^{197}Au + ^{18}O at 159 MeV in the two-dimensional deformation-parameter space shown in Fig. 5.14. Hydrodynamical transport coefficients in Werner-Wheeler approximation were used in this calculation together with the assumption that the rotational angular momentum is perpendicular to the fission axis and the probability P_I for an angular momentum $I\hbar$ is

$$P_I \propto \begin{cases} 2I + 1 & \text{if } I \le 70 \\ 0 & \text{else,} \end{cases}$$

neglecting in this context the small angular momentum carried by the odd proton of ^{215}Fr [282]. The figure shows results for an angular momentum of $60\hbar$ of rigid rotation and two different values for the viscosity constant η_0. Increasing viscosity slows down the fission motion, as expected. The evaporation of light particles is seen to enhance the long tails of the distribution and therefore increases the average fission time.

The evolution of the prescission multiplicities of neutrons, protons and α particles with time is shown in Fig. 5.19 for the same system and with the inclusion of all rotational states up to $70\hbar$. The enhancement of the multiplicities with increasing friction is clearly seen. Also shown in the figure is the

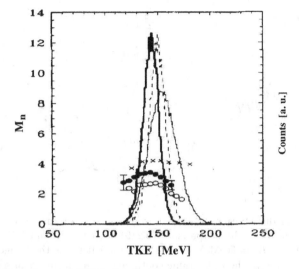

Fig. 5.20. Experimental TKE distribution (thin histogram) measured in coincidence with the prescission neutron multiplicities M^n (crosses) for $^{215}\text{Fr}^*$. The results of the calculation of these quantities with $\eta_0 = 0.015\,\text{TP}$ is shown by the dashed-line histogram and the open circles. The thick-line histogram and the full circles represent results of a model calculation with $\eta_0 = 0.06$ TP (after Ref. [282]).

time dependence of the average temperature of those fission fragments which passed the scission line until this time.

In Fig. 5.20 the calculated correlation between the TKE distribution of the fission fragments and the coincident neutron prescission multiplicity is plotted for two different viscosity constants. It is compared with experimental data from Refs. [635,636]. To match the experimental average value of the TKE one would prefer the smaller η_0, in agreement with the results shown in Fig. 5.13. However the neutron multiplicity is then almost a factor of 2 smaller than the experimental value. Also the width of the TKE distribution is almost a factor of 2 smaller than the experimental result. The mechanism which leads to the width of the distribution of center-of-mass distances between the fragments at scission, shown in Fig. 5.14, causes mainly the distribution of TKE. One may therefore argue that the energy surface along the cut of the fission valley with the scission line is too stiff. This can have a variety of reasons, e.g. the use of a sharp-surface mass-formula in the scission region, where the smoother finite-range mass-formula may be more appropriate or an inadequate shape parametrization around scission. An increase of the number of collective variables also leads to an increase of the widths. Unfortunately, the problem was sometimes fixed by redefining the scission configuration in a rather ad hoc manner by means of a critical neck radius at which Langevin

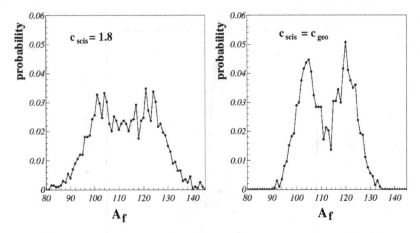

Fig. 5.21. Mass distributions obtained from solving two-dimensional Langevin equations in c and α in the shape class (2.219) with $h = 0$ and initial condition (5.103), $I = 60$, and $E^* = 56$ MeV for ^{227}Pa. In the left frame the Langevin process is stopped at $c = 1.8$, in the right frame continued to the geometrical scission point (from Ref. [604]).

dynamics is supposed to stop [40, 600, 637, 638]. Such strategies obviously introduce additional fudge parameters into the model.

Choosing the elongation and the asymmetry as collective variables Schmitt et al. [604] obtained the mass distributions from fission of the compound nucleus ^{227}Pa with an excitation energy of 56 MeV and rotational angular momentum $I = 60\hbar$ shown in Fig. 5.21. The (c, h, α)-shape parametrization (2.219) was used and h was kept constant at zero. In the left frame the Langevin process was stopped at $c = 1.8$, corresponding to a neck size of approximately 4 fm, in the right frame the calculation was continued to the geometrical scission point. The drastic difference in the distributions is obvious. Temperature-dependent shell corrections were added to the liquid-drop free energy, hydrodynamical inertial parameters and wall-and-window friction were used in these calculations.

As an example for results obtained in three-dimensional Langevin calculations we show the probability of fission events of ^{260}Rf as function of TKE and fragment mass-number as contour plot in the upper frame of Fig. 5.22 from the work of Karpov et al. [600]. In the lower frame experimental data from Ref. [639] are given for comparison. The calculation was done in the (c, h, α) shape class, α being mapped for numerical convenience on a coordinate $\eta = (3/8)\alpha c^3$ which approaches $(A_1 - A_2)/(A_1 + A_2)$ towards scission. The finite-range liquid-drop model was used for the driving force. Hydrodynamical values in Werner-Wheeler approximation were employed for the inertial tensor, and the wall-and-window formula with a reduction factor 0.1 for the friction tensor.

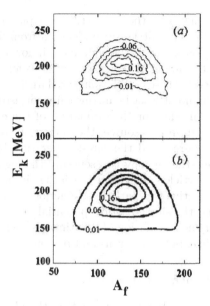

Fig. 5.22. Yield of fragments from fission of ^{260}Rf with $E^* = 74.2$ MeV as function of TKE and fragment mass. The numbers on the contour lines give the yield in percent, the total yield being normalized to 200%. The lower frame gives experimental results from [639] (from Ref. [600]).

The evolution of the system between contact and formation of a compound state in fusion-fission reactions is of particular interest in attempts to produce superheavy nuclei [640]. Despite considerable uncertainties in understanding the details of the dynamics in this phase, as discussed above, several attempts have been made to describe it in terms of Langevin equations in various shape parametrizations and with specially fitted friction parameters. A rather recent overview of these calculations was given by Abe et al. [641].

All calculations using the Langevin process are based on a number of ad hoc assumptions:

- choice of the shape class and identification of the "relevant" collective variables,
- definition of initial conditions or assumptions on the dynamics of the transient phase between contact and compound nucleus formation,
- shape dependence of the level-density parameter in the free energy and in the emission probabilities for light particles,
- the temperature and shape-dependence of shell, pairing, and vibrational corrections to Bethe's Thomas-Fermi approximation of the nuclear level density are rarely considered, even when the excitation energy is small enough to warrant their use; for an exception see ref. [642],
- shape and temperature dependence of the transport parameters.

The Langevin calculations have therefore still an exploratory character and limited predictive power. Very little can be learned from fitting just one parameter like the reduction factor of the wall-and-window friction to a limited set of experimental data, taking all other assumptions for granted.

The input data, used in these calculations, are largely of a phenomenological nature. But they should at least be intrinsically consistent. Unfortunately, often the level density, needed for the evaluation of the evaporation rate of light particles, is taken from one source, the entropy in the Langevin equations from a different source, and the zero-temperature binding energy has still another origin. But the entropy is essentially the logarithm of the level density and the single-particle model from which the binding energy is derived is based on the same effective two-particle interaction from which the binding energy, the entropy and the single-particle potential are obtained. The latter should be the same for the calculation of Strutinsky's shell corrections and in the transmission coefficients for neutrons and protons.

5.2.2 The Fokker-Planck equation

Instead of generating the probability $w(\mathbf{q}, \mathbf{p}, t)$ in the phase space of the collective variables numerically from bunches of fission trajectories, one may seek to derive a differential equation for $w(t)$ and obtain solutions for a given initial probability $w(t = t_0)$. Volume elements in phase space will refer in the following to the contravariant components of the coordinates $d\{q^i\}$ and the covariant components of the momenta $d\{p_i\}$, in short-hand notation $d\mathbf{q}$ and $d\mathbf{p}$. The volume element $d\mathbf{q}d\mathbf{p}$ is therefore a scalar. The probability density $w(\mathbf{q}, \mathbf{p}, t)$ must then also be a scalar. In some treatments of transport theory covariance of all equations with respect to general transformations in phase space is required [643, 644]. This is also the case in the textbook by Risken [590], which is recommended for an in-depth discussion of the Fokker-Planck equation and of its solution techniques. In the present context this global covariance would be somewhat artificial. Since coordinates and momenta are connected in our case by Newtonian equations, a transformation in coordinate space induces the inverse transformation in momentum space.

The Kramers-Moyal expansion

One introduces the conditional probability $P(\mathbf{q}, \mathbf{p}; t + \tau | \mathbf{q}'\mathbf{p}'; t)$ to find the system at time $t + \tau$ at the phase-space point $(\mathbf{q}, \mathbf{p})$ if it was at time t at the point $(\mathbf{q}', \mathbf{p}')$. In terms of this transition probability the equation of motion for $w(t)$, called Chapman-Kolmogorov equation, is

$$w(\mathbf{q}, \mathbf{p}, t + \tau) = \int P(\mathbf{q}, \mathbf{p}; t + \tau | \mathbf{q}', \mathbf{p}'; t) w(\mathbf{q}', \mathbf{p}', t) d\mathbf{q}' d\mathbf{p}' . \qquad (5.106)$$

For small τ one expects the transition probability P to be peaked at $\mathbf{q} = \mathbf{q}'$ and $\mathbf{p} = \mathbf{p}'$ since points in phase space far away from the point $(\mathbf{q}, \mathbf{p})$ should not influence the situation there in a very short time. One is therefore led to evaluate the integral in Eq. (5.106) by a moment expansion. In the Kramers-Moyal expansion one introduces new integration variables $\mathbf{Q} = \mathbf{q} - \mathbf{q}'$ and $\mathbf{P} = \mathbf{p} - \mathbf{p}'$ and expands the integrand in a Taylor series with respect to the new variables $\mathbf{Q}$ and $\mathbf{P}$ up to terms of second order

$$w(\mathbf{q}, \mathbf{p}, t + \tau) = \int \left([Pw] - \frac{\partial [Pw]}{\partial q^i} Q^i - \frac{\partial [Pw]}{\partial p_i} P_i + \frac{1}{2} \frac{\partial^2 [Pw]}{\partial q^i \partial q^j} Q^i Q^j \right.$$
$$\left. + \frac{\partial^2 [Pw]}{\partial q^i \partial p_j} Q^i P_j + \frac{1}{2} \frac{\partial^2 [Pw]}{\partial p_i \partial p_j} P_i P_j \right) d\mathbf{Q} d\mathbf{P} , \tag{5.107}$$

where the abbreviation $[Pw] = P(\mathbf{q} + \mathbf{Q}, \mathbf{p} + \mathbf{P}, t + \tau | \mathbf{q}, \mathbf{p}; t) w(\mathbf{q}, \mathbf{p}, t)$ was used. Since $\mathbf{Q}$ and $\mathbf{P}$ do not explicitly depend on $\mathbf{q}$ and $\mathbf{p}$, one can write

$$\frac{\partial^2 [Pw]}{\partial q^i \partial q^j} Q^i Q^j = \frac{\partial}{\partial q^i} \left(\frac{\partial [Pw] Q^j}{\partial q^j} \right) Q^i ,$$

where the right-hand side is a properly covariant expression, which is therefore also true for the expression with the second partial derivative on the left side of the equation. Introducing the moments

$$1 = \int P(\mathbf{q} + \mathbf{Q}, \mathbf{p} + \mathbf{P}; t + \tau | \mathbf{q}, \mathbf{p}; t) d\mathbf{Q} d\mathbf{P},$$

$$\mathcal{M}_{Q_i} = \int P(\mathbf{q} + \mathbf{Q}, \mathbf{p} + \mathbf{P}; t + \tau | \mathbf{q}, \mathbf{p}; t) Q^i d\mathbf{Q} d\mathbf{P},$$

$$\mathcal{M}_{Q_i Q_j} = \int P(\mathbf{q} + \mathbf{Q}, \mathbf{p} + \mathbf{P}; t + \tau | \mathbf{q}, \mathbf{p}; t) Q^i Q^j d\mathbf{Q} d\mathbf{P},$$

and analogous moments $\mathcal{M}_{P^i}$, $\mathcal{M}_{P^i P^j}$, and $\mathcal{M}_{Q_i P^j}$, Eq. (5.107) becomes

$$w(\mathbf{q}, \mathbf{p}, t + \tau) = w(\mathbf{q}, \mathbf{p}, t) - \frac{\partial (w \mathcal{M}_{Q_i})}{\partial q^i} - \frac{\partial (w \mathcal{M}_{P_i})}{\partial p_i}$$
$$+ \frac{1}{2} \left(\frac{\partial^2 (w \mathcal{M}_{Q_i Q_j})}{\partial q^i \partial q^j} + \frac{\partial^2 (w \mathcal{M}_{P^i P^j})}{\partial p_i \partial p_j} + 2 \frac{\partial^2 (w \mathcal{M}_{Q_i P^j})}{\partial q^i \partial p_j} \right) . \tag{5.108}$$

In order to associate a concrete meaning with the preceding formalism we relate the Kramers-Moyal moments to the coefficients of the Langevin equations. In its standard form Kramers' transport theory is assumed to describe systems at fixed temperature. We will therefore not use Eqs. (5.97), but assume that the temperature is constant. Integrating the first two Langevin equations (5.95) and (5.96) over a short time interval τ one obtains

$$q^i(t + \tau) - q^i(t) = \tau (m^{-1})^{ij} p_j(t)$$

$$p_i(t + \tau) - p_i(t) = \tau K_i^{(\text{det})} + \sum_j g_{ij} \int_t^{t+\tau} \Gamma_j(t') dt' ,$$

where

$$K_i^{(\text{det})} = -(\partial_{q^i} F + (1/2)\partial_{q^i}(m^{-1})^{jk} p_j p_k + \eta_{ij}(m^{-1})^{jk} p_k) \tag{5.109}$$

is the deterministic force and where the Ito discretization of the stochastic integral was used. These equations yield the following averages over $\mathbf{q}(t+\tau)$ and $\mathbf{p}(t+\tau)$ at fixed $\mathbf{q}(t)$ and $\mathbf{p}(t)$

$$\mathcal{M}_{Q_i} = \langle q^i(t+\tau) - q^i(t) \rangle = \tau(m^{-1})^{ij} p_j(t) , \tag{5.110}$$

$$\mathcal{M}_{P^i} = \langle p_i(t+\tau) - p_i(t) \rangle = \tau K_i^{(\text{det})} + \sum_j g_{ij} \int \langle \Gamma_j(t') \rangle dt' = \tau K_i^{(\text{det})} , \tag{5.111}$$

$$\begin{aligned}
\mathcal{M}_{Q_i Q_j} &= \langle (q^i(t+\tau) - q^i(t))(q^j(t+\tau) - q^j(t)) \rangle \\
&= \tau^2 (m^{-1})^{ik}(m^{-1})^{jl} p_k(t) p_l(t) ,
\end{aligned} \tag{5.112}$$

$$\begin{aligned}
\mathcal{M}_{Q_i P^j} &= \langle (q^i(t+\tau) - q^i(t))(p_j(t+\tau) - p_j(t)) \rangle \\
&= \tau^2 (m^{-1})^{ik} p_k(t) K_j^{(\text{det})} ,
\end{aligned} \tag{5.113}$$

$$\begin{aligned}
\mathcal{M}_{P^i P^j} &= \langle (p_i(t+\tau) - p_i(t))(p_j(t+\tau) - p_j(t)) \rangle \\
&= \tau^2 K_i^{(\text{det})} K_j^{(\text{det})} + \sum_{kl} g_{ik} g_{jl} \int_t^{t+\tau} dt' \int_t^{t+\tau} dt'' \langle \Gamma_k(t') \Gamma_l(t'') \rangle \\
&= \tau^2 K_i^{(\text{det})} K_j^{(\text{det})} + 2\tau \sum_k g_{ik} g_{jk} ,
\end{aligned} \tag{5.114}$$

where $\langle \Gamma_j(t) \rangle = 0$ and Eq. (5.90) were used. Inserting these results into Eq. (5.108), dividing the equation by τ, and letting τ go to zero, one obtains the Fokker-Planck equation

$$\dot{w} = -\frac{\partial((m^{-1})^{ij} p_j w)}{\partial q^i} - \frac{\partial(K_i^{(\text{det})} w)}{\partial p_i} + \frac{\partial^2(D_{ij} w)}{\partial p_i \partial p_j} \tag{5.115}$$

with the diffusion tensor $D_{ij} = \sum_k g_{ik} g_{jk}$.

Frequently fission dynamics is assumed to be "overdamped", i.e. the inertial term $\dot{p}$ in the Langevin equations (5.96) is neglected, compared with the friction term $\eta_{ij}(m^{-1})^{jk} p_k$. Introducing the contravariant inverse friction tensor $(\eta^{-1})^{ij}$, one obtains one set of first-order differential equations in time, instead of two sets

$$\dot{q}^j = (\eta^{-1})^{ji}(-\partial_{q^i} F + \sum_k g_{ik} \Gamma_k) .$$

In this case the Kramers-Moyal moments are given by

$$\mathcal{M}_{Q_i} = -\tau(\eta^{-1})^{ij}\partial_{q^j}F \,,$$

$$\mathcal{M}_{Q_iQ_j} = \tau(\eta^{-1})^{il}[\tau\partial_{q^l}F\partial_{q^k}F + 2\sum_r g_{lr}g_{kr}](\eta^{-1})^{jk}$$

and all other moments are of the order τ^2. Eq. (5.108) therefore yields

$$\dot{w} = \frac{\partial[(\eta^{-1})^{ij}(\partial_{q^j}F)w]}{\partial q^i} + \frac{\partial^2[(\eta^{-1})^{ik}D_{kl}(\eta^{-1})^{jl}w]}{\partial q^i\partial q^j} \,, \qquad (5.116)$$

which is called Smoluchowski equation.

In order to allow for a change of the temperature along each Langevin trajectory according to the Langevin equation (5.97), one may introduce the temperature as an additional stochastic variable in the probability w. The Kramers-Moyal expansion yields then a few additional terms in the resulting Fokker-Planck equation.

To account for particle emission during the fission process, Strumberger, Dietrich, and Pomorski added the vector $\mathcal{N} = \{n^i_{\alpha\beta}\}$ to the arguments of $w(\mathbf{q}, \mathbf{p}; t)$ [645, 646], where $n^i_{\alpha\beta}$ are the integer-valued numbers ≥ 0 of evaporated particles of type i emitted in the energy interval α and the angular momentum interval β. The quantity $w(\mathbf{q}, \mathbf{p}, \mathcal{N}; t)d\mathbf{q}\,d\mathbf{p}$ is therefore the probability that the system is at time t in the phase-space cell $d\mathbf{q}\,d\mathbf{p}$ around the the point $(\mathbf{q}\,\mathbf{p})$ in phase space and has emitted $n^i_{\alpha\beta}$ particles of type i into the energy and angular momentum bins α and β, respectively. A change of the value of the function w for a given value, say 1, of one of the arguments $n^{i_0}_{\alpha_0\beta_0}$ during a short interval τ can have two reasons: either the cell (i_0, α_0, β_0) was empty and a particle is emitted into this cell with the rate $r^{i_0}_{\alpha_0\beta_0}(\mathbf{q}, \mathbf{p}, ..., n^{i_0}_{\alpha_0\beta_0}{=}0, ...)$ out of an initial configuration with $n^{i_0}_{\alpha_0\beta_0} = 0$, whose probability is $w(..., n^{i_0}_{\alpha_0\beta_0}{=}0, ...; t)$; this increases the value of w, or a second particle is emitted into the same cell out of a configuration with $n^{i_0}_{\alpha_0\beta_0} = 1$ in which case $w(..., n^{i_0}_{\alpha_0\beta_0}{=}1, ...; t)$ is reduced since such events belong now to the argument $n^{i_0}_{\alpha_0\beta_0} = 2$. By this reasoning the master equation

$$\dot{w}(..., n^i_{\alpha\beta}, ...; t) = r^i_{\alpha\beta}(..., n^i_{\alpha\beta}{-}1, ...)w(..., n^i_{\alpha\beta}{-}1, ...; t)$$
$$-r^i_{\alpha\beta}(..., n^i_{\alpha\beta}, ...)w(..., n^i_{\alpha\beta}, ...; t)$$

is obtained if all other variables of w would be kept fixed. Allowing all variables to change, Strumberger et al. [645, 646] derived the generalized Fokker-Planck equation

$$\dot{w} = -\frac{\partial((m^{-1})^{ij}p_jw)}{\partial q^i} - \frac{\partial(K^{(\mathrm{det})}_iw)}{\partial p_i} + \frac{\partial^2(D_{ij}w)}{\partial p_i\partial p_j}$$
$$+ \sum_{i,\alpha,\beta}\left[r^i_{\alpha\beta}(..., n^i_{\alpha\beta}{-}1, ...)w(..., n^i_{\alpha\beta}{-}1, ...; t)\right.$$
$$\left.-r^i_{\alpha\beta}(..., n^i_{\alpha\beta}, ...)w(..., n^i_{\alpha\beta}, ...; t)\right]. \qquad (5.117)$$

In this expression $w(..., n_{\alpha\beta}^i - 1, ...; t)$ is defined to vanish if $n_{\alpha\beta}^i - 1 = 0$. For a given vector $\mathcal{N}$ one can calculate the mass ΔM_{nucl}, the charge ΔZ, the energy ΔE, and the angular momentum ΔJ, carried away by the emitted particles. The force K_i, the transport matrices m_{ij} and D_{ij}, and the emission rates $r_{\alpha\beta}^i$ depend on these quantities besides their dependence on $\mathbf{q}$ and the temperature T. In this way the evaporation and the collective motion are dynamically coupled, although they are treated in Eq. (5.117) as statistically independent.

We introduced the Langevin equations on a strictly phenomenological basis. Some authors [647] prefer to derive them from the Fokker-Planck equation (5.115). Assuming that there are stochastic, first-order equations of motion for the phase-space variables $\mathbf{q}$ and $\mathbf{p}$ of the form

$$\dot{q}^i = (h_1(\mathbf{q}, \mathbf{p}))^i + \sum_j (g_1(\mathbf{q}, \mathbf{p}))_j^i (\Gamma_1(t))_j \tag{5.118}$$

$$\dot{p}_i = (h_2(\mathbf{q}, \mathbf{p}))_i + \sum_j (g_2(\mathbf{q}, \mathbf{p}))_{ij} (\Gamma_2(t))_j, \tag{5.119}$$

we will show that the Fokker-Planck equation (5.115) implies that the functions $\mathbf{h}(\mathbf{q}, \mathbf{p})$ and $\mathbf{g}(\mathbf{q}, \mathbf{p})$ have just the form which leads to the Langevin equations (5.95) and (5.96). In Eqs. (5.118) and (5.119) it is assumed that the $\Gamma_i(t)$ are Gaussian-distributed random numbers with

$$\langle (\Gamma_\nu(t))_i \rangle = 0 \quad \text{and} \quad \langle (\Gamma_\nu(t))_i (\Gamma_\mu(t'))_j \rangle = 2\delta_{\nu\mu}\delta_{ij}\delta(t - t'), \quad \nu, \mu = 1, 2.$$

For a sufficiently small time step τ the equations of motion (5.118) and (5.119) yield

$$q^i(t + \tau) - q^i(t) =$$
$$\tau(h_1(\mathbf{q}, \mathbf{p}))^i + \sum_j (g_1(\mathbf{q}, \mathbf{p}))_j^i \int_t^{t+\tau} (\Gamma_1(t'))_j dt' + \mathcal{O}(\tau^{3/2}) \tag{5.120}$$

and

$$p_i(t + \tau) - p_i(t) =$$
$$\tau(h_2(\mathbf{q}, \mathbf{p}))_i + \sum_j (g_2(\mathbf{q}, \mathbf{p}))_{ij} \int_t^{t+\tau} (\Gamma_2(t'))_j dt' + \mathcal{O}(\tau^{3/2}). \tag{5.121}$$

From the Fokker-Planck equation follows for the Kramers-Moyal moments

$$\lim_{\tau \to 0} \frac{1}{\tau} \mathcal{M}_{Q_i} = (m^{-1})^{ij} \langle p_j \rangle, \qquad \lim_{\tau \to 0} \frac{1}{\tau} \mathcal{M}_{P^i} = (m^{-1})^{ij} K^{(\text{det})},$$

$$\lim_{\tau \to 0} \frac{1}{\tau} \mathcal{M}_{P^i P^j} = D_{ij}, \qquad \lim_{\tau \to 0} \frac{1}{\tau} \mathcal{M}_{Q_i Q_j} = \lim_{\tau \to 0} \frac{1}{\tau} \mathcal{M}_{Q_i P^j} = 0.$$

Using these results and inserting the expressions (5.120) and (5.121) into Eqs. (5.110)-(5.114) yields

$$(h_1(\mathbf{q}, \mathbf{p}))^i = (m^{-1}(\mathbf{q}))^{ij} p_j, \qquad (h_2(\mathbf{q}, \mathbf{p}))_i = K_i^{(\text{det})}(\mathbf{q}, \mathbf{p}),$$
$$(g_1(\mathbf{q}, \mathbf{p}))^i_j = 0, \qquad \sum_k (g_2)_{ik}(g_2)_{jk} = D_{ij}(\mathbf{q}),$$

which completes the proof.

Kramers' rate formula

For a system, caught in a potential minimum and leaking out over a barrier, Kramers gave an approximate, stationary solution of the Fokker-Planck equation (5.115) in the one-dimensional case [510]. In his approach it is assumed that the transport coefficients do not depend on the coordinates and that the temperature is neither time-dependent nor shape-dependent. Kramers' result was generalized to the multi-dimensional case by Weidenmüller and J.-S. Zhang [634] with the same assumptions on the temperature and the transport coefficients. Below we follow essentially the derivation of the rate formula given by these authors.

We seek a stationary solution of Eq. (5.115) under the assumption that the system is essentially in thermal equilibrium in the ground-state potential-minimum at $\mathbf{q} = 0$ and is only slightly disturbed by the outflow over the barrier. The equilibrium distribution at temperature T is $w_0(\mathbf{q}, \mathbf{p}, T) = e^{-H/T}$, where $H(\mathbf{q}, \mathbf{p}, T) = (1/2)(m^{-1})^{ij} p_i p_j + F(\mathbf{q}, T)$ is the conservative part of the Hamiltonian and F is the free energy. For the disturbed distribution the ansatz

$$w(\mathbf{q}, \mathbf{p}) = f(\mathbf{q}, \mathbf{p}) w_0(\mathbf{q}, \mathbf{p}) \tag{5.122}$$

is made and inserted into the Fokker-Planck equation, making use of the assumed shape independence of m_{ij} and η_{ij}. A stationary solution has to satisfy the equation

$$0 = \dot{w} = -(m^{-1})^{ij} p_j \frac{\partial w}{\partial q^i} - \frac{\partial (K_i^{(\text{det})} w)}{\partial p_i} + D_{ij} \frac{\partial^2 w}{\partial p_i \partial p_j}. \tag{5.123}$$

One then gets the following equation for the correction function $f(\mathbf{q}, \mathbf{p})$

$$(m^{-1})^{ij} p_j \frac{\partial f}{\partial q^i} - \left[\partial_{q^i} F - \eta_{ij}(m^{-1})^{jl} p_l \right] \frac{\partial f}{\partial p_i} - T \eta_{ij} \frac{\partial^2 f}{\partial p_i \partial p_j} = 0, \tag{5.124}$$

where Eqs. (5.91) and (5.109) (neglecting the $\mathbf{q}$-dependence of the inertial tensor) have been used. The boundary conditions for $f(\mathbf{q}, \mathbf{p})$ are $f = 1$ for $\mathbf{q} = 0$ and $f = 0$ for $q^i \to \infty$.

In order to obtain the current density at the barrier $\mathbf{q} = \mathbf{q}_{\text{sd}}$

$$j^i(\mathbf{q}_{\mathrm{sd}}) = (m^{-1})^{ij} \int p_j w(\mathbf{q}_{\mathrm{sd}}, \mathbf{p}) d\mathbf{p}, \qquad (5.125)$$

a solution of Eq. (5.124) in the vicinity of $\mathbf{q}_{\mathrm{sd}}$ is needed. Expanding $\partial_{q^i} F(\mathbf{q})$ to linear order in $q^i - q^i_{\mathrm{sd}}$ and using $\partial_{q^i} F(\mathbf{q}_{\mathrm{sd}}) = 0$, one obtains the differential equation

$$(m^{-1})^{ij} p_j \frac{\partial f}{\partial q^i} + \Big[(m^{-1})^{jk} \eta_{ij} p_k - (q^j - q^j_{\mathrm{sd}}) F_{|ij}(\mathbf{q}_{\mathrm{sd}}) \Big] \frac{\partial f}{\partial p_i}$$
$$= T\eta_{ij} \frac{\partial^2 f}{\partial p_i \partial p_j}, \qquad (5.126)$$

whose coefficients are linear in q^i and p_i. We abbreviate $\partial_{q^i} \partial_{q^j} F(\mathbf{q}_{\mathrm{sd}})$ by $F_{|ij}$. As in Kramers' one-dimensional case, the differential operator on the left-hand side of Eq. (5.126) will be shown to depend on a linear combination y of the vectors q^i and p_i

$$y = a^i p_i - b^i m_{ij}(q^j - q^j_{\mathrm{sd}}) \qquad (5.127)$$

only, where a^i and b^i are constant vectors. Using the relations $\partial f / \partial q^i = -b^j m_{ji} \partial f / \partial y$ and $\partial f / \partial p_i = a^i \partial f / \partial y$, Eq. (5.126) becomes

$$\Big[-p_j b^j + \{ (m^{-1})^{jk} \eta_{ij} p_k - (q^j - q^j_{\mathrm{sd}}) F_{|ij}(\mathbf{q}_{\mathrm{sd}}) \} a^i \Big] \frac{df}{dy}$$
$$= T a^i \eta_{ij} a^j \frac{d^2 f}{dy^2}. \qquad (5.128)$$

If the differential operator on the left side of this equation depends only on y, the square bracket must be a multiple $-h$ of y, identically in p_i and $q^i - q^i_{\mathrm{sd}}$. This leads to the system of linear equations for a^i and b^i

$$(m^{-1})^{ij} \eta_{kj} a^k + h a^i - b^i = 0, \qquad (5.129)$$
$$F_{|ij}(\mathbf{q}_{\mathrm{sd}}) a^j + h m_{ij} b^j = 0. \qquad (5.130)$$

We introduce the matrices

$$\eta^i_k = (m^{-1})^{ij} \eta_{jk}, \quad F^i_k = (m^{-1})^{ij} F_{|jk}$$

and the unit matrix δ^i_k. Multiplying Eq. (5.130) with $(m^{-1})^{kj}$, Eqs. (5.129) and (5.130) can be written

$$(\eta^i_k + h\delta^i_k) \, a^k - \delta^i_k b^k = 0,$$
$$F^i_k \qquad a^k + h\delta^i_k b^k = 0. \qquad (5.131)$$

These equations have a solution if, and only if, h satisfies the secular equation

$$\det \begin{pmatrix} \eta^i_k + h\delta^i_k & -\delta^i_k \\ F^i_k & h\delta^i_k \end{pmatrix} = \det(h^2 \delta^i_k + h\eta^i_k + F^i_k) = 0. \qquad (5.132)$$

It was shown in Ref. [634] that there is always one non-negative solution h if the matrix m_{ij} is positive definite and the matrix $F_{|ij}(\mathbf{q}_{sd})$ has one negative and $n-1$ positive eigenvalues, which is the case for a first order saddle-point $\mathbf{q}_{sd}$. It is convenient to choose the normalization of a^i such that

$$a^i \eta_{ij} a^j = h/(2T).$$ (5.133)

The differential equation (5.128) then reads

$$-y\frac{df}{dy} = \frac{1}{2}\frac{d^2 f}{dy^2}.$$ (5.134)

Note that the transformation to the variable y in Eq. (5.128) is only possible if the expansion of F around $\mathbf{q}_{sd}$ is restricted to the second derivative.

The solution $f(y)$ of Eq. (5.134) with the boundary conditions $f(-\infty) = 0$ and $f(\infty) = 1$ is

$$f(y) = \frac{1}{\sqrt{\pi}}\int_{-\infty}^{y} e^{-\xi^2} d\xi = \frac{1}{2}\operatorname{erfc}(-y).$$

The main contribution to the integral comes from the area around $\xi = 0$. Therefore the upper limit of the integral corresponding to the ground state has been shifted to infinity without introducing a large error. It will be shown below that the vector a^i has the direction of the fission current. Therefore y approaches $-\infty$ when $\mathbf{q}$ goes to $+\infty$, i.e. to the region far to the right of the saddle point. Again, the details have a negligible effect on $f(y)$. In order to calculate the current (5.125) we use the identity

$$p_i \exp[-(m^{-1})^{lj} p_l p_j/(2T)] = -T m_{ij}\partial_{p_j} \exp[-(m^{-1})^{lk} p_l p_k/(2T)]$$

and obtain from Eq. (5.122) after partial integration

$$j^i(\mathbf{q}_{sd}) = a^i T \exp\left\{\left[-F(\mathbf{q}_{sd}) - \frac{1}{2}F_{|ij}(\mathbf{q}_{sd})(q^i - q^i_{sd})(q^j - q^j_{sd})\right]\Big/T\right\}$$
$$\times \int f'(y)\exp\left(-\frac{(m^{-1})^{lk} p_l p_k}{2T}\right) d\mathbf{p}$$ (5.135)

in the vicinity of $\mathbf{q}_{sd}$. The integral is an n-dimensional Gaussian integral and can be evaluated analytically

$$\int \exp\left\{-[a^i p_i - b^i m_{ij}(q^j - q^j_{sd})]^2 - \frac{(m^{-1})^{lk} p_l p_k}{2T}\right\} d\mathbf{p} =$$
$$\left(\frac{(2\pi T)^n}{\det C^{ij}}\right)^{1/2}\exp\left(-\frac{b^i b^j}{1 + 2Ta^2}m_{il}m_{jk}(q^l - q^l_{sd})(q^k - q^k_{sd})\right)$$

in terms of the scalar $a^2 = a^i m_{ij} a^j$ and the tensor

$$C^{ij} = (m^{-1})^{ij} + 2Ta^i a^j .$$ (5.136)

We also used its inverse

$$(C^{-1})_{ij} = m_{il}\left(\delta^l_j - \frac{2Ta^l m_{jk}a^k}{1 + 2Ta^2}\right)$$

to obtain this result. Inserting it into Eq. (5.135) and introducing the symmetric tensor

$$\Psi_{ij} = F_{|ij}(\mathbf{q}_{\mathrm{sd}}) + \frac{2T}{1 + 2Ta^2} m_{ik}b^k m_{jl}b^l ,$$

the current density becomes

$$j^i = Ta^i\left(\frac{(2\pi T)^n}{\pi \det C^{ij}}\right)^{1/2} e^{-[2F(\mathbf{q}_{\mathrm{sd}})+\Psi_{ij}(q^i-q^i_{\mathrm{sd}})(q^j-q^j_{\mathrm{sd}})]/(2T)} .$$ (5.137)

The vector a^i is seen to have the direction of the current.

Using Eqs. (5.131) and (5.133), Ψ_{ij} can be rewritten as

$$\Psi_{ij} = F_{|ij}(\mathbf{q}_{\mathrm{sd}}) - \frac{F_{|il}a^k a^l F_{|kj}}{a^l F_{|lk}a^k} .$$

This form shows that $\Psi_{ij}a^j = 0$, i.e. a^i is eigenvector of the matrix Ψ with eigenvalue 0. The remaining $n-1$ eigenvectors are orthogonal to a^i and span the hyperplane perpendicular to the fission direction at the saddle point. Integration of the norm of the current-density vector, $\sqrt{j^i j_i} = \sqrt{j^i m_{ik} j^k}$, over this hyperplane yields the total current I. In terms of the $n-1$ nonvanishing eigenvalues λ_i, $i = 2, \ldots n$ of Ψ the integration yields

$$I = Ta\left(\frac{(2\pi T)^n}{\pi \det C^{ij}}\right)^{1/2} e^{-F(\mathbf{q}_{\mathrm{sd}})/T} \int e^{-\sum_{i=2}^n \lambda_i \xi_i^2/(2T)} \prod_{i=2}^n d\xi_i$$

$$= (2\pi T)^n \left(\frac{T}{2(\prod_{i=2}^n \lambda_i) \det C^{ij}}\right)^{1/2} \frac{a}{\pi} e^{-F(\mathbf{q}_{\mathrm{sd}})/T} ,$$ (5.138)

where the ξ_i are the $q^i - q^i_{\mathrm{sd}}$ in the eigenrepresentation of the matrix Ψ. Using again Eqs. (5.131) and (5.133) one can show that

$$\det C^{ij} = \frac{1 + 2Ta^2}{\det m_{ij}} = -\frac{2Ta^i F_{|ij}(\mathbf{q}_{\mathrm{sd}})a^j}{\hbar^2 \det m_{ij}} .$$

In Ref. [634] the identity

$$\prod_{i=2}^n \lambda_i = \frac{\det F_{|ij}(\mathbf{q}_{\mathrm{sd}})}{a^i F_{|ij}(\mathbf{q}_{\mathrm{sd}})a^j} a^2$$

was proved. With these results the fission current becomes

$$I = \frac{h}{2\pi} \left(\frac{\det m_{ij}}{|\det F_{|ij}(\mathbf{q}_{sd})|} \right)^{1/2} (2\pi T)^n \, e^{-F(\mathbf{q}_{sd})/T} . \tag{5.139}$$

To obtain the fission rate, I has to be divided by the normalization integral around the ground-state minimum at $\mathbf{q} = 0$

$$N = \int e^{-[(m^{-1}(0))^{ij} p_i p_j - 2F(\mathbf{q})]/(2T)} d\mathbf{q} d\mathbf{p}$$

$$= [(2\pi T)^n \det m_{ij}(0)]^{1/2} \int e^{-F(\mathbf{q})/T} d\mathbf{q} .$$

We evaluate $F(\mathbf{q})$ in the integrand around the minimum to second order in q^i: $F(\mathbf{q}) = F(0) + (1/2)F_{|ij}(0)q^i q^j$ and obtain

$$N = (2\pi T)^n \left(\frac{\det m_{ij}(0)}{\det F_{|ij}(0)} \right)^{1/2} e^{-F(0)/T} . \tag{5.140}$$

With Eq. (5.139) the rate is therefore

$$r_{\text{Kramers}} = \frac{I}{N} = \frac{h(\mathbf{q}_{sd})}{2\pi} \left(\frac{\det m_{ij}(\mathbf{q}_{sd}) \det F_{|ij}(0)}{|\det F_{|ij}(\mathbf{q}_{sd})| \det m_{ij}(0)} \right)^{1/2} e^{-F_{\text{barr}}/T} \tag{5.141}$$

with the temperature-dependent barrier $F_{\text{barr}} = F(\mathbf{q}_{sd}) - F(0)$. Although we assumed that the matrices m_{ij} and η_{ij} (the latter appears in the definition of h) are shape-independent in the vicinity of the ground state and the saddle point, we take at least their variation over large distances along the fission path into account by using the mass at ground-state deformation in Eq. (5.140) and the saddle-point values of m_{ij} and η_{ij} in Eq. (5.139). If one also allows for a change of the temperature along the fission path, one obtains the rate formula (5.104), where h is called Kramers frequency, $h = \omega_K$.

In the one-dimensional case Eq. (5.104) simplifies to

$$r_{\text{Kramers}} = \frac{T_{sd}}{T_0} \frac{\omega_K}{2\pi} \frac{\omega_0}{\omega_{sd}} e^{-[F(q_{sd};T_{sd})/T_{sd} - F(q_0;T_0)/T_0]} , \tag{5.142}$$

where

$$\omega_0 = \left(\frac{F''(q_0;T_0)}{m(q_0)} \right)^{1/2} \quad \text{and} \quad \omega_{sd} = \left(-\frac{F''(q_{sd};T_{sd})}{m(q_{sd})} \right)^{1/2} \tag{5.143}$$

are the oscillator frequencies in the ground-state minimum and in the inverted osculating parabola at the saddle point, respectively, and the Kramers frequency is

$$h = \omega_K = \sqrt{\omega_{sd}^2 + \left(\frac{\eta(q_{sd})}{2m(q_{sd})} \right)^2} - \frac{\eta(q_{sd})}{2m(q_{sd})} . \tag{5.144}$$

Unfortunately, no distinction is usually made between T_0 and T_{sd}.

In Fig. 5.17 the horizontal full line is Kramers' rate in the two-dimensional case. The one-dimensional rate with the same potential, mass and friction parameters is shown by the dash-dotted line. The rate is seen to increase with an increasing number of degrees of freedom. The histogram comes from two-dimensional Langevin equations. It approaches the two-dimensional Kramers rate fairly well in this case. The rate formula (5.104) was derived by Fröbrich and Tillack [626] from a path-integral formulation of the problem. They obtained r_{Kramers} in the classical limit. They also got a rate formula in which quantum corrections are taken into account approximately. The result is shown in Fig. 5.17 by the dotted line. In their quantal calculation however no allowance was made for the zero-point energy in the fission degree of freedom ϵ_0, which would increase the barrier height by ϵ_0 and compensate partly the upwards shift shown in Fig. 5.17.

In view of the harmonic approximations of the free energy around the minimum and the saddle point, the corresponding shift of the upper and lower limits of the integral in Eq. (5.128) to $\pm\infty$, and the rather crude way of accounting for the shape dependence of the transport parameters and the temperature, it is not surprising that the limit of stationary rate, achieved in calculations with the Langevin equations, is not matched exactly by the Kramers rate formula with the same dimension and the same mass, friction, and free energy parameters. Deviations of up to 20% have been found [626].

Error estimates for the accuracy of Kramers' formula in one dimension were discussed by Edholm and Leimar [648] in terms of the higher derivatives of the free energy $[F'''(q)]^2/[F''(q)]^3$ and $F^{\text{iv}}(q)/[F''(q)]^2$ for $q = 0$ and $q = q_{\text{sd}}$. In the derivation of Kramers' formula it was assumed that the scission point lies sufficiently far to the right of the saddle point that the back-scattering process is only negligibly inhibited by scission [649]. It was also assumed that there is only one barrier. For decay over a multiple-barrier landscape Gontchar and Fröbrich [650] derived a rate formula under the condition that the minima and saddle points lie sufficiently far apart from each other so that the harmonic approximations around all extrema are justified.

In the limit of vanishing friction Kramers' frequency becomes in the one-dimensional case $h = \omega_K = \omega_{\text{sd}}$ and the rate is

$$\lim_{\eta \to 0} r_{\text{Kramers}} = \frac{\omega_0}{2\pi} e^{-F_{\text{barr}}/T} . \tag{5.145}$$

Comparison of this equation with the Bohr-Wheeler formula (4.9) shows that they differ by a factor $T/(\hbar\omega_0)$ – apart from the barrier F_{barr} in the exponential, which is a difference of free energies in Eq. (5.145) and a difference of potentials in Eq. (4.9). One could have obtained the same preexponential factor in the Bohr-Wheeler and the Kramers formulae if the level density of the initial system $\rho(E^*)$ in Eq. (4.5) would be written as a folding product of the (constant) level density of the fission degree of freedom $(\hbar\omega_0)^{-1}$ and the level density of the remaining system $\rho_{\text{int}} = d(e^S)/dE^*$

$$\rho(E^*) = \frac{1}{\hbar\omega_0} * \rho_{\text{int}} = \int_0^{E^*} \rho_{\text{int}}(E^* - \epsilon)d\epsilon/(\hbar\omega_0) \approx e^{S(E^*)}/(\hbar\omega_0).$$

It may appear natural to separate out the fission degree of freedom when calculating ρ_{int}. However, the form (4.9) is generally preferred with the argument that the quadrupole mode can be recognized spectroscopically as a collective mode at best for the three lowest quantum states whose amplitude does by far not reach to the saddle point. It would therefore not make sense to give this degree of freedom a special treatment in the level density of the compound state. This argument undermines of course the basis for the Langevin or Fokker-Planck treatment of fission altogether. The issue has been discussed controversially since the early days of the transition-state theory [382,649,651].

Approximate solution techniques for the Fokker-Planck equation

In early applications of the Fokker-Planck equation in nuclear fission theory the global moment-expansion technique was used, see e.g. the review article by Adeev and Pashkevich [637]) and work cited there. In this approach averaged coordinates and momenta are introduced

$$\overline{\mathbf{q}}(t) = \int \mathbf{q}\, w(\mathbf{q}, \mathbf{p}, t)\, d\mathbf{q}d\mathbf{p}, \quad \overline{\mathbf{p}}(t) = \int \mathbf{p}\, w(\mathbf{q}, \mathbf{p}, t)\, d\mathbf{q}d\mathbf{p} \qquad (5.146)$$

and the mass and friction tensors $m_{ij}(\mathbf{q})$ and $\eta_{ij}(\mathbf{q})$ in the Fokker-Planck equation (5.115) are replaced by their values at $\overline{\mathbf{q}}$, $m_{ij}(\overline{\mathbf{q}})$ and $\eta_{ij}(\overline{\mathbf{q}})$, respectively. One further approximates the free energy around $\overline{\mathbf{q}}$ by its quadratic expansion $\partial_{q^i} F(\mathbf{q}) \approx \partial_{q^i} F(\mathbf{q})|_{\mathbf{q}=\overline{\mathbf{q}}} + C_{ij}(\overline{\mathbf{q}})(q^j - \overline{q}^j)$ with $C_{ij} = \partial_{q^i}\partial_{q^j} F(\mathbf{q})|_{\mathbf{q}=\overline{\mathbf{q}}}$. For the sake of simplicity we will neglect the term $\partial_{q^i}(m^{-1})^{jk}$; see Ngô and Hofmann [652] for relaxing this latter assumption. The deterministic force (5.109) in the Fokker-Planck equation is then given by

$$K_i^{(\text{det})} = -\left[\partial_{q^i} F(\overline{\mathbf{q}}) + C_{ij}(\overline{\mathbf{q}})(q^j - \overline{q}^j) + \eta_{ij}(\overline{\mathbf{q}})(m^{-1})^{jk}p_k\right]. \qquad (5.147)$$

With these approximations the Fokker-Planck equation becomes linear in $\mathbf{q}$ and $\mathbf{p}$

$$\dot{w} = -(m^{-1})^{ij}p_j \frac{\partial w}{\partial q^i} + \left[\partial_{q^i} F + C_{ij}(q^j - \overline{q}^j)\right] \frac{\partial w}{\partial p_i}$$

$$+ \eta_{ij}(m^{-1})^{jk} \frac{\partial p_k w}{\partial p_i} + T\eta_{ij} \frac{\partial^2 w}{\partial p_i \partial p_j}. \qquad (5.148)$$

Multiplying this equation by $\mathbf{q}$ or $\mathbf{p}$ and integrating over all phase-space variables, one obtains, after partial integration, Newton's equations of motions for $\overline{\mathbf{q}}$ and $\overline{\mathbf{p}}$

$$\dot{\overline{q}}^i = (m^{-1})^{ij}\overline{p}_j$$
$$\dot{\overline{p}}_i = -\partial_{q^i} F - \eta_{ij}(m^{-1})^{jk}\overline{p}_k, \qquad (5.149)$$

where we assumed that w vanishes for $|\mathbf{q}| \to \infty$ and $|\mathbf{p}| \to \infty$ sufficiently fast.

For the three variance matrices

$$S_{q^i q^j} = \int (q^i - \overline{q}^i)(q^j - \overline{q}^j)\, w\, d\mathbf{q}d\mathbf{p}\,, \quad S_{p_i p_j} = \int (p_i - \overline{p}_i)(p_j - \overline{p}_j)\, w\, d\mathbf{q}d\mathbf{p}\,,$$

$$S_{q^i p_j} = \int (q^i - \overline{q}^i)(p_j - \overline{p}_j)\, w\, d\mathbf{q}d\mathbf{p}$$

a set of ordinary, coupled differential equations can be derived. Multiplying Eq. (5.148) with $(q^i - \overline{q}^i)(q^j - \overline{q}^j)$ and integrating over the whole phase space one obtains $\dot{S}_{q^i q^j}$. Similarly the time derivatives of the other two variances are calculated. The result is

$$\dot{S}_{q^i q^j} = (m^{-1})^{il} S_{q^j p_l} + (m^{-1})^{jl} S_{q^i p_l}$$
$$\dot{S}_{q^i p_j} = (m^{-1})^{il} S_{p_l p_j} - C_{jl} S_{q^l q^i} - (m^{-1})^{lk} \eta_{lj} S_{q^i p_k} \qquad (5.150)$$
$$\dot{S}_{p_i p_j} = -C_{il} S_{q^l p_j} - C_{jl} S_{q^l p_i} - (m^{-1})^{lk} \big[\eta_{il} S_{p_k p_j} + \eta_{jl} S_{p_i p_k} \big]$$
$$\qquad + \eta_{ij} T\,.$$

Observables O are typically functions of the phase space variables $O = f(\mathbf{q}, \mathbf{p})$. Their mean value and variance are obtained by

$$\overline{O}(t) = \int f(\mathbf{q}, \mathbf{q}) w(\mathbf{p}, \mathbf{q}; t)\, d\mathbf{q}\, d\mathbf{p}$$

and

$$\Delta O = \int \big[f(\mathbf{q}, \mathbf{p}) - \overline{O} \big]^2 w(\mathbf{q}, \mathbf{p}; t)\, d\mathbf{q}\, d\mathbf{p}$$

with

$$w(\mathbf{q}, \mathbf{p}; t) = \pi^{-n} \left(\det \left| \begin{matrix} S_{q^i q^j} & S_{q^i p_j} \\ S_{q^j p_i} & S_{p_i p_j} \end{matrix} \right| \right)^{1/2} \exp \Big\{ - S_{q^i q^j}(q^i - \overline{q}^i)(q^j - \overline{q}^j)$$
$$- 2 S_{q^i p_j}(q^i - \overline{q}^i)(p_j - \overline{p}_j) - S_{p_i p_j}(p_i - \overline{p}_i)(p_j - \overline{p}_j) \Big\}\,.$$

The scheme was first applied to the asymmetry degree of freedom of strongly necked-in configurations in quasifission reactions [625, 652–654] and used by Pomorski and Hofmann [655] and Adeev, Gontchar et al. [656,657] to describe fission dynamics.

For diffusion problems involving only one collective variable there are several additional approaches to solve the Fokker-Planck equation numerically. The simplest procedure, for which library routines are available [658], is to solve it on a grid in phase space and time. This was done for example in Refs. [647,659,660]. Scheuter and Hofmann devised a solution technique, which they called propagator method [661]. It is constructed to give more accurate results than the global moment-expansion method. A variety of solution methods are also described in Ref. [590], Chaps. 5, 6, and in particular 10, including expansions of the distribution function in eigenfunctions of Hermite operators related to the Fokker-Planck operator on the r.h.s. of Eq. (5.115). Most of these methods can in principle be used also for multidimensional systems. However, for more than one-dimensional problems it appears that it is easier to solve the equivalent Langevin equations.

5.2.3 Werner-Wheeler approximation of the hydrodynamical inertial tensor

In nonadiabatic fission studies the inertial tensor has mostly been calculated in the hydrodynamical model. In this model the inertia of fission dynamics is identified with the inertia of an irrotational, incompressible liquid drop. In terms of the classical velocity field $\mathbf{v}(\mathbf{r})$ and the (constant) mass density ρ_m the kinetic energy is

$$E_{\text{kin}} = \frac{1}{2}\rho_m \int \mathbf{v}^2(\mathbf{r})d^3r\,, \qquad (5.151)$$

which we want to write in the form (5.83) in terms of the collective velocities $\dot{q}_i$. The velocity field of an ideal liquid can be derived from a velocity potential ϕ, which satisfies the Laplace equation with Dirichlet boundary conditions on the nuclear surface

$$\left.\begin{array}{l} \Delta\phi = 0 \\ \partial_n\phi = \dot{n}(\mathbf{r};\mathbf{q}) = \dot{q}^i\partial_{q^i}n(\mathbf{q}), \end{array}\right\} \qquad (5.152)$$

where $n(\mathbf{r};\mathbf{q})$ is the displacement field of the boundary in the direction of the surface normal. The velocity field is therefore a linear function of the collective velocities $\dot{q}^i$ and the kinetic energy becomes a quadratic form of the collective velocities $E_{\text{kin}} = (1/2)m_{ij}\dot{q}^i\dot{q}^j$. According to a theorem by Kelvin (see Ref. [662] Sec. 45) the flow field of an ideal liquid leads to the smallest possible kinetic energy, connected with given boundary velocities $\dot{\mathbf{q}}$. One obtains therefore a lower limit for the inertia in this hydrodynamical model.

Exact, analytical solutions of the boundary value problem are only known in rather restricted shape classes, see for instance Ref. [662], Sec. 382 or Refs. [30, 663, 664]. Also expansions in terms of complete sets of basis functions have been used in special cases [665]. The velocity field is however almost always calculated in the Werner-Wheeler approximation [82, 666]. For axially symmetric shapes the velocity has a radial component v_ρ and a component in the axial direction v_z. The Werner-Wheeler approximation consists in the assumption that v_z does not depend on the radial coordinate ρ and can be represented by a vector field $a_i(z;\mathbf{q})$

$$v_z = a_i(z;\mathbf{q})\,\dot{q}^i \qquad (5.153)$$

and that v_ρ is proportional to ρ and can be represented by a second vector field $b_i(z;\mathbf{q})$

$$v_\rho = (\rho/\rho_{\text{surf}})\,b_i(z;\mathbf{q})\,\dot{q}^i\,, \qquad (5.154)$$

where $\rho_{\text{surf}}(z;\mathbf{q})$ is the value of ρ on the nuclear surface at position z. Since the velocity field shall be sourceless, $\partial_\rho(\rho v_\rho) + \rho\partial_z v_z = 0$, one obtains

$$b_i = -(1/2)\rho_{\text{surf}}\, \partial_z a_i \, . \qquad (5.155)$$

Insertion of this result into Eq. (5.151) leads to the expression

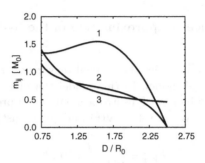

Fig. 5.23. Werner-Wheeler mass-tensor components $m_{r_{\text{neck}} r_{\text{neck}}}$, curve 1, $m_{r_{\text{neck}} D}$, curve 2, and m_{DD}, curve 3 for symmetric fission as function of the center-of-mass distance D of the two halves of the system and the neck radius r_{neck}. The calculation was based on the Cassinian ovaloid shape-parametrization with parameters ϵ and α_4, which were mapped on the variables D and r_{neck}. Lengths are measured in units of the radius R_0 of the spherical compound nucleus, masses in units of the total mass (after Ref. [280]).

$$m_{ij} = \pi \rho_m \int_{z_{\min}}^{z_{\max}} \left(a_i a_j + \frac{1}{8}\rho_{\text{surf}}^2\, a_i' a_j' \right) \rho_{\text{surf}}^2(z; \mathbf{q})\, dz \qquad (5.156)$$

for the inertial tensor, where $a_i' = \partial_z a_i$. If the center-of-mass coordinate $z_{\text{c.m.}}$ changes in time in the chosen shape parametrization, the spurious velocity

$$\dot{z}_{\text{c.m.}} = \pi \frac{\rho_m}{M_{\text{total}}}\, \dot{q}^i \partial_{q^i} \int_{z_{\min}}^{z_{\max}} z \rho_{\text{surf}}^2(z; \mathbf{q}) dz$$

has to be subtracted from v_z in Eq. (5.153) before inserting a_i into Eq. (5.156).

The Werner-Wheeler approximation implies that a slice of the fluid enclosed between two planes perpendicular to the axis keeps its volume and retains plane surfaces while moving. We define in particular the volumes

$$V^+(z; \mathbf{q}) = \pi \int_{z}^{z_{\max}} \rho_{\text{surf}}^2(z'; \mathbf{q})\, dz' \qquad (5.157)$$

and

$$V^-(z; \mathbf{q}) = \pi \int_{z_{\min}}^{z} \rho_{\text{surf}}^2(z'; \mathbf{q})\, dz' \, , \qquad (5.158)$$

whose total (convective) time derivatives vanish

$$\frac{d}{dt}V^{\pm}(z;\mathbf{q}) = \partial_z V^{\pm}\dot{z} + \partial_{q^i}V^{\pm}\dot{q}^i = 0\,. \tag{5.159}$$

Inserting Eq. (5.157) yields

$$\dot{z} = \frac{1}{\rho_{\text{surf}}^2(z;\mathbf{q})}\,\dot{q}^i\partial_{q^i}\int_z^{z_{\max}}\rho_{\text{surf}}^2(z';\mathbf{q})\,dz'$$

and with Eq. (5.153) one obtains an explicit expression for the vector $a_i(z;\mathbf{q})$ in terms of the shape function $\rho_{\text{surf}}(z;\mathbf{q})$

$$a_i(z;\mathbf{q}) = \frac{1}{\rho_{\text{surf}}^2(z;\mathbf{q})}\,\partial_{q^i}\int_z^{z_{\max}}\rho_{\text{surf}}^2(z';\mathbf{q})\,dz'\,. \tag{5.160}$$

A similar result can be derived with the use of $V^-(z;\mathbf{q})$.

The Werner-Wheeler field is incompressible, but in general not irrotational. The kinetic energy $E_{\text{kin}} = (1/2)m_{ij}\dot{q}^i\dot{q}^j$ is therefore equal to or larger than the kinetic energy of an ideal fluid. Only for spheroidal deformations of arbitrary eccentricity and small-amplitude quadrupole oscillations is the Werner-Wheeler velocity irrotational [82]. It should however be stressed that in some important limiting cases the Werner-Wheeler mass tensor has very desirable properties: It is translational invariant, it becomes infinite for the asymmetry degree-of-freedom when the neck radius vanishes [446], since the flux between the emerging fragments must vanish in the hydrodynamical model when the neck vanishes. It yields the reduced mass in the distance coordinate $r_{\text{c.m.}}$ between the centers of mass of the fragments beyond scission and the inertia connected with the neck radius ρ_{neck} approaches zero when the neck vanishes as illustrated in Fig. 5.23.

Pomorski and Bartel generalized the Werner-Wheeler approximation to nonaxial nuclei of the shape class (2.220) [275]. They retain the ansatz (5.153) and introduce instead of Eq. (5.154) for the x and y components of the velocity field the ansatz

$$v_x = x[\alpha_i(z;\mathbf{q}) + \beta_i(z;\mathbf{q})]\dot{q}^i$$
$$v_y = y[\alpha_i(z;\mathbf{q}) - \beta_i(z;\mathbf{q})]\dot{q}^i\,.$$

Vanishing of the divergence of the velocity field leads to the equation $\alpha_i(z;\mathbf{q}) = -(1/2)\partial_z a_i(z;\mathbf{q})$ instead of Eq. (5.155). The quantity β_i is derived from the condition that the x and y components of the velocity field at the surface should coincide with the surface velocity, expressed by $\dot{\mathbf{q}}$. The condition (5.159) determines again $a_i(z;\mathbf{q})$ as function of the surface shape. In the coordinates (2.220) one obtains

$$a_i(z;\mathbf{q}) = -\frac{1}{\tilde{\rho}_{\text{surf}}^2(z;\mathbf{q})}\int_z^{z_{max}}\frac{\partial\tilde{\rho}_{\text{surf}}^2(z';\mathbf{q})}{\partial q^i}dz'\,, \tag{5.161}$$

where $\tilde{\rho}_{\text{surf}}^2$ is defined in Eq. (2.221). In terms of the quantity

$$A_i^\rho(z;\mathbf{q}) = \frac{1}{2}\tilde{\rho}_{\text{surf}}^2 \left(\frac{\partial \tilde{\rho}_{\text{surf}}^2}{\partial q^i} + \frac{\partial \tilde{\rho}_{\text{surf}}^2}{\partial z} a_i \right)$$

the mass tensor becomes

$$m_{ij} = \pi \rho_m \begin{cases} \displaystyle\int_{z_{min}}^{z_{max}} \tilde{\rho}_{\text{surf}}^2 \left(\frac{1}{2(1-\eta^2)^{1/2}} A_i^\rho A_j^\rho + a_i a_j \right) dz \ , & q^i, q^j \in \{c,h,\alpha\}, \\[4mm] \dfrac{\eta}{4(1-\eta^2)^{3/2}} \displaystyle\int_{z_{min}}^{z_{max}} A_j^\rho \tilde{\rho}_{\text{surf}}^4 \, dz \ , & q^i = \eta, \ q^j \in \{c,h,\alpha\}, \\[4mm] \dfrac{1}{8(1-\eta^2)^{3/2}} \displaystyle\int_{z_{min}}^{z_{max}} \tilde{\rho}_{\text{surf}}^4 \, dz \ , & q^i = q^j = \eta \, . \end{cases}$$

Inertial parameters for adiabatic, collective motion, as discussed in Secs. 5.1.1 and 5.1.2 can be larger than the liquid-drop values by factors up to 5. The following qualitative reason has been given for this discrepancy [87]: The last occupied single-particle orbits contribute to the total density a pattern of nodal lines and nodal surfaces which undergoes rapid changes as function of the deformation. These changes are connected with relatively large mass transport when the deformation parameters change only by small amounts, which leads to rapidly changing, large contributions to the collective inertia. Even though these fluctuations of the inertia are smeared out by the residual interactions, in particular of the pairing type, a large contribution remains.

5.2.4 The friction tensor

A variety of mechanisms have been proposed to describe the coupling of the collective degrees of freedom to the "intrinsic" system, leading to widely different friction tensors. Historically the oldest approach is to treat an excited nucleus as a classical, viscous liquid drop. The microscopic basis for the dynamical properties of a classical fluid is the picture of interacting molecules, whose mean free path is very short compared to the length scale of the flow field. However, in the range of nuclear excitation energies, we consider in this book, the mean-free-path of nucleons is larger than the radius of a fissile nucleus. This was already stressed by Hill and Wheeler [87]. They therefore surmised that fission motion is organized by the interaction of the surface with almost free nucleons. In classical, macroscopic physics this behavior is characteristic of a Knudsen gas [667]. Groß [668] considered a Knudsen gas in a cylinder, closed by a movable piston and calculated the energy exchange of the gas with the piston as function of the velocity of the piston.

Extending the idea of the piston model to a more realistic geometry, Swiatecki and his collaborators [669] considered a classical Knudson gas enclosed in the nuclear surface, which changes its shape along the fission path, transferring energy to the gas. In this model the surface acts as an agent moved from outside as in the cranking model, discussed in Sec. 5.1.1.

Several variants of particle-hole excitation models have also been proposed, where the energy exchange between the time-dependent mean field and the gas of quasiparticles is studied [670–672]. They differ in the assumed mechanism by which the energy, taken out of the collective motion, is eventually thermalized. All friction models, however, are of the cranking type, i.e. are lacking selfconsistency on the microscopic level.

Some authors assume that the width of giant resonances is due to the same damping mechanism as seen in fission [673–675]. It should however be stressed that giant resonances get their collective character by the residual two-particle interaction, rather than the surface of the single-particle potential as in fission [676, 677]. It is therefore not at all certain that the two types of collective modes have the same damping mechanism.

Hydrodynamical viscosity

In a laminar flow of a viscous, incompressible fluid energy is converted from the flow field into heat at a rate, given by Rayleigh's dissipation function F (see e.g. Ref. [662] §329)

$$-\dot{E} = 2F = \eta_0 \int \Phi(\mathbf{r}) d^3 r$$

with the viscosity constant η_0 and

$$\Phi(\mathbf{r}) = \sum_{ij} \partial_{x_i} v_j(\mathbf{r}) [\partial_{x_i} v_j(\mathbf{r}) + \partial_{x_j} v_i(\mathbf{r})]$$

in terms of the Cartesian components v_i of the velocity field and Cartesian coordinates x_i. This expression can be rewritten in a coordinate-independent form [82]

$$\Phi = \nabla^2 \mathbf{v}^2 + (\mathrm{rot}\,\mathbf{v})^2 - 2\mathrm{div}\,(\mathbf{v} \times \mathrm{rot}\,\mathbf{v}). \qquad (5.162)$$

Inserting the Werner-Wheeler expression (5.153)-(5.155) for the velocity field into Eq. (5.162) gives in cylindrical coordinates [82]

$$-\dot{E} = \eta_{ij} \dot{q}^i \dot{q}^j$$

$$= \eta_0 \pi \int_{z_{\min}}^{z_{\max}} \left[3a_i' a_j' + \frac{1}{8}\rho^2(z;\mathbf{q}) a_i'' a_j'' \right] \rho^2(z;\mathbf{q}) dz\, \dot{q}^i \dot{q}^j. \qquad (5.163)$$

in terms of the first and second derivatives of the field $a_i(z;\mathbf{q})$, Eq. (5.160), with respect to z. For a flow field which satisfies Eqs. (5.151) the volume integral over $\Phi(\mathbf{r})$ can be converted into a surface integral

$$\dot{E} = -\eta_0 \int (\mathrm{grad}\,v^2) \cdot d\boldsymbol{\sigma}.$$

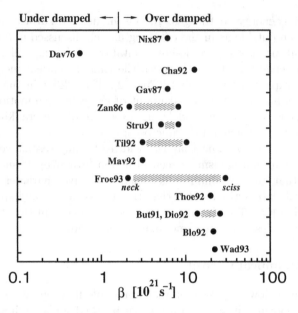

Fig. 5.24. Experimentally determined reduced dissipation coefficients $\beta = \eta_{ij}(m^{-1})^{ij}$. The fits are made in the framework of one-body and two-body friction models. The references are: Nix87 [678], Dav76 [82], Cha92 [679], Gav87 [615], Zan86 [613], Stru91 [646], Til92 [282], Mav92 [633], Froe93 [593], Thoe92 [680], But91 [681], Dio92 [682], Blo92 [683], Wad93 [684] (after Ref. [685]).

In order to obtain the correct velocity field for a viscous, classical fluid one would have to solve at least approximately the (linearized) Navier-Stokes equations with a free surface (see e.g. Hasse [686]). However, we do not even have a classical fluid with a short mean-free-path length. That the expression (5.163) was nevertheless so widely used seems to be due to the convenient employment of the parameter η_0 as a fudge factor to fit data. Unfortunately there is a tendency to fit some set of data by just one "dissipation parameter", say $\beta = \eta_{ij}(m^{-1})^{ij}$. Not surprisingly, widely differing values of this parameter were obtained with different sets of data and different additional assumptions in the dynamical models, used in these fits. A collection of examples is shown in Fig. 5.24. Note the logarithmic scale on the abscissa of the figure.

One-body friction, classical approach

The wall formula

For compact shapes the one-body dissipation mechanism of the "wall" formula was proposed in Ref. [669] by working out in detail the concept of the nuclear Doppler effect, introduced by Hill and Wheeler [87]: Assume a Knudson gas with an isotropic velocity distribution, characterized by a distribution function

$\rho_0 f(v)$, where ρ_0 is the (number) density and f is the normalized velocity distribution, $\int f(v)d^3v = 1$. The number of particles in the velocity interval v_z, $v_z + dv_z$ is given by

$$g(v_z)dv_z = \rho_0 dv_z \int f(v)dv_\phi v_\rho dv_\rho.$$

Introducing $v = \sqrt{v_z^2 + v_\rho^2}$ as new integration variable instead of v_ρ, yields

$$g(v_z) = 2\pi\rho_0 \int_{v_z}^{\infty} vf(v)dv.$$

Therefore

$$dg(v_z)/dv_z = -2\pi\rho_0 v_z f(v_z). \tag{5.164}$$

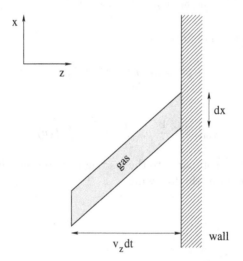

Fig. 5.25. Geometry of the wall friction.

All particles which hit the surface element $dxdy$ of a plane, fixed wall in the x,y-plane during the time interval dt lie in the oblique column with the ground surface $dxdy$ and height $v_z dt$, as sketched in Fig. 5.25. The number of particles in the column is $dN_0 = v_z dt dx dy g(v_z)$. When the wall moves with the velocity $\dot{n}$ with respect to the bulk of the gas in the positive z direction, this number becomes

$$dN = dxdydt(v_z - \dot{n})g(v_z). \tag{5.165}$$

Each particle transfers the momentum $2M_{\mathrm{nucl}}(v_z - \dot{n})$ during the collision if the wall is ideally elastic and the total momentum transfer during dt on the surface element $dxdy$ of the wall is $2M_{\mathrm{nucl}}\, dxdydt \int dv_z(v_z - \dot{n})^2 g(v_z)$. The momentum transfer per time and per surface element is the pressure p

$$p = 2M_{\mathrm{nucl}} \int_{\dot{n}}^{\infty} (v_z - \dot{n})^2 g(v_z)dv_z \, .$$

We perform a partial integration (assuming that $g(v_z)$ falls off sufficiently fast for large v_z so that $\lim_{v_z \to \infty} v_z^3 g(v_z) = 0$) and insert the relation (5.164). The pressure then becomes

$$p = \frac{4\pi}{3}\rho_m \int_{\dot{n}}^{\infty} (v_z - \dot{n})^3 v_z f(v_z)dv_z \, ,$$

where $\rho_m = M_{\mathrm{nucl}}\, \rho_0$ is the mass density. This expression is rewritten as a series in increasing powers of $\dot{n}$

$$p = \frac{1}{3}\rho_m \overline{v^2} - \rho_m \overline{v}\dot{n} + \rho_m \dot{n}^2 - \frac{1}{3}\rho_m (\overline{1/v})\dot{n}^3 + p_{\mathrm{corr}} \, , \qquad (5.166)$$

where the averages are defined by

$$\overline{v^n} = 4\pi \int_0^{\infty} v^{n+2} f(v)dv \qquad (5.167)$$

and

$$p_{\mathrm{corr}} = -\frac{4\pi}{3}\rho_m \int_0^{\dot{n}} (v - \dot{n})^3 v f(v)dv$$

is of the order $(\dot{n}/v)^5$ compared to the leading term on the r.h.s. of Eq. (5.166) and will be neglected in the following.

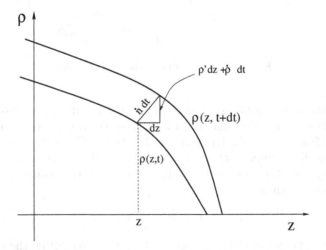

Fig. 5.26. Geometrical relations used for the derivation of Eq. (5.170)

When the wall is displaced by δn in the normal direction, the energy transferred from the surface element $dx\,dy$ to the gas is $\delta E = -p\delta n\,dx\,dy$. Dividing by δt and integrating over the whole surface of the nucleus yields the energy transfer from the nuclear surface to the gas per unit time

$$\dot{E} = -\oint p\dot{n}\,d\sigma\,. \tag{5.168}$$

Insertion of the expansion (5.166) of p into this equation gives

$$\dot{E} = -\frac{1}{3}\rho_m\overline{v^2}\oint\frac{\delta n d\sigma}{\delta t} + \rho_m\overline{v}\oint \dot{n}^2 d\sigma + \mathcal{O}(\dot{n}^3/v^3)\,.$$

For volume-conserving deformations δn the integral in the first term on the r.h.s. of this equation vanishes. The wall formula has therefore the simple form

$$\dot{E} = \rho_m\overline{v}\oint \dot{n}^2 d\sigma = \rho_m\overline{v}\oint \frac{dn}{dq^i}\frac{dn}{dq^j}\,d\sigma\,\dot{q}^i\dot{q}^j\,. \tag{5.169}$$

Implicitly the assumption is made that the wall is rigid in the sense that it can compensate a local momentum transfer on one side of the surface with the same momentum transfer on the opposite side without being deformed.

For an axially symmetric shape, given by a shape function $\rho(z;\mathbf{q})$ in cylindrical coordinates, $\dot{n}$ becomes

$$\dot{n} = \mathbf{v}_s\cdot\hat{n} = \dot{\rho}\left[1 + \left(\frac{\partial\rho}{\partial z}\right)^2\right]^{-1/2} \tag{5.170}$$

where $\mathbf{v}_s$ is the velocity of the surface and $\hat{n}$ the unit vector in the direction of the surface normal, pointing outwards. To see this one can obtain from Fig. 5.26 the relation

$$\dot{n}dt = dz\sqrt{\left(\dot{\rho}\frac{dt}{dz} + \rho'\right)^2 + 1}\,, \tag{5.171}$$

where $\dot{\rho}$ is the time derivative of ρ and $\rho' = \partial\rho/\partial z$. The slope of the surface normal with respect to the z-axis is $-1/\rho'$. One therefore gets the equation

$$\rho'dz + \dot{\rho}dt = -\frac{1}{\rho'}dz\,,$$

from which follows

$$\dot{\rho}\frac{dt}{dz} = -\frac{1}{\rho'}(1 + \rho'^2)\,.$$

Insertion into Eq. (5.171) yields Eq. (5.170). It is convenient to rewrite the latter equation in the form

$$\dot{n} = \dot{q}^i\rho\frac{\partial\rho}{\partial q^i}\left[\rho^2 + \left(\rho\frac{\partial\rho}{\partial z}\right)^2\right]^{-1/2}\,,$$

Using this result the friction tensor η_{ij} becomes [687]

$$\eta_{ij}^{(\text{wall})} = \rho_m \bar{v} \oint \frac{dn}{dq^i} \frac{dn}{dq^j} d\sigma$$

$$= \frac{\pi}{2} \rho_m \bar{v} \int_{z_{\min}}^{z_{\max}} dz \frac{\partial(\rho^2)}{\partial q^i} \frac{\partial(\rho^2)}{\partial q^j} \left[\rho^2 + \frac{1}{4} \left(\frac{\partial(\rho^2)}{\partial z} \right)^2 \right]^{-1/2}. \quad (5.172)$$

The average $\bar{v}$ can be calculated in the grand canonical ensemble with the temperature $T = 1/\beta$ and the chemical potential $\mu \approx \epsilon_f = (1/2) M_{\text{nucl}}^* v_f^2$ if $\mu\beta \gg 1$, where ϵ_f and v_f are the Fermi energy and the Fermi velocity, respectively,

$$\bar{v} = \frac{4\pi \int v^3 [1 + \exp(\epsilon - \mu)\beta]^{-1} dv}{4\pi \int v^2 [1 + \exp(\epsilon - \mu)\beta]^{-1} dv}.$$

Using the Sommerfeld expansion (3.8) and writing $\epsilon = (1/2) M_{\text{nucl}}^* v^2$, the numerator and the denominator can be expanded in powers of $1/(\mu\beta)$. Up to second order in the expansion parameter one obtains

$$\bar{v} = \frac{3}{4} v_f \left[1 + \frac{1}{12} \left(\frac{\pi}{\epsilon_f \beta} \right)^2 \right]. \quad (5.173)$$

The temperature-dependent term is small and neglected in applications of the wall formula. In Thomas-Fermi approximation the factor in front of the integral in the wall formula (5.169) can therefore be written

$$\rho_m \bar{v} = \frac{\hbar k_f^4}{2\pi^2} \quad (5.174)$$

in terms of the Fermi wave number $k_f = v_f/\hbar$. In Ref. [688] the numerical value 1.026×10^{-22} MeV s fm^{-4} was suggested for the constant (5.174).

From the derivation of the wall formula it is clear that $\dot{n}$ is the velocity of the surface with respect to the bulk of the gas, assumed to be in thermal equilibrium. When applied to a translating or rotating nucleus, $\dot{n}$ has to be measured in the co-moving, co-rotating frame of reference. For configurations close to scission it appears natural to associate the average velocity of the particles in each of the two nascent fragments with the center-of-mass velocity of the fragment $\dot{D}_\nu$, $\nu = L, R$. The friction tensor then becomes [621]

$$\eta_{ij}^{(\text{wall 2})} = \frac{\pi}{2} \rho_m \bar{v} \left(\int_{z_{\min}}^{z_N} I_L(z) dz + \int_{z_N}^{z_{\max}} I_R(z) dz \right), \quad (5.175)$$

where

$$I_\nu = \left(\frac{\partial(\rho^2)}{\partial q^i} + \frac{\partial(\rho^2)}{\partial z} \frac{\partial D_\nu}{\partial q^i} \right) \left(\frac{\partial(\rho^2)}{\partial q^j} + \frac{\partial(\rho^2)}{\partial z} \frac{\partial D_\nu}{\partial q^j} \right) \left(\rho^2 + \left[\frac{\partial(\rho^2)}{2\partial z} \right]^2 \right)^{-1/2}$$

with the smallest neck radius at z_N and D_L and D_R are the centers-of-mass distances of the nuclear volume to the left and to the right of z_N, respectively.

The assumption of a reflection of particles on a sharp surface would require an infinite rectangular potential box. Since actual single-particle potentials are smooth and finite, one expects that the momentum transfer is smeared out over a surface layer, whose thickness is of the order of the surface thickness. Randrup and Swiatecki [688] therefore introduced a surface-peaked function $Y_{\mathrm{fric}}(\mathbf{r})$ in close analogy to the surface-peaked function $g(\mathbf{r}) = \mathcal{E}(\mathbf{r}) - c_1\rho(\mathbf{r})$ introduced in Sec. 2.2.1. The one-dimensional integral over Y_{fric} in the direction of the surface normal and the higher surface moments are assumed to be independent of the nuclear shape. The leptodermous expansion (2.128) then leads to the expansion

$$\dot{E} = \int Y_{\mathrm{fric}}(\mathbf{r})d^3r = \int Y_{\mathrm{fric}}(z)dz \cdot \int d\sigma + \int Y_{\mathrm{fric}}(z)zdz \cdot \int \kappa d\sigma$$

+terms involving higher moments of Y_{fric},   (5.176)

where z is in the direction of the surface normal and κ is twice the mean curvature.

The window formula

In analogy to the proximity potential, which had to be added to the leptodermous expansion terms of the mass formula, also a proximity friction was postulated in Ref. [688]. It would account for the momentum transfer between the potential wall of one of the nascent fragments and the particles of the other fragment around the scission configuration. This mechanism should be particularly important in the entrance channel of a heavy-ion fusion reaction. Unfortunately, the idea was no longer pursued in the framework of the wall formula. Only in the linear response treatment of friction was it considered in Refs. [624, 689, 690].

Instead, an older suggestion by Swiatecki [691] was further developed, in Ref. [669] which lead to the "window" formula: Assume a nuclear shape in the vicinity of the scission configuration, where the two nascent fragments A and B have already acquired a relative center-of-mass velocity $\dot{\mathbf{D}}$. There may still be a hole of size $\Delta\sigma$ in the single-particle potential of the neck region through which particles can cross from A to B and carry their momentum and angular momentum from A to B and vice versa. In the average this leads to a braking of the dynamics in the $\mathbf{D}$-degree of freedom if the transfered momentum is thermally equilibrated in the new environment of the transfered particle. We will assume this to be the case and postpone a discussion of the justification to the end of this section.

For a quantitative discussion of the effect we use a coordinate system in which the bulk of the gas in the container A is at rest. The window shall be in the x, y plane and the z direction, perpendicular to the window, shall point from A to B. The change of momentum of the gas in the container A per unit time has three contributions:

- The first contribution comes from particles which leave A. As in the case of the wall formula there are $dN = \Delta\sigma dt(v_z - \dot{n})g(v_z)$ particles reaching the hole of size $\Delta\sigma$ during time dt. Each particle carries the momentum $M_{\text{nucl}}\mathbf{v}$ out of A. After averaging in the x, y-plane, only the z-component v_z of $\mathbf{v}$ remains. This leads to a drag force in the $-z$ direction

$$F_{AB} = -\int_{\dot{n}}^{\infty} \frac{dN}{dt} M_{\text{nucl}} v_z dv_z$$

$$= -M_{\text{nucl}}\Delta\sigma \int_{\dot{n}}^{\infty} (v_z - \dot{n}) v_z g(v_z) dv_z . \qquad (5.177)$$

Partial integration and the use of Eq. (5.164) yields

$$F_{AB} = -2\pi\rho_m\Delta\sigma \int_{\dot{n}}^{\infty} \left(\frac{1}{3}v_z^3 - \frac{\dot{n}}{2}v_z^2\right) v_z f(v_z) dv_z .$$

Neglecting terms of order $(\dot{n}/v)^3$, the lower limit of the integral can be moved to 0. Introducing the moments $\overline{v^n}$, defined in Eq. (5.167), one finds

$$F_{AB} = -\frac{1}{2}\rho_m\Delta\sigma \left(\frac{1}{3}\overline{v^2} - \frac{\dot{n}}{2}\overline{v}\right) . \qquad (5.178)$$

- The second contribution is connected with the flow of particles from B to A. To calculate the number of particles which cross the window from B to A, we introduce the velocity $\mathbf{v}'$ with respect to the moving frame of B. The number of particles which reach the hole during time dt from within B is $dN = dt\Delta\sigma(v_z' + \dot{n} - \dot{D}_z)g(v_z')$, where $\dot{D}_z$ is the z-component of the relative velocity $\dot{\mathbf{D}}$ of B with respect to A. Each particle transfers a momentum $M_{\text{nucl}}(-v_z'\hat{z} + \dot{\mathbf{D}}) = M_{\text{nucl}}[(\dot{D}_z - v_z')\hat{z} + \dot{D}_\rho\hat{\rho}]$, where $\hat{z}$ is the unit vector in the z-direction and $\hat{\rho}$ the unit vector in the direction of the component of $\dot{\mathbf{D}}$ perpendicular to $\hat{z}$. The force on A connected with the momentum transfer from B to A is

$$\mathbf{F}_{BA} = M_{\text{nucl}}\Delta\sigma \int (v_z' + \dot{n} - \dot{D}_z)(-v_z'\hat{z} + \dot{D}_z\hat{z} + \dot{D}_\rho\hat{\rho})g(v_z')dv_z'$$

Using the same transformations which lead from Eq. (5.177) to Eq. (5.178), neglecting quadratic terms in dn, dD_z, and dD_ρ, one finds

$$\mathbf{F}_{BA} = \frac{1}{2}\rho_m\Delta\sigma \left[-\frac{1}{3}\overline{v^2}\hat{z} + \frac{1}{2}\overline{v}[(2\dot{D}_z - \dot{n})\hat{z} + \dot{D}_\rho\hat{\rho}]\right] . \qquad (5.179)$$

- There is a net force on the gas resulting from collisions with the open surface $\Sigma_A - \Delta\sigma$ of the container A

$$\mathbf{F}_{\text{surf}} = -\int_{\Sigma_A - \Delta\sigma} p_{\text{stat}} d\boldsymbol{\sigma} \qquad (5.180)$$

with the static pressure $p_{\text{stat}} = (1/3)\rho_m\overline{v^2}$ (see e.g. Eq. (5.166)).

Adding the contributions (5.178)-(5.180) one obtains a total drag force

$$\mathbf{F}_A = \frac{\rho_m}{2} \left[-\frac{2\overline{v^2}}{3} \left(\int_{\Sigma_A - \Delta\sigma} d\boldsymbol{\sigma} + \Delta\sigma\hat{z} \right) + \Delta\sigma\overline{v} \left(\dot{D}_z\hat{z} + \frac{\dot{D}_\rho}{2}\hat{\rho} \right) \right].$$

The first term in the bracket vanishes since it amounts to an integral of the momentum transfer between gas and container A over its whole surface. The second term gives rise to a rate of change of the energy of the gas in A because of the presence of B

$$\dot{E} = \mathbf{F}_A\dot{\mathbf{D}} = \frac{1}{2}\Delta\sigma\rho_m\overline{v} \left(\dot{D}_z^2 + \frac{1}{2}\dot{D}_\rho^2 \right)$$

$$= \frac{1}{2}\Delta\sigma\rho_m\overline{v} \left(\frac{\partial D_z}{\partial q^i}\frac{\partial D_z}{\partial q^j} + \frac{1}{2}\frac{\partial D_\rho}{\partial q^i}\frac{\partial D_\rho}{\partial q^j} \right) \dot{q}^i\dot{q}^j. \quad (5.181)$$

In applications to heavy-ion fusion and deep-inelastic reactions the z-component of $\mathbf{F}_A$ contributes to the radial friction force, the ρ-component to the tangential friction.

We have so far assumed that the potential is a rectangular box and the mean one-sided flux of nucleons $n_0 = \rho_m\overline{v}$ through the window is that in bulk matter. In order to take the more realistic diffuse surface of the potential into account, Randrup [692] first considered two juxtaposed plane surfaces with a distance s between half density surfaces and calculates the local flux $n(s)$ in Thomas-Fermi approximation (without gradient terms) with the Seyler-Blanchard two-body interaction [179]. Using the same argument which leads from the interaction per unit area $e(s)$ to the proximity force $V(s)$ between two spherical nuclei with half-density radii R_1 and R_2 in Eq. (1.8) one obtains instead of the factor $\Delta\sigma\rho_m\overline{v} = \Delta\sigma n_0$ in Eq. (5.181) the expression

$$n_0\Delta\sigma \rightarrow \int n\,d\sigma \approx 2\pi\overline{R}\int_s^\infty n(s')ds',$$

where $n(s)$ plays the role of $e(s)$ in Eq. (1.8), $\overline{R} = R_1R_2/(R_1 + R_2)$ is the reduced radius, and $s = D - R_1 - R_2$ is the closest distance between the two surfaces. Randrup tabulates the dimensionless functions $\psi(\zeta) = n(b\zeta)/n_0$ of the dimensionless variable $\zeta = s/b$, where b is the surface diffuseness and the integral

$$\Psi(\zeta) = \int_\zeta^\infty \psi(\zeta')d\zeta'.$$

In terms of the latter function one has

$$n_0\Delta\sigma \rightarrow 2\pi\overline{R}n_0b\Psi(s/b), \quad (5.182)$$

which is to be inserted into Eq. (5.181). The standard value for the diffuseness b is 1 fm. Randrup also gives a closed-form approximation for $\Psi(\zeta)$

$$\Psi(\zeta) = \begin{cases} 1.4 - \zeta & \text{for } \zeta \leq -0.4 \\ 1.6 - 0.5\zeta - (1.8/\pi) \sin\left(\frac{\zeta+0.4}{3.6}\pi\right) & \text{for } -0.4 \leq \zeta \leq 3.2 \\ 0 & \text{for } \zeta \geq 3.2 \,. \end{cases}$$

Because of the Thomas-Fermi approximation used by Randrup, the function Ψ falls off too fast for large ζ. An improved expression for Ψ was reported by Ko [693], which takes also tunneling through the potential barrier between the nuclei into account.

If one considers a change of the asymmetry in time, there is an induced net mass-current in the neck. By a similar mechanism as discussed above it is connected with an energy transfer from the walls to the gas, which acts as a drag force in the asymmetry degree of freedom. The rate of energy transfer is given by

$$\dot{E} = \frac{16}{9}\rho_m \bar{v} \frac{1}{\Delta\sigma} \dot{V}_1^2 = \frac{16}{9}\rho_m \bar{v} \frac{1}{\Delta\sigma} \frac{\partial V_1}{\partial q^i} \frac{\partial V_1}{\partial q^j} \dot{q}^i \dot{q}^j \,, \tag{5.183}$$

where V_1 is the volume on one side of the window. For a derivation we refer to Swiatecki [285]. Adding the contributions from Eqs. (5.181) and (5.183) to the friction tensor one finds

$$\eta_{ij}^{(\text{window})} =$$
$$\frac{1}{2}\rho_m \bar{v} \left[\Delta\sigma \left(\frac{\partial D_z}{\partial q^i} \frac{\partial D_z}{\partial q^j} + \frac{1}{2} \frac{\partial D_\rho}{\partial q^i} \frac{\partial D_\rho}{\partial q^j} \right) + \frac{32}{9} \frac{1}{\Delta\sigma} \frac{\partial V_1}{\partial q^i} \frac{\partial V_1}{\partial q^j} \right] , \tag{5.184}$$

where the correction (5.182) has not yet been inserted in the first term. The "wall and window" friction tensor $\eta_{ij}^{(\text{w+w})} = \eta_{ij}^{(\text{wall 2})} + \eta_{ij}^{(\text{window})}$ should account for both, the energy transfer to the gas because of collisions with the wall and because of momentum flux between A and B. It applies to a situation where a drift velocity between the gas in A and in B exists.

The wall-and-window formula

One expects a smooth transition between the regime in which the wall formula (5.169) applies and the part of the fission or fusion path where the wall-and-window friction-tensor $\eta_{ij}^{(\text{w+w})}$ should be used. Nix and Sierk proposed the phenomenological ansatz [694]

$$\eta_{ij}^{\text{total}} = [1 - c(\mathbf{q})]\,\eta_{ij}^{(\text{w+w})} + c(\mathbf{q})\,\eta_{ij}^{(\text{wall})} \,. \tag{5.185}$$

The function $c(\mathbf{q})$ is somewhat arbitrary. In Ref. [694] it was chosen as

$$c = \sin^2\left(\frac{\pi}{2}x\right)$$

with $x = \rho_{\text{neck}}^2 / \min(\rho_1^2, \rho_2^2)$, where ρ_1 and ρ_2 are the transverse semi-axes of the two outer ellipsoids in the shape parametrization of three quadratic surfaces (2.226) and ρ_{neck} is the neck radius. The latter is taken to be zero for separated fragments. Feldmeier [695] proposed a modified form for the function $c(\mathbf{q})$

$$c = \begin{cases} 0 & \text{for} \quad 0 \leq x \leq 0.8 \\ \sin^2\left(\frac{x-0.8}{0.2}\frac{\pi}{2}\right) & \text{for} \quad 0.8 \leq x \leq 1 \\ 1 & \text{for} \quad x \geq 1. \end{cases}$$

Still another variant for $c(\mathbf{q})$, which contains additional parameters to be fitted to individual nuclei, was reported by Błocki [286].

The energy transferred to the gas is not automatically thermalized. Since particle-particle collisions play no role in a Knudson gas, the authors of the one-body friction model [669] argued that successive collisions with the wall would randomize the momentum transfered to individual nucleons if the surface is sufficiently "irregular". In more precise terms: because the model in which the wall formula was derived is a three-dimensional, time-dependent billiard, one can say it is assumed that this billiard has a chaotic dynamical behavior. Błocki, Shi, and Swiatecki investigated this numerically for (axially symmetric) multipole vibrations of a spherical cavity [696]. For multipolarity larger than 2 the wall-formula prediction is fairly well reproduced.

Błocki et al. [697] introduced a measure μ(shape) for the degree of chaoticity of a three-dimensional, axially symmetric billiard of given shape. For that purpose the authors determined numerically the largest Lyapunov coefficient λ for a trajectory starting inside the billiard box by comparing the deviation $\delta(t)$ in phase space from a neighboring trajectory with an initial small deviation $\delta_0 = \delta(0)$. For chaotic dynamics one expects the trajectories to deviate exponentially with time, $\delta(t) \sim \exp(\lambda t)$, for large time. To determine λ a procedure of Benettin et al. [698] was used, based on the relation

$$\lambda = \lim_{\delta_0 \to 0} \lim_{k \to \infty} \frac{1}{k\tau} \sum_{i=1}^{k} \ln \frac{\delta(\tau)}{\delta_0},$$

which converges reasonably well with increasing time $t = k\tau$ (typically $t > 10^4\, R_0/v$). A trajectory is counted as chaotic if $\lambda > 10^{-3}$. By a uniform sampling of initial values from the phase space (the volume of the billiard times the momentum space with a constraint on the kinetic energy and the angular momentum projection on the symmetry axis) one determines the ratio μ of chaotic trajectories to the total number of trajectories. A billiard with a completely chaotic dynamics would have $\mu = 1$, regular behavior is characterized by $\mu = 0$. The motion of a particle in a static billiard of spherical shape with multipole deformations of order l yields μ-values between 0 and 1. They approach 1 with increasing multipole order and increasing amplitude. For a spheroidal cavity the Hamiltonian is separable the motion is therefore regular and $\mu = 0$. These findings are in close analogy to the time-dependent multipole deformations considered in Ref. [696].

Chaudhuri and Pal calculated chaoticity coefficients $\mu(c)$ in the (c,h,α) shape class (2.219) for $h = \alpha = 0$ [699]. They found $\mu(c)$ to increase monoto-

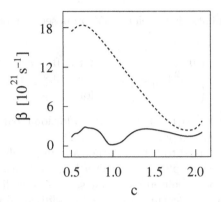

Fig. 5.27. Dissipation parameter $\beta(c)$. Full line: including chaoticity reduction, dashed line: original wall formula (from Ref. [699]).

nously from the spherical shape at $c = 1$, where it is of course 0, to the scission configuration around $c = 1.8$, confirming qualitatively a result for a pure quadrupole mode obtained in Ref. [697]. The authors used $\mu(c)$ as a shape-dependent reduction factor of the classical wall formula. The resulting dissipation parameter $\beta = \mu(c)\eta^{(\text{wall } 2)}(c)/m(c)$ – with Werner-Wheeler mass parameter $m(c)$ – is shown in Fig. 5.27. Chaos-related arguments to reduce the window friction are based on the implicit assumption that there can be many successive collisions with the wall before its shape has changed appreciably. This may not be realistic.

Use of one-body friction to describe fission and fusion-fission data in the framework of the Fokker-Planck or the Langevin equations appears to require a substantially reduced wall friction than given by the classical formulae (5.169) and (5.175). The simplest way to fix the problem is the introduction of a shape-independent "reduction" factor k_s. Values for k_s in the range $0.2 < k_s < 0.5$ were reported from fits of a variety of data [638]. This however spoils the main advantage of the wall-and-window formalism: that it does not involve any fudge parameter.

A variant of the classical wall formula was suggested by Sierk, Koonin, and Nix [700]. The authors argue that it is not realistic to assume that the bulk of the gas of nucleons remains at rest for any motion of the surface, except translations and rotations. Instead of the integrand $\dot{n}^2$ in the surface integral of the wall formula (5.169) they suggest therefore to take the square of the difference between the normal velocity of the surface $\dot{n}$ and the normal component of a laminar flow field $\mathbf{v}_{\text{lam}}$ in "some" normal distance λ from the surface. For the reference flow field $\mathbf{v}_{\text{lam}}$ the Werner-Wheeler field (5.153), (5.154) is used. Expanding the component of this field in the direction of the surface normal to linear order in the distance from the surface

$$(v_{\text{lam}})_n = \dot{n} + \lambda \frac{\partial (v_{\text{lam}})_n}{\partial n},$$

the modified window formula becomes

$$\dot{E} = \rho_m \overline{v} \lambda^2 \oint \left(\frac{\partial (v_{\text{lam}})_n}{\partial n} \right)^2 d\sigma \,, \tag{5.186}$$

where λ was fitted to fission data with the result $\lambda \approx 1.7$ fm. The shape-dependence of the modified wall formula is different from that of the classical wall formula and resembles qualitatively the shape-dependence of the hydrodynamical, two-body friction. In this variant of the wall formula a nucleon-nucleon scattering length of the order of λ is indirectly introduced in conflict with the original assumption of a large mean-free-path length in nuclei of low temperature.

One-body friction, quantal approach

Nuclear Doppler effect, superfluid case

For rather low excitation energies ($T < 1$ MeV) Yannouleas, Dworzecka, and Griffin [701] considered a reduction of one-body dissipation because of superfluidity effects. They treated nuclear collective motion in the two-fluid model with a core of a superfluid liquid and an ideal gas of a few excited quasiparticles. In this model the mean field is produced by the superfluid core. It changes its shape on a slower time scale than the motion of the quasiparticles. The superfluid component has zero entropy and all excitation energy resides in the quasiparticle component, which is assumed to be in thermal equilibrium.

For an estimate of the wall friction in this approach the BCS model with Thomas-Fermi level density was used, which we discussed in Sec. 3.6. Since we are here interested in a system with time-reflection symmetry, $n_k = n_{\overline{k}}$ and $\mu = 0$ in Eq. (3.86). Neglecting the term proportional to $G/2$ in Eq. (3.87), the quasiparticle energy becomes

$$\mathcal{E}_k = \sqrt{\tilde{\epsilon}_k^2 + \Delta^2(T)} \tag{5.187}$$

in terms of the temperature-dependent gap $\Delta(T)$ and the single-particle energy $\tilde{\epsilon}_k$ with respect to the chemical potential λ. In the bulk, where the mean field is constant,

$$\tilde{\epsilon}_k = \frac{\hbar^2 k^2}{2M_{\text{nucl}}} - \lambda \,, \tag{5.188}$$

where $\hbar k$ is the momentum of the quasiparticle. Its velocity is

$$\mathbf{v}_k = \frac{1}{\hbar} \text{grad}_k \mathcal{E}_k = \frac{\hbar}{M_{\text{nucl}}} \frac{\tilde{\epsilon}_k}{\mathcal{E}_k} \mathbf{k} \,. \tag{5.189}$$

Since $\tilde{\epsilon}_k$ can be positive or negative depending on $\epsilon_k = \hbar^2 k^2/(2M_{\text{nucl}})$ being larger or smaller than λ, $\mathbf{v}$ and $\mathbf{k}$ can have the same direction (particle-like quasiparticles) or the opposite direction (hole-like quasiparticles).

The temperature-dependent gap $\Delta(T)$ follows from the gap equation (3.88). With the Thomas-Fermi level density for either neutrons or protons

$$g(\epsilon) = \frac{\mathcal{V}}{2\pi^2}\left(\frac{2M_{\mathrm{nucl}}}{\hbar^2}\right)^{3/2}\epsilon^{1/2} \tag{5.190}$$

the gap equation becomes

$$\frac{2}{G\mathcal{V}} = \frac{1}{(2\pi)^3}\left(\frac{2M_{\mathrm{nucl}}}{\hbar^2}\right)^{3/2}\int_{\lambda-\omega}^{\lambda+\omega}\frac{\tanh(\beta\mathcal{E}/2)}{\mathcal{E}(\epsilon)}\epsilon^{1/2}d\epsilon\,, \tag{5.191}$$

where $\mathcal{V}$ is the nuclear volume. For a more detailed discussion of the Thomas-Fermi BCS formalism we refer to Ref. [702]. There also the important relations

$$T_{\mathrm{crit}} = \frac{e^C}{\pi}\Delta(T=0) \tag{5.192}$$

with the Euler constant $C = 0.577$ and

$$\ln\frac{\Delta(T=0)}{\Delta(T)} = 2I_1(\beta\Delta) \tag{5.193}$$

with

$$I_1(y) = \int_0^\infty\frac{1}{1+e^{\sqrt{x^2+y^2}}}\frac{dx}{\sqrt{x^2+y^2}} \tag{5.194}$$

are derived.

The distribution of quasiparticles with energy $\mathcal{E}_k$ and temperature $T = 1/\beta$ is $f(\mathcal{E}_k) = (1+e^{\beta\mathcal{E}_k})^{-1}$. The authors of Ref. [701] choose a coordinate system in which the surface element $dxdy$ is at rest and the outwards pointing normal in the z-direction. The distribution of particles hitting the wall is therefore Galilei-shifted by the velocity $-\dot{n}$ in this coordinate system and has the distribution $f(\mathcal{E}_k - \dot{n}\hbar k_z)$, where terms of second order in $\beta\hbar k_z\dot{n}$ are neglected. The ensemble of scattered particles is assumed to be in thermal equilibrium as in the classical wall formula and it is therefore characterized by the distribution $f(\mathcal{E}_k + \dot{n}\hbar k_z)$. In analogy to Eq. (5.165) the number of quasiparticles hitting the surface element $dxdy$ during time dt is given by

$$dN = \frac{4}{(2\pi)^3}\int d^3k v_z f(\mathcal{E}_k - \dot{n}\hbar k_z)[1 - f(\mathcal{E}_k + \dot{n}\hbar k_z)]dxdydt\,. \tag{5.195}$$

The factor 4 in front of the integral accounts for the spin and isospin degrees of freedom, the factor $(2\pi)^3$ comes from proper normalization of momentum space in Thomas-Fermi approximation. The integration range is different for particle-like and hole-like quasiparticles

$$\begin{matrix} k_f < k < \infty & 0 \le \theta < \pi/2 & \text{particle-like} \\ 0 \le k \le k_f & \pi/2 \le \theta < \pi & \text{hole-like,} \end{matrix} \tag{5.196}$$

where θ is the angle between the direction of $\mathbf{k}$ and the z-axis. The term $f(\mathcal{E}_k - \dot{n}\hbar k_z)$ counts the probability for a quasiparticle with energy $\mathcal{E}_k$, the term $[1 - f(\mathcal{E}_k + \dot{n}\hbar k_z)]$ gives the probability that the scattering state is empty.

We can rewrite the expression

$$f(\mathcal{E}_k - \dot{n}\hbar k_z)[1 - f(\mathcal{E}_k + \dot{n}\hbar k_z)] =$$
$$\frac{e^{\beta\mathcal{E}_k}}{(1 + e^{\beta\mathcal{E}_k})^2} \frac{e^{\beta\mathcal{E}_k + \beta\hbar k_z \dot{n}}}{e^{\beta\mathcal{E}_k}} \frac{1 + e^{\beta\mathcal{E}_k}}{1 + e^{\beta\mathcal{E}_k + \beta\hbar k_z \dot{n}}} \frac{1 + e^{\beta\mathcal{E}_k}}{1 + e^{\beta\mathcal{E}_k - \beta\hbar k_z \dot{n}}}$$

and multiply numerator and denominator of each of the four factors on the r.h.s. of this equation by $e^{-\beta\mathcal{E}_k}$. The first two factors yield $-df(\mathcal{E}_k)/d(\beta\mathcal{E}_k)$ and $e^{\beta\hbar k_z \dot{n}}$, respectively. Taking into account that $\mathcal{E}_k/T \equiv \beta\mathcal{E}_k \geq \beta\Delta \gg 1$ and therefore neglecting $e^{-\beta\mathcal{E}_k}$ compared to 1, the last two factors yield 1. Using Eq. (5.189), one has $v_z = \pm|d\mathcal{E}_k/(\hbar dk)| \cos\theta$, where the upper sign refers to particle-like quasiparticles, the lower to hole-like states. Inserting these results into Eq. (5.195), one obtains

$$dN = \mp\frac{4}{(2\pi)^3 \beta\hbar} \int d^3k \cos\theta \left|\frac{d\mathcal{E}_k}{dk}\right| \frac{df}{d\mathcal{E}_k} e^{\beta\hbar k_z \dot{n}} dxdydt .$$

The momentum transfer for each collision is $2\hbar k_z$ and therefore the pressure becomes

$$p = \mp\frac{8}{(2\pi)^3 \beta} \int d^3k \, k \cos^2\theta \left|\frac{d\mathcal{E}_k}{dk}\right| \frac{df}{d\mathcal{E}_k} e^{\beta\hbar k_z \dot{n}}$$
$$= \mp\frac{2}{\pi^2\beta} \int dk \, k^3 \left|\frac{d\mathcal{E}_k}{dk}\right| \frac{df(\mathcal{E}_k)}{d\mathcal{E}_k} \int d\theta \sin\theta \cos^2\theta e^{\beta\hbar k \cos\theta \dot{n}} . \quad (5.197)$$

As in Eq. (5.168) the energy dissipation rate is obtained by multiplying the pressure p with $\dot{n}$ and integrating over the whole surface. Expanding the exponential $e^{\beta\hbar k \cos\theta \dot{n}}$ in Eq. (5.197) in a power series, the first term gives a vanishing contribution to the surface integral because of volume conservation (as in the derivation of Eq. (5.169)). We therefore obtain in the lowest non-vanishing order of $\dot{n}$

$$\dot{E} = \frac{2\hbar}{\pi^2} \left[\int_0^{k_f} dk k^4 \left|\frac{d\mathcal{E}_k}{dk}\right| \frac{df(\mathcal{E}_k)}{d\mathcal{E}_k} \int_{\pi/2}^{\pi} d\theta \sin\theta \cos^3\theta \right.$$
$$\left. - \int_{k_f}^{\infty} dk k^4 \left|\frac{d\mathcal{E}_k}{dk}\right| \frac{df(\mathcal{E}_k)}{d\mathcal{E}_k} \int_0^{\pi/2} d\theta \sin\theta \cos^3\theta \right] \oint \dot{n}^2 d\sigma ,$$

where contributions of hole-like and particle-like quasiparticles are added with their appropriate integration limits (5.196). The angular integration yields $\mp(1/4)$ and the dissipation rate can be written

$$\dot{E} = -\frac{\hbar}{2\pi^2} \int_0^\infty dk k^4 \left|\frac{d\mathcal{E}_k}{dk}\right| \frac{df(\mathcal{E}_k)}{d\mathcal{E}_k} \oint \dot{n}^2 d\sigma \,.$$

Using the relation $|d\mathcal{E}_k/dk|dk = |d\mathcal{E}_k/d\tilde{\epsilon}_k|d\tilde{\epsilon}_k$ and Eq. (5.188) one obtains after partial integration

$$\dot{E} = \frac{4M_{\mathrm{nucl}}^2}{\pi^2\hbar^3} \int_{-\lambda}^\infty (\tilde{\epsilon} + \lambda) f(\tilde{\epsilon}) d\tilde{\epsilon} \oint \dot{n}^2 d\sigma \,. \tag{5.198}$$

To evaluate the energy integral we neglect terms of the order $e^{-\beta\lambda}$ and get

$$\int_{-\lambda}^\infty (\tilde{\epsilon} + \lambda) f(\tilde{\epsilon}) d\tilde{\epsilon} \approx \int_{-\infty}^\infty (\tilde{\epsilon} + \lambda) f(\tilde{\epsilon}) d\tilde{\epsilon} = 2\lambda \int_0^\infty f(\tilde{\epsilon}) d\tilde{\epsilon} \equiv I_2 \,, \tag{5.199}$$

where the symmetry $f(\tilde{\epsilon}) = f(-\tilde{\epsilon})$ was used, which follows from $\mathcal{E}_k(\tilde{\epsilon}) = \mathcal{E}_k(-\tilde{\epsilon})$. Introducing the new integration variable $y = \sqrt{(\tilde{\epsilon}/\Delta)^2 + 1}$ one obtains

$$I_2 = 2\lambda\Delta \int_1^\infty \frac{y}{\sqrt{y^2 - 1}} \frac{dy}{1 + e^{\beta\Delta y}} = \begin{cases} \lambda\sqrt{2\pi\Delta/\beta} & \beta\Delta \gg 1 \,,^1 \\ (2\ln 2)\lambda/\beta & \beta\Delta \to 0 \,.^2 \end{cases} \tag{5.200}$$

(The result for I_2 reported in Ref. [701] does not agree with these limits). With the mass density $\rho_m = 2M_{\mathrm{nucl}}k_F^3/(3\pi^2)$, the mean velocity $\overline{v} = 3\hbar k_f/(4M_{\mathrm{nucl}})$, and the chemical potential $\lambda = \hbar^2 k_f^2/(2M_{\mathrm{nucl}})$ the energy dissipation rate (5.198) can be written as the wall formula with a reduction factor k_s

$$\dot{E} = k_s \rho_m \overline{v} \oint \dot{n}^2 d\sigma \,, \tag{5.201}$$

where $k_s = 2\lambda^{-2} I_2$. If there were no pairing correlations, $\Delta = 0$ and $k_s = (4\ln 2)(\lambda\beta)^{-1}$ for all temperatures $1/\beta$. In the case of finite pairing, k_s is smaller than this expression for all $T < T_{\mathrm{crit}}$.

Linear response approach to friction

In another approach the adiabatic, linear response theory of Sec. 5.1.1 was extended to finite temperatures. For that purpose it is useful to rewrite matrix elements of the form $\langle l; q | \partial_q \hat{H} | 0; q \rangle$ between the ground state and a particle-hole state l, consisting of a hole state h and a particle state p, as $\langle h | \partial_q \hat{H} | p \rangle$ for

---

[1] If $\beta\Delta \gg 1$ and $y \geq 1$, one has $[1 + \exp(\beta\Delta y)]^{-1} \approx \exp(-\beta\Delta y)$ and the resulting Laplace integral yields $I_2 = 2\lambda\Delta K_1(\beta\Delta)$ (Ref. [703], part 1, formula 2.41). Inserting the first term of the asymptotic expansion for the modified Hankel function K_1, yields the indicated limit.

[2] For small $\beta\Delta$ the new integration variable $z = \beta\Delta y$ is introduced and one uses the result $\int_0^\infty dz[1 + \exp z]^{-1} = \ln 2$.

a single particle shell-model Hamiltonian $\hat{H}$. Particle-hole excitation energies are given by $E_l - E_0 = \epsilon_p - \epsilon_h \equiv \epsilon_{hp}$. In terms of the expansion coefficients $a_l(q(t)) \equiv a_{hp}(q(t))$ of Eq. (5.1) and the fermion distribution-function $f(\epsilon) = (1 + e^{\beta(\epsilon - \lambda)})^{-1}$ the rate of energy dissipation out of the collective motion can be written

$$
\dot{E} = -\sum_{ph} f(\epsilon_h)[1 - f(\epsilon_p)]\epsilon_{hp}\frac{d|a_{ph}|^2}{dt}
$$

$$
= -\sum_{ph} f(\epsilon_h)[1 - f(\epsilon_p)]\epsilon_{hp}a_{ph}\dot{a}_{ph} + c.c. , \tag{5.202}
$$

where the Pauli factor $f(\epsilon_h)[1 - f(\epsilon_p)]$ gives the probability that the initial single-particle state ϵ_h is occupied and the final state ϵ_p unoccupied. The sum $\sum_{hp}$ runs in both indices over all single-particle states. Inserting the result (5.4) and using the relation (5.8) $\langle h|\partial_q|p\rangle = \langle h|\partial_q\hat{H}|p\rangle/\epsilon_{hp}$, one obtains

$$
\dot{E} = -\dot{q}\sum_{ph} f(\epsilon_h)[1 - f(\epsilon_p)]\langle p|\partial_q\hat{H}|h\rangle_t
$$

$$
\times \int_{t_0}^{t} dt'\dot{q}(t')\frac{\langle h|\partial_q\hat{H}|p\rangle_{t'}}{\epsilon_{hp}(t')} e^{(i/\hbar)\int_{t'}^{t}\epsilon_{hp}(t'')dt''} + c.c. \tag{5.203}
$$

Replacing the integration variable t' by $t + t'$ and moving t_0 to $-\infty$, the integral can be written

$$
\int_{-\infty}^{0} dt'\dot{q}(t + t')\frac{\langle h|\partial_q\hat{H}|p\rangle_{t+t'}}{\epsilon_{hp}(t + t')} \exp\left(\frac{i}{\hbar}\int_{t+t'}^{t}\epsilon_{hp}(t'')dt''\right) . \tag{5.204}
$$

The integral (5.204) is called "memory integral" because the integrand requires knowledge of the system for all times prior to t.

If we neglect the time-dependence of the matrix element, of ϵ_{hp}, and of $\dot{q}$, the result is

$$
\dot{E} = -\dot{q}^2\sum_{ph} f(\epsilon_h)[1 - f(\epsilon_p)]\left[\frac{|\langle h|\partial_q\hat{H}|p\rangle|^2}{\epsilon_{hp}}\int_{-\infty}^{0} e^{-(i/\hbar)\epsilon_{hp}t'}dt' + c.c.\right] . \tag{5.205}
$$

In Ref. [704] and in later publications of the same authors the expression $f(\epsilon_h) - f(\epsilon_p)$ appears in Eq. (5.205), instead of the Pauli factor $f(\epsilon_h)[1 - f(\epsilon_p)]$. Since the operator $\partial_q\hat{H}$ is Hermitian, the square bracket in this equation is odd when h and p are exchanged. Therefore the contribution of the $f(\epsilon_p)$ term to $\dot{E}$ is equal to the contribution of the $f(\epsilon_h)$ term. The interpretation of this modified form of the response in Ref. [704] is therefore that excitations are only allowed from occupied states, but in the final states no Pauli blocking is considered.

The integral in Eq. (5.205) yields $2\pi\hbar\delta(\epsilon_{hp})$. If there is a gap between the last occupied and the first unoccupied state, $\epsilon_{hp} > 0$ for all particle-hole states and therefore $\dot{E}$ would vanish. This rather unphysical result must be due to inappropriate implicit or explicit approximations made in the derivation of Eq. (5.205).

Various proposals have been made to avoid the appearance of the δ function $\delta(\epsilon_{hp})$ from the time integral in Eq. (5.205). It has been argued that the particle-hole excitations have a finite lifetime τ and that the particle-hole energies ϵ_{hp} should be replaced by $\epsilon_{hp} + i\Gamma_{hp}$ with $\hbar/\Gamma = \tau$ [704]. One then obtains

$$\dot{E} \equiv -\dot{q}^2\eta(q) = -\dot{q}^2 \sum_{ph} f(\epsilon_h)[1 - f(\epsilon_p)]|\langle h|\partial_q\hat{H}|p\rangle|^2 \frac{4\hbar\epsilon_{hp}\Gamma_{hp}}{(\epsilon_{hp}^2 + \Gamma_{hp}^2)^2} \quad (5.206)$$

with the friction coefficient η. Several semiempirical expressions for the temperature dependence of the width Γ have been suggested. Bertsch [705] estimated the state-independent width $\Gamma = (280\text{MeV})(T/\epsilon_f)^2$ for collisions with thermal excitations in a Fermi liquid. Hofmann proposed the state-dependent expression

$$\Gamma_k = \frac{1}{\Gamma_0} \frac{(\epsilon_k - \lambda)^2 + \pi^2 T^2}{1 + c^{-2}[(\epsilon_k - \lambda)^2 + \pi^2 T^2]} , \quad (5.207)$$

where ϵ_k may be either a particle or a hole state and $\Gamma_0 \approx 33$ MeV, $c \approx 20$ MeV, see in Ref. [674] Eq. 6.76. Ivanyuk and Pomorski [706] used Nörenberg's thermal equilibration time [707]

$$\tau_{\text{therm}} = (33 \text{ MeV}^2)\frac{1 + \exp[-(\epsilon_k - \lambda)/T]}{(\epsilon_k - \lambda)^2 + \pi^2 T^2} \sec^{-21} \quad (5.208)$$

to get a width $\Gamma_k = \hbar/\tau_{\text{therm}}$. The expression (5.208) is obtained from the assumption that it takes in the average 3 collisions to thermally equilibrate a particle or a hole with a mean collision time in a Fermi liquid

$$\tau = \hbar\Lambda\frac{1 + \exp[-(\epsilon_k - \lambda)/T]}{(\epsilon_k - \lambda)^2 + \pi^2 T^2} , \quad \text{for} \quad |\epsilon_k - \lambda|/\lambda \ll 1,$$

where Λ is a cutoff energy [708], which Bertsch [709] adapted to the case of heavy-ion collisions. From his result Nörenberg obtained the prefactor in Eq. (5.208).

Gross [624] suggested to take the dependence of the matrix element $\langle h|\partial_q\hat{H}|p\rangle_{t+t'}$ on t' in the integral (5.204) into account. The theory was formulated only for radial friction in the entrance channel of deep-inelastic reactions. In this case the collective variable q is the center-of-mass distance D. Assuming that the velocity $\dot{D}$ and the excitation energy ϵ_{hp} are constant during the rather short interaction time in these reactions, the matrix element splits off approximately a factor $\exp(t'\dot{D}/a)$ if $(R_1 \gg a, R_2 \gg a)$, where a is the diffuseness of the transition density [624, 690]. Similar to Eq. (5.206) the friction coefficient η is then given by

$$\eta(D) = \sum_{hp} f(\epsilon_h)[1 - f(\epsilon_p)]|\langle h|\partial_D V|p\rangle|^2 \frac{4\hbar^2(\dot{D}/a)}{\epsilon_{hp}^2 + (\hbar\dot{D}/a)^2}, \qquad (5.209)$$

where $\partial_D \hat{H} = \partial_D V$ is used. Since we neglect shape changes of the two colliding nuclei, the matrix element $|\langle h|\partial_D V|p\rangle|^2$ consists of a contribution from the moving potential V_A of the nucleus A producing particle-hole excitations in the nucleus B and a term where the roles of A and B are exchanged. The radial friction formfactor obtained from Eq. (5.209) depends sensitively on the choice of the radii and the diffuseness of the potentials and transition densities of nuclei A and B. It is only weakly temperature-dependent and can be even larger than the temperature-independent window-friction formfactor [690], in contrast to the result reported in Ref. [706].

It should be stressed that in all friction models discussed so far a locally isotropic momentum distribution of the nucleons is assumed, even at the surface and no shell effects are considered in the level density, i.e. we restricted the discussion to the Thomas-Fermi approximation. This may be particularly unrealistic at the surface compared to a HF or HFB description.

Time-dependent shell model

In the linear response treatment of friction only one hit of a nucleon with the wall is taken into account. The further fate of the nucleon is not considered. (It would take the next order of the time-dependent perturbation theory to describe two successive collisions with the wall.) Therefore a numerical solution of the time-dependent Schrödinger equation for a gas of either noninteracting [710] or paired nucleons [711–714] in an externally cranked single-particle potential was considered a more realistic model for energy dissipation in large-scale collective motion. The method was called nonadiabatic, independent particle (or quasiparticle) model [30]. It was applied to the dynamics of fission between the outer saddle point of actinides and scission [710, 711, 713] and to deep-inelastic heavy-ion collisions [714, 715]. No temperature dependence was considered in these models.

One may be tempted to take the difference

$$\Delta E = \langle \mathbf{q}(t)|\hat{H}_\mathbf{q}|\mathbf{q}(t)\rangle - \langle 0; \mathbf{q}(t=0)|\hat{H}_\mathbf{q}|0; \mathbf{q}(t=0)\rangle$$

as the intrinsic excitation energy at time t, where $|\mathbf{q}(t)\rangle$ is the solution of the time-dependent Schrödinger equation (5.2) and $|0; \mathbf{q}(t=0)\rangle$ the HF ground state at $t = 0$. However, ΔE contains the collective kinetic energy and "genuine" intrinsic excitations as well. Note that in the linear response theory one can distinguish between virtual excitations, leading to the cranking expression for the collective mass, and real excitations, assumed – probably not always with sufficient reason – to be irreversibly taken out of the collective motion. But this separation is only meaningful in first order of the perturbation theory in the adiabatic basis. A satisfactory distinction between collective energy, including collective kinetic energy, and "dissipated" energy is in principle not possible in a model built only from time-reversible operators. When Boneh and Fraenkel [710] found, for instance, a ΔE at scission almost equal to the gain in potential energy between saddle and scission points, one cannot conclude that the motion is overdamped without knowing the contribution of the collective kinetic energy to ΔE. An approximate separation of the collective kinetic energy was proposed by Glas and Mosel [714], applicable to the entrance channel of deep-inelastic and fusion reactions.

In particular when pairing correlations are taken into account, the dissipation of energy is often identified with Hill and Wheeler's slipping mecha-

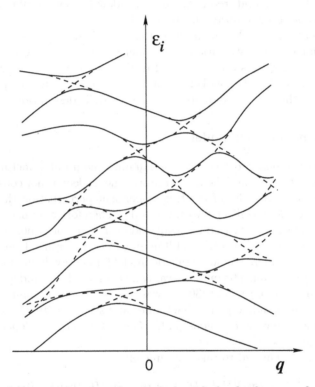

Fig. 5.28. Schematic drawing of a single-particle level scheme as function of a deformation parameter q in the single-particle potential. Full lines are adiabatic levels. Dashed lines connect diabatic levels across avoided crossing points (after Wilets, Ref. [30]).

nism [87]. In the schematic drawing of a single-particle level scheme along the fission path of Fig. 5.28 the adiabatic levels are shown by full lines, the diabatic ones, which are allowed to cross, by dashed lines. In a near-crossing situation the nodal structure of the corresponding wave function changes slowly as function of the deformation along the diabatic levels, whereas drastic changes of this structure occur along the adiabatic lines. (For methods of numerical construction of a diabatic level scheme we refer in the general case to Delos [716] and for fusion reactions in the framework of the two-center shell model to the paper [717] by Ayik and Nörenberg). Landau [718], Zener [719], and Stückelberg [720] derived the closed expression

$$P = \exp\left\{ -\frac{2\pi |\langle a|\hat{H}'|b\rangle|^2}{\hbar |\dot{q}\partial_q(\epsilon_a(q) - \epsilon_b(q))|} \right\} \tag{5.210}$$

for the probability that a particle in an occupied level a before the crossing point jumps to an unoccupied level b while the deformation variable q changes with constant velocity $\dot{q}$ across the region of the crossing point. The single-particle operator $\hat{H}'$ is responsible for the repulsion of the adiabatic states in a near-crossing situation by coupling the two diabatic states. The formula is derived under the two conditions that (1) the diabatic levels depend linearly on q in a wider surrounding of the crossing point and that (2) subsequent crossings are sufficiently far apart that they can be treated as isolated events. These conditions are sometimes only poorly fulfilled along the fission path of actinides. A remarkable feature of the probability (5.210) is its nonanalytic dependence on the velocity $\dot{q}$. Shifting gradually an increasing number of nucleons in the diabatic scheme of Fig. 5.28 above the Fermi level was called in Ref. [87] slippage.

Schütte and Wilets considered pairing correlations in the framework of the BCS theory in an externally cranked mean-field potential and derived a formula analogous to Eq. (5.210) for the probability that a pair of quasiparticles continues to occupy the pair of states with single-particle energy $\epsilon_i(q)$ when these states cross the chemical potential λ from below [712]

$$P_i = \exp\left\{ -\frac{\pi \Delta^2}{|\dot{q}\partial_q(\epsilon_i(q) - \lambda)|} \right\}. \tag{5.211}$$

The formula is derived under the same conditions as Eq. (5.210) and in addition, the gap Δ was assumed to be independent of the deformation q. Numerical calculations showed that in more than 90% of all two-quasiparticle excitations the two quasiparticles remained paired in the diabatic state and its time-reversed partner [712]. To estimate the energy ΔE exclusively due to the slippage mechanism along the fission path with a paired wave function, Schütte et al. [713] considered a model Hamiltonian for independent quasiparticles

$$\hat{H} = \sum_i \mathcal{E}_i \hat{\alpha}_i^+ \hat{\alpha}_i \,,$$

where the quasiparticle energies $\mathcal{E}_i$ are defined in Eq. (2.51) and the quasiparticle operators $\hat{\alpha}_i$ in Eq. (2.38). The authors restricted the Hilbert space for the dynamics of the system to the BCS ground state $|BCS\rangle$ and $2n$ paired quasiparticle states ($n = 1, 2 \ldots$). Counting the number of states j crossing the chemical potential along the fission path until time t and associating a squared amplitude $|c_j|^2 = P_j$ with these two-quasiparticle states, the dissipated energy was identified with $\Delta E = 2\sum_j \mathcal{E}_j |c_j|^2$. Allowance was also made in Ref. [713] for crossings of an occupied state j with an unoccupied quasiparticle state j' and a partial population of this state at the expense of $|c_j|^2$. A comparison of the dissipated energy as a function of time with the wall-and-window dissipation mechanism and with hydrodynamical viscosity shows that the slippage mechanism leads to considerably smaller friction than the full one-body mechanism and to somewhat larger friction than hydrodynamical dissipation with a friction coefficient fitted to observed total kinetic energies of the fission fragments.

Another way of separating intrinsic excitation from collective kinetic energy was proposed by Nörenberg in his "dissipative diabatic dynamics" (DDD) model [670]. He postulates the existence of an irrotational collective velocity field and defines the phase of all time-dependent, diabatic single-particle states so that they include the collective velocity potential [717,721]. From the energy expectation-value, calculated with these states, one gets the same collective mass tensor (5.11) as from the cranking model. Unlike other time-dependent shell-model approaches the DDD model contains an element of irreversibility by allowing a statistical decay of excited, diabatic particle-hole states towards a thermal distribution with a decay time τ, given by Eq. (5.208). If the typical time of the collective motion is similar or shorter than τ, the DDD model yields a retarded friction force of the form $\int_{t_0}^t \dot{\mathbf{q}}(t')\xi(t,t')dt'$ with a memory kernel $\xi(t,t') = c(t')\exp[-\int_{t'}^t dt''/\tau(t'')]$ [707]. The DDD model was used to describe the entrance channel of fusion and deep-inelastic reactions [723–725].

In order to justify the wall formula in a quantal description of nuclear dynamics, Błocki, Skalski and Swiatecki [726] used a spherical cavity with periodic multipole vibrations as the potential in the time-dependent Schrödinger equation. They filled 112 neutrons into the lowest single-particle states of the spherical box and propagated them in time over one period of a multipole vibration. The energy transferred to the particles was calculated as function of time. After one period this energy remained in general below the wall-formula prediction and also below the linear response estimate of Ref. [672]. Irregularities were observed for very slow vibrations, i.e. in the adiabatic limit, which could be traced back to a single avoided level crossing in the range of the chosen amplitude. This should however not be relevant for fission of excited compound nuclei where the effect of single Landau-Zener transitions is washed out by the finite-temperature of the system. In fact, the wall-and-window mechanism is supposed to apply to large-scale collective motion and to average over several avoided level crossings of the single-particle level scheme along the fission path.

References

1. O. Hahn, F. Straßmann, Naturwiss. **27**, 11 (1939); Naturwiss. **27**, 89 (1939)
2. O. R. Frisch, Physics Today **20**, 43 (Nov. 1967)
3. E. Fermi, E. Amaldi, B. Pontecorvo, F. Rasetti, E. Segrè, Ricerca scientifica **5**, 282 (1934)
4. E. Fermi, in *Nobel Lectures, Physics, 1922-1941* (Elsevier, Amsterdam, 1965) pp. 414
5. I. Curie, P. Savitch, J. Phys. et le Radium **8**, 385 (1937); J. Phys. et le Radium **9**, 355 (1938)
6. L. Meitner, O. Hahn, F. Straßmann, Z. Phys. **106**, 249 (1937)
7. I. Noddack, Angewandte Chemie **47**, 653 (1934)
8. H. Menke, G. Herrmann, Radiochimica Acta **16**, 119 (1971)
9. O. Hahn, *Erlebnisse und Erkenntnisse* (Econ, Düsseldorf, 1975)
10. P. Brix, Phys. Blätter **45**, 2 (1989)
11. A. Hardy, L. Sexl, *Lise Meitner* (Rowohlt, Hamburg, 2002)
12. H. Rechenberg, in *Die großen Physiker,* edited by K. v. Meÿenn (C. H. Beck, München, 1997), vol. **2**, pp. 210
13. R. Reed, *The Path to Nuclear Fission, the Story of Lise Meitner, Otto Hahn* (Rosemarie Reed Productions Ltd., 2005)
14. C. F. v. Weizsäcker, Z. Phys. **96**, 431 (1935)
15. N. Bohr, Nature **137**, 344 (1936)
16. N. Bohr, F. Kalckar, Kgl. Danske Vid. Selskab, Math. Phys. Medd. **14** (10), (1937)
17. L. Meitner, O. R. Frisch, Nature **143**, 239 (1939)
18. O. R. Frisch, Nature **143**, 276 (1939)
19. H. v. Halban, F. Joliot, L. Kowarski, Nature **143**, 470, 680 (1939)
20. G. N. Flerov, K. A. Petrzhak, Phys. Rev. **58**, 89 (1940); Journ. of Phys. (USSR) **3**, 275 (1940)
21. E. Feenberg, Phys. Rev. **55**, 504 (1939)
22. J. Frenkel, Phys. Rev. **55**, 987 (1939); Journ. of Phys. (USSR) **1**, 125 (1939)
23. C. F. v. Weizsäcker, Naturwiss. **27**, 133 (1939)
24. N. Bohr, J. A. Wheeler, Phys. Rev. **56**, 426 (1939).
25. H. Pelzer, E. Wigner, Z. phys. Chemie B **15**, 445 (1932)
26. E. Wigner, Trans. Faraday Soc. **34**, 29 (1938)
27. S. Flügge, Naturwiss. **27**, 402 (1939)

H. J. Krappe and K. Pomorski, *Theory of Nuclear Fission,*
Lecture Notes in Physics 838, DOI: 10.1007/978-3-642-23515-3,
© Springer-Verlag Berlin Heidelberg 2012

28. *50 Years with Nuclear Fission* eds. J. W. Behrens, A. D. Carlson (Amer. Nucl. Society, La Grange Park, Ill. 1989) vol. **1** pp. 2-91.

29. H. A. Bethe, R. F. Bacher, Rev. Mod. Phys. **8**, 82 (1936)

30. L. Wilets, *Theories of Nuclear Fission* (Clarendon, Oxford, 1964)

31. V. G. Nossoff, *Proc. of the Intern. Conf. on the Peaceful Uses of Atomic Energy, Geneva 1955* (UN, New York, 1956) vol. **2**, pp. 205

32. W. J. Swiatecki, Phys. Rev. **101**, 651 (1956); Phys. Rev. **104**, 993 (1956)

33. K. T. R. Davies, A. J. Sierk, J. Comp. Phys. **18**, 311 (1975)

34. S. Cohen, W. J. Swiatecki, Ann. Phys. (N.Y.) **22**, 406 (1963)

35. J. R. Nix, Nucl. Phys. **A130**, 241 (1969)

36. U. L. Businaro, S. Gallone, Nuovo Cim. **1**, 629 (1955); Nuovo Cim. **1**, 1277 (1955); Nuovo Cim. **5**, 315 (1957)

37. W. D. Myers, W. J. Swiatecki, Nucl. Phys. **81**, 1 (1966)

38. Lord Rayleigh, Proc. Lond. Math. Soc. **10**, 4 (1878); Proc. Roy. Soc. **29**, 71 (1879)

39. J. Plateau, *Statique experimentale et théorique des liquides soumis aux seules forces moléculaires* (Gauthier-Villars, Paris, 1873)

40. U. Brosa, S. Großmann, A. Müller, Phys. Rep. **197C**, 167 (1990)

41. J. Błocki, J. Randrup, W. J. Swiatecki, C. F. Tsang, Ann. Phys. (N.Y.) **105**, 427 (1977)

42. G. A. Pik-Pichak, J. Exper. Theor. Phys. (USSR) **34**, 341 (1958), [Sov. Phys. JETP **7**, 238 (1958)]

43. G. A. Pik-Pichak, J. Exper. Theor. Phys. (USSR) **42**, 1294 (1962), [Sov. Phys. JETP **15**, 897 (1962)]; J. Exper. Theor. Phys. (USSR) **43**, 1701 (1962), [Sov. Phys. JETP **16**, 1201 (1963)]; Yad. Fiz. **31**, 98 (1980), [Sov. J. Nucl. Phys. **31**, 52 (1980)]

44. S. Cohen, F. Plasil, W. J. Swiatecki, Ann. Phys. (N.Y.) **82**, 557 (1974)

45. J. R. Hiskes, University of California Lawrence Radiation Laboratory Report UCRL-9275 (1960)

46. R. Beringer, W. J. Knox, Phys. Rev. **121**, 1195 (1961)

47. D. Ward, R. M. Diamond, W. J. Swiatecki, R. M. Clark, M. Cromaz, M. A. Deleplanque, P. Fallon, A. Goergen, G. J. Lane, I. Y. Lee, A. O. Macchiavelli, W. Myers, F. S. Stephens, C. E. Svensson, K. Vetter, Phys. Rev. C **66**, 024317 (2002)

48. P. Appell, *Mécanique rationelle* (Gauthier-Villars, Paris, 1932) vol. **4**, part 1, chapter 8

49. L. Lichtenstein, *Gleichgewichtsfiguren rotierender Flüssigkeiten* (Springer, Berlin, 1933)

50. S. Chandrasekhar, *Ellipsoidal Figures of Equilibrium* (Yale University, Yale, 1969)

51. J. Meyer-ter-Vehn, Physik und Didaktik **4**, 317 (1979)

52. W. J. Swiatecki, Nucl. Phys. **A574**, 233c (1994)

53. B. Derjaguin, Kolloid Z. **69**, 155 (1934)

54. B. V. Deryaguin, N. V. Churaev, V. M. Muller, *Surface Forces* (Plenum, New York, 1987)

55. R. Bass, Phys. Lett. **47B**, 139 (1973)

56. J. Wilczyński, *Proc. Third IAEA Symposium on the Physics, Chemistry of Fission, Rochester, 1973* (IAEA, Vienna, 1974), vol. **2**, pp. 269; Nucl. Phys. **A216**, 386 (1974)

57. J. Randrup, W. J. Swiatecki, C. F. Tsang, University of California Lawrence Berkeley Laboratory, Annual Report 1973, LBL-2366, pp. 143
58. J. Błocki, W. J. Swiatecki, Ann. Phys. (N.Y.) **132**, 53 (1981)
59. R. Bradley, Phil. Mag. (7) **13**, 853 (1932)
60. Lord Rayleigh, Phil. Mag. (5) **14**, 187 (1882)
61. J. Zeleny, Phys. Rev. **10**, 1 (1917)
62. D. Duft, T. Achtzehn, R. Müller, B. A. Huber, T. Leisner, Nature **421**, 128 (2003)
63. J. Zeleny, Phys. Rev. **2** (1914) 69; Proc. Camb. Phil. Soc. **18** 71 (1915)
64. J. J. Nolan, Proc. Roy. Irish Acad. **37** 28 (1925)
65. W. A. Macky, Proc. Roy. Soc. A **133** 565 (1931)
66. C. T. R. Wilson, G. I. Taylor, Proc. Camb. Phil. Soc. **22**, 728 (1925); G. E. Taylor, A. D. McEwan, J. Fluid Mech. **22**, 1 (1965)
67. V. V. Pashkevich, H. J. Krappe, J. Wehner, Z. Phys. D **40**, 338 (1997)
68. D. Duft, H. Lebius, H. Huber, B. A. Guet, T. Leisner, Phys. Rev. Lett. **89**, 0884503 (2002)
69. H. J. Krappe, Z. Phys. D **23**, 269 (1992)
70. V. V. Pashkevich, Nucl. Phys. **A169**, 275 (1971)
71. C.F. Tsang, in *Proc. Intern. Coll. on Drops, Bubbles*, eds. D. J. Collins, M. S. Plesset, M. M. Saffren (California Institute of Technology, Jet Propulsion Laboratory, 1974), vol. **1**, pp. 85; PHD Thesis, University of California, Berkeley 1969, report UCRL-18899
72. A. Frohn, N. Roth, *Dynamics of Droplets*, (Springer, Berlin, 2000) Chap. 7
73. T. Leisner, private communication
74. M. Orme, Phys. of Fluids A **3** (1991) 2936; Prog. Energy Combustion Sci. **23**, 65 (1997)
75. S. Flügge, Ann. d. Phys. (Leipzig) **39**, 373 (1941)
76. M. Fierz, Helv. Phys. Acta, **16**, 365 (1943)
77. S. Flügge, K. Woeste, Sitzungsberichte der Marburger Akad. der Wiss. **75**, 46 (1952)
78. D. L. Hill, in *Proc. of the Second UN Intern. Conf. on the Peaceful Uses of Atomic Energy, Geneva 1958* (UN, Geneva, 1958), vol. **15**, pp. 244
79. C. T. Alonso, University of California, Berkeley 1973, reports LBL 2366, pp. 133, LBL 4000, pp. 157
80. H. H. Tang, C. Y. Wong, J. Phys. A **7**, 1038 (1974)
81. R. W. Hasse, Ann. Phys. (N.Y.) **93**, 68 (1975)
82. K. T. R. Davies, A. T. Sierk, J. R. Nix, Phys. Rev. C **13**, 2385 (1976)
83. A. R. Edmonds, *Angular Momentum in Quantum Mechanics* (Princeton University, Princeton, 2nd edition 1960).
84. A. Bohr, Mat. Fys. Medd. Dan. Vid. Selsk. **26** (14) (1952)
85. A. Bohr, B. R. Mottelson: *Nuclear Structure* (Benjamin, New York 1974) vol. **2**
86. G. Herzberg, *Molecular Spectra, Molecular Structure* (van Nostrand, New York, 1950)
87. D. L. Hill, J. A. Wheeler, Phys. Rev. **89**, 1102 (1953)
88. O. Haxel, J. H. D. Jensen, H. E. Suess, Phys. Rev. **75**, 1766 (1949)
89. M. Goeppert-Mayer, Phys. Rev. **75**, 1969 (1949)
90. J. Rainwater, Phys. Rev. **79**, 432 (1950); Rev. Mod. Phys. **48**, 385 (1976)
91. A. Bohr, Phys. Rev. **81**, 134 (1951)
92. S. G. Nilsson, Mat. Fys. Medd. Dan. Vid. Selsk. **29** (16) (1955)

93. S. G. Nilsson, C. F. Tsang, A. Sobiczewski, Z. Szymanski, S. Wycech, C. Gustafson, I. L. Lamm, P. Möller, B. Nilsson, Nucl. Phys. **A131**, 1 (1969)
94. J. Bardeen, L. N. Cooper, J. R. Schrieffer, Phys. Rev. **108**, 1175 (1957)
95. A. Bohr, B. R. Mottelson, D. Pines, Phys. Rev. **110**, 936 (1958)
96. M. Bolsterli, E. O. Fiset, J. R. Nix, L. Norton, Phys. Rev. C **5**, 1050 (1972)
97. S. T. Belyaev, Mat. Fys. Medd. Dan. Vid. Selsk. **31** (11) (1959)
98. D. R. Bés, Z. Szymański, Nucl. Phys. **28**, 42 (1961)
99. V. M. Strutinsky, Yad. Fiz. (USSR) **3** 614 (1966), [Sov. J. Nucl. Phys. **3**, 449 (1966)]; Nucl. Phys. **A95**, 420 (1967)
100. P. Möller, S. G. Nilsson, J. R. Nix, Nucl. Phys. **A229**, 292 (1974)
101. I. Ragnarsson, S. G. Nilsson, R. K. Sheline, Phys. Rep. **45C**, 1 (1978)
102. V. M. Strutinsky, Nucl. Phys. **A122**, 1 (1968)
103. M. Brack, J. Damgaard, A. S. Jensen, H. C. Pauli, V. M. Strutinsky, C. Y. Wong, Rev. Mod. Phys. **44**, 320 (1972)
104. S. M. Polikanov, V. A. Druin, V. A. Karnaukhov, V. L. Mikheev, A. A. Pleve, N. K. Skobelev, V. G. Subbotin,G. M. Ter-Akopian, V. A. Fomichev, J. Exper. Theor. Phys. (USSR) **42**, 1464 (1961), [Sov. Phys. JETP **15**, 1062 (1962)]
105. P. Möller, S. G. Nilsson, Phys. Lett. **31B**, 283 (1970)
106. R. Vandenbosch, J. R. Huizenga, *Nuclear Fission* (Academic, New York, 1973).
107. Yu. A. Muzychka, V. V. Pashkevich, V. M. Strutinsky, Yad. Fiz. (USSR) **8**, 716 (1969), [Sov. J. Nucl. Phys. **8**, 417 (1969)]
108. P. Ring, P. Schuck, *The Nuclear Many-Body Problem* (Springer, Heidelberg, 1980).
109. P. Hohenberg, W. Kohn, Phys. Rev. **136**, B864 (1964)
110. W. Kohn, L. J. Sham, Phys. Rev. **140**, A1133 (1965)
111. R. M. Dreizler, E. K. U. Gross, *Density Functional Theory* (Springer, Heidelberg, 1990).
112. I. Zh. Petkov, M. V. Stoitsov, *Nuclear Density Functional Theory* (Clarendon Press, Oxford, 1991).
113. B. D. Serot, J. D. Walecka, Adv. Nucl. Phys. **16**, 1 (1986)
114. M. Brack, R. Bhaduri, *Semiclassical Physics* (Adisson-Wesley, Reading, NY, 1997).
115. M. Bender, P. H. Heenen, P. G. Reinhard, Rev. Mod. Phys. **75**, 121 (2003)
116. M. Baranger, Phys. Rev. **122**, 992 (1961)
117. N.N. Bogolyubov, Sov. Phys. JETP **7**, 41 (1958), [Sov. Phys. Usp. **67**, 236 (1959)]
118. D. Gogny, in *Nuclear Self Consistent Fields*, eds. G. Ripka, M. Porneuf (North Holland, Amsterdam, 1975), pp. 333.
119. S.J. Krieger, P. Bonche, H. Flocard, P. Quentin, M. Weiss, Nucl. Phys. **A517**, 275 (1990)
120. K. Sieja, A. Baran, Phys. Rev. C **68**, 044308 (2003)
121. G. Audi, A. H. Wapstra, C. Thibault, Nucl. Phys. **A729**, 337 (2003)
122. W. Satuła, J. Dobaczewski, W. Nazarewicz, Phys. Rev. Lett. **81**, 3599 (1998)
123. P. Möller, J. R. Nix, Nucl. Phys. **A536**, 20 (1992)
124. J. A. Sheikh, P. Ring, E. Lopes, R. Rossignoli, Phys. Rev. C **66**, 044318 (2002)
125. H. J. Lipkin, Ann. Phys. (N.Y.) **9**, 272 (1960)
126. Y. Nogami, Phys. Rev. **134**, B313 (1964)
127. H. C. Pradhan, Y. Nogami, J. Law, Nucl. Phys. **A201**, 357 (1973)
128. J. Dobaczewski, W. Nazarewicz, Phys. Rev. C **47**, 2418 (1993)

129. H. J. Krappe, S. Fadeev, Nucl. Phys. **A645**, 559 (1999)
130. D. G. Madland, J. R. Nix, Nucl. Phys. **A476**, 1 (1988)
131. R. G. Seyler, C. H. Blanchard, Phys. Rev. **124**, 227 (1961)
132. M. Brack, C. Guet, H.-B. Håkansson, Phys. Rep. **123C**, 275 (1985)
133. B. Grammaticos, A. Voros, Ann. Phys. (N.Y.) **123**, 359 (1979)
134. J. Bartel, K. Bencheikh, Eur. Phys. J. **A14**, 179 (2002)
135. X. Campi, S. Stringari, Nucl. Phys. **A337**, 313 (1980)
136. S. Großmann, unpublished manuscript.
137. J. Bartel, M. Brack, M. Durand, Nucl. Phys. **A445**, 263 (1985)
138. J. Bartel, P. Quentin, M. Brack, C. Guet, and H.-B. Håkansson, Nucl. Phys. **A386**, 79 (1982)
139. J. D. Walecka, Ann. Phys. (N.Y.) **83**, 491 (1974)
140. P. Ring, Prog. Part. Nucl. Phys. **37**, 193 (1996)
141. J. Boguta, A. R. Bodmer, Nucl. Phys. **A292**, 413 (1977)
142. P. Ring, Y. K. Gambhir, G. A. Lalazissis, Comp. Phys. Comm. **105**, 77 (1997)
143. V. Blum, G. Lauritsch, J. A. Maruhn, P. G. Reinhard, J. Comp. Phys. **100**, 364 (1992)
144. R. Brockmann, Phys. Rev. C **18**, 1510 (1978)
145. A. Bouyssy, J. F. Mathiot, N. van Giai, S. Marcos, Phys. Rev. C **36**, 380 (1987)
146. P.-G. Reinhard, M. Rufa, J. Maruhn, W. Greiner, J. Friedrich, Z. Phys. **A323**, 13 (1986)
147. G. A. Lalazissis, M. M. Sharma, P. Ring, Y. K. Gambhir, Nucl. Phys. **A608**, 202 (1996)
148. M. Bender, K. Rutz, P.-G. Reinhard, J. A. Maruhn, W. Greiner, Phys. Rev. C **60**, 034304 (1999)
149. T. Bürvenich, M. Bender, J. A. Maruhn, P.-G. Reinhard, Phys. Rev. C **69**, 014307 (2004)
150. M. Bender, K. Rutz, P.-G. Reinhard, J. A. Maruhn, Eur. Phys. J. **A7**, 467 (2000)
151. F. Tondeur, S. Goriely, J. M. Pearson, M. Onsi, Phys. Rev. C **62**, 024308 (2000)
152. S. Goriely, F. Tondeur, J. M. Pearson, At. Data Nucl. Data Tables **77**, 311 (2001)
153. R. E. Peierls, J. Yoccoz, Proc. Phys. Soc. (London) **A70**, 381 (1957)
154. H. D. Zeh, Z. Phys. **188**, 361 (1965)
155. R. Dreizler, P. Federman, B. Giraud, E. Osnes, Nucl. Phys. **A113**, 145 (1968)
156. H. Flocard, N. Onishi, Ann. Phys. (N.Y.) **254**, 275 (1996)
157. A. Góźdź, Phys. Lett. **152B**, 281 (1985)
158. A. Góźdź, K. Pomorski, M. Brack, E. Werner, Nucl. Phys. **A442**, 26 (1985)
159. A. Góźdź, K. Pomorski, Nucl. Phys. **A451**, 1 (1986)
160. B. F. Bayman, Nucl. Phys. **15**, 33 (1960)
161. K. Dietrich, H. J. Mang, J. H. Pradal, Phys. Rev. **135**, B22 (1964)
162. J. W. Negele, D. Vautherin, Phys. Rev. C **5**, 1472 (1972)
163. J. S. Bell, T. H. R. Skyrme, Phil. Mag. **1**, 1055 (1956)
164. T. H. R. Skyrme, Nucl. Phys. **9**, 615 (1958)
165. S. O. Bäckmann, A. D. Jackson, J. Speth, Phys. Lett. **56B**, 209 (1975)
166. M. Beiner, H. Flocard, N. van Giai, P. Quentin, Nucl. Phys. **A238**, 29 (1975)
167. J. Dobaczewski, H. Flocard, J. Treiner, Nucl. Phys. **A422**, 103 (1983)
168. E. Chabanat, P. Bonche, P. Haensel, J. Meyer, R. Schaeffer, Nucl. Phys. **A635**, 231 (1998); ibid. **A643**, 441 (1998), Erratum.

169. P.-G. Reinhard, H. Flocard, Nucl. Phys. **A584**, 467 (1995)

170. M. Samyn, S. Goriely, P.-H. Heenen, J. M. Pearson, F. Tondeur, Nucl Phys. **700** (2002) 142; Phys. Rev. C **66**, 024326 (2002)

171. D. Vautherin, D. M. Brink, Phys. Rev. C **5**, 626 (1972)

172. J. C. Slater, Phys. Rev. **81**, 385 (1951)

173. M. J. Giannoni, P. Quentin, Phys. Rev. **21**, 2076 (1980)

174. G. Audi, A. H. Wapstra, Nucl. Phys. **A595**, 409 (1995)

175. M. Samyn, S. Goriely, M. Bender, J.M. Pearson, Phys. Rev. C **70**, 044309 (2004)

176. D. Gogny, *Proc. Int. Conf. on Nucl. Phys., Munich, 1973* eds. J. de Boer, H. J. Mang (North Holland, Amsterdam, 1973) vol. **1**, pp. 48

177. D. M. Brink, E. Bocker, Nucl. Phys. **91**, 1 (1967)

178. J. Decharge, D. Gogny, Phys. Rev. C **21**, 1568 (1980)

179. R. G. Seyler, C. H. Blanchard, Phys. Rev. **131**, 355 (1963)

180. W. D. Myers, W. J. Swiatecki, Ann. Phys. (N.Y.) **55**, 395 (1969)

181. W. D. Myers, W. J. Swiatecki, Nucl. Phys. **A601**, 141 (1996)

182. W. D. Myers, W. J. Swiatecki, Ann. Phys. (N.Y.) **84**, 186 (1974)

183. G. Süssmann, Z. Phys. **A274**, 145 (1975)

184. J. Friedrich, N. Voegler, Nucl. Phys. **A373**, 192 (1982)

185. A. Sommerfeld, Z. Phys. **47**, 1 (1927)

186. J. Damgaard, H. C. Pauli, V. V. Pashkevich, V. M. Strutinsky, Nucl. Phys. **A137**, 432 (1969)

187. Y. H. Chu, B. K. Jennings, M. Brack, Phys. Lett. **68B**, 407 (1977)

188. H. J. Krappe, in *Heavy Ions, Nuclear Structure, Proc. 14th Mikołajki Summer School of Physics 1981*, eds. B. Sikora, Z. Wilhelmi, Nuclear Science Research Conference Series, vol. **5** (Harwood Academic, New York 1984) pp. 197

189. J. Friedrich, N. Voegler, P.-G. Reinhard, Nucl. Phys. **A459**, 10 (1986)

190. R. D. Woods, D. S. Saxon, Phys. Rev. **95**, 577 (1954)

191. S. Ćwiok, J. Dudek, W. Nazarewicz, J. Skalski, T. Werner, Comp. Phys. Comm. **46**, 379 (1987).

192. J. Blomquist, S. Wahlborn, Ark. Fys. **16**, 545 (1960)

193. E. Rost, Phys. Lett. **26B**, 184 (1968)

194. V. A. Chepurnov, Yad. Fiz. **5**, 955 (1967), [Sov. J. Nucl. Phys. **6**, 696 (1968)]

195. W. D. Myers, Nucl. Phys. **A145**, 387 (1969)

196. H. C. Pauli, Phys. Rep. **7**, 35 (1973)

197. P. Möller, J. R. Nix, W. D. Myers, W. J. Swiatecki, At. Data Nucl. Data Tables **59**, 185 (1995)

198. K. T. R. Davies, R. L. Becker, J. Math. Phys. **19**, 2207 (1978)

199. H. Iwe, Z. Phys. **A304**, 347 (1982)

200. W. Hasse, W. D. Myers, *Geometrical Relationships of Macroscopic Nuclear Physics* (Springer, Heidelberg, 1988)

201. G. A. Korn, T. M. Korn, *Mathematical Handbook for Scientists, Engineers* (McGraw-Hill, New York, 1968) Sec. 17.3

202. A. Erdélyi, *Asymptotic Expansions* (Dover, New York 1956) Chap. 1

203. H. J. Krappe, J. R. Nix, A. J. Sierk, Phys. Rev. C **20**, 992 (1979)

204. H. J. Krappe, J. Comp. Phys. **50**, 499 (1983)

205. K. T. R. Davies, J. R. Nix, Phys. Rev. C **14**, 1977 (1976)

206. N. W. Ashcroft, N. D. Mermin, *Solid State Physics* (Saunders College, Philadelphia, 1981)

207. F. Bloch, Z. Phys. **57**, 545 (1929)
208. W. D. Myers, At. Data Nucl. Data Tables **17**, 411 (1976)
209. E. Wigner, Phys. Rev. **51**, 947 (1937)
210. W. D. Myers, W. J. Swiatecki, Nucl. Phys. **A612**, 249 (1997)
211. P. Möller, J. R. Nix, Nucl. Phys. **A272**, 502 (1976)
212. W. D. Myers, *Droplet Model of Atomic Nuclei* (IFI/Plenum, New York, 1977)
213. W. Satuła, R. Wyss, Phys. Lett. **393B**, 1 (1997)
214. W. Satuła, D. J. Dean, J. Gary, S. Mizutori, W. Nazarewicz, Phys. Lett. **407B**, 103 (1997)
215. P. Möller, J. R. Nix, Nucl. Phys. **A361**, 117 (1981); At. Data Nucl. Data Tables **26**, 165 (1981)
216. T. Johansson, S. G. Nilsson, Z. Szymanski, Ann. Phys. (Paris) **5**, 377 (1970)
217. P. Möller, J. R. Nix, W. J. Swiatecki, Nucl. Phys. **A492**, 349 (1989)
218. G. Audi, *Midstream Atomic Mass Evaluation, 1989*, unpublished.
219. M. S. Antony, *Nuclide Chart, Strasbourg 2002* (Impressions François, Haguenau, 2002)
220. L. L. Foldy, Phys. Rev. **83**, 397 (1951)
221. K. Pomorski, J. Dudek, Int. J. Mod. Phys. **E13**, 107 (2004)
222. W. D. Myers, W. J. Swiatecki, Nucl. Phys. **A536**, 61 (1992)
223. G. Audi, A. H. Wapstra, Nucl. Phys. **A565**, 1 (1993)
224. W. D. Myers, W. J. Swiatecki, Ark. Fys. **36**, 343 (1967)
225. H.J. Krappe, J.R. Nix, in *Proc. 3rd IAEA Symposium on the Physics, Chemistry of Fission, Rochester, New York, 1973* (IAEA, Vienna, 1974) vol. **1**, pp. 159
226. W. M. Howard, P. Möller, At. Data Nucl. Data Tables **25**, 219 (1980)
227. P. Möller, J. R. Nix, At. Data Nucl. Data Tables **39**, 213 (1988)
228. K. Pomorski, J. Dudek, Phys. Rev. C **67**, 044316 (2003)
229. A. H. Wapstra, G. Audi, R. Hoekstra, At. Data. Nucl. Data Tables **39**, 281 (1988)
230. P. Möller, A. J. Sierk, A. Iwamoto, Phys. Rev. Lett. **92**, 072501 (2004)
231. A. E. S. Green, *Nuclear Physics* (McGraw-Hill, New York, 1955) pp. 185, 250
232. D. N. Delis, Y. Blumenfeld, D. R. Bowman, N. Colonna, K. Hanold, K. X. Jing, M. C. Meng, G. F. Peaslee, G. J. Wozniak, L. G. Moretto, Nucl. Phys. **A534**, 403 (1991)
233. K. X. Jing, L. G. Moretto, A. C. Veeck, N. Colonna, I. Lhenry, K. Tso, K. Hanold, W. Skulski, Q. Sui, G. J. Wozniak, Nucl. Phys. **A645**, 203 (1999)
234. P. Möller, W. D. Myers, W. J. Swiatecki, J. Treiner, in *Proc. 7th Int. Conf. on Nuclear Masses and Fundamental Constants, Darmstadt-Seeheim Germany, 1984* (Schriftenreihe "Wissenschaft und Technik" vol. 26) ed. O. Klepper (Technische Hochschule Darmstadt, 1984) pp. 457
235. P. Möller, W. D. Myers, W. J. Swiatecki, J. Treiner, At. Data Nucl. Data Tables, **39**, 225 (1988)
236. G. G. Bunatian, V. M. Kolomietz, V. M. Strutinsky, Nucl. Phys. **A188**, 225 (1972)
237. K. Dietrich, in *The Structure of Nuclei, Lectures presented at the International Course on Nuclear Theory, ICTP, Trieste, 1971* (IAEA, Vienna, 1972) pp. 373
238. M. Abramowitz, I. A. Stegun (eds.) *Handbook of Mathematical Functions* (Nat. Bureau of Standards, Washington DC, 5th printing, 1966)
239. M. Brack, H. C. Pauli, Nucl. Phys. **A207**, 401 (1973)
240. V. M. Strutinsky, F. A. Ivanyuk, Nucl. Phys. **A255**, 405 (1975)

241. J. Dudek, W. Nazarewicz, P. Olanders, Nucl. Phys. **A420**, 285 (1984)

242. H.J. Krappe, U. Wille, in *Proc. 2nd IAEA Symposium on the Physics, Chemistry of Fission, Vienna, 1969* (IAEA, Vienna, 1970) pp. 197

243. Y. Yariv, T. Ledergerber, H. C. Pauli, Z. Phys. **A278**, 225 (1976)

244. V. M. Strutinsky, F. A. Ivanyuk, V. V. Pashkevich, Yad. Fis. **31**, 88 (1980), [Sov. J. Nucl. Phys. **31**, 46 (1980)]

245. F. Tondeur, Journ. Phys. (Paris) **33**, 825 (1972)

246. F. A. Ivanyuk, V. M. Strutinsky, Z. Phys. **A286**, 291 (1978)

247. F. A. Ivanyuk, V. M. Strutinsky, Z. Phys. **A290**, 107 (1979)

248. F. A. Ivanyuk, V. M. Strutinsky, Z. Phys. **A293**, 337 (1979)

249. F. A. Ivanyuk, Z. Phys. **A316**, 233 (1984)

250. K. Pomorski, Phys. Rev. C **70**, 044306 (2004)

251. V. M. Strutinsky, Pramāṇa **33**, 21 (1989)

252. A. B. Migdal, V. P. Krainov, *Approximation Methods in Quantum Mechanics* (Benjamin, Reading, Mass. 1969) Sec. 3.2

253. D. H. E. Gross, Phys. Lett. **42B**, 41 (1972)

254. R. K. Bhaduri, C. K. Ross, Phys. Rev. Lett. **27**, 606 (1971)

255. B. K. Jennings, Nucl. Phys. **A207** 538 (1973)

256. A. K. Dutta, J.-P. Arcoragi, J. M. Pearson, R. Behrmann, F. Tondeur, Nucl. Phys. **A458**, 77 (1986)

257. F. Tondeur, A. K. Dutta, J. M. Pearson, R. Behrmann, Nucl. Phys. **A470**, 93 (1987)

258. Y. Aboussir, J. M. Pearson, A. K. Dutta, F. Tondeur, Nucl. Phys. **A549**, 155 (1992)

259. Y. Aboussir, J. M. Pearson, A. K. Dutta, F. Tondeur, At. Data Nucl. Data Tables **61**, 127 (1995)

260. A. Mamdouh, J. M. Pearson, M. Rayet, F. Tondeur, Nucl. Phys. **A644**, 389 (1998); **A679**, 337 (2001)

261. W. Younes, D. Gogny, Phys. Rev. C **80**, 054313 (2009)

262. H. Flocard, P. Quentin, A. K. Kerman, D. Vautherin, Nucl. Phys. **A203**, 433 (1973)

263. J. F. Berger, M. Girod, D. Gogny, Nucl. Phys. **A502**, 85c (1989)

264. M. Warda, J. L. Egido, L. M. Robledo, K. Pomorski, Phys. Rev. C **66**, 014310 (2002)

265. B. Giraud, J. LeTourneux, S. K. M. Wong, Phys. Lett. **32B**, 23 (1970)

266. V. M. Strutinsky, J. Exper. Theor. Phys. (USSR) **42**, 1571 (1962), [Sov. Phys. JETP **15**, 1091 (1964)]; J. Exper. Theor. Phys. (USSR) **45**, 1891, 1900 (1963), [Sov. Phys. JETP **18**, 1298, 1305 (1964)]

267. V. M. Strutinsky, N. Ya. Lyashchenko, N. A. Popov, J. Exper. Theor. Phys. **43**, 584 (1962), [Sov. Phys. JETP **16**, 418 (1963)]; Nucl. Phys. **46**, 639 (1963)

268. F. A. Ivanyuk, K. Pomorski, Phys. Rev. C **79**, 054327 (2009)

269. F. Ivanyuk, K. Pomorski, Int. J. Mod. Phys. **19**, 514 (2010)

270. F. Ivanyuk, Int. J. Mod. Phys. **18**, 879 (2009)

271. J. N. P. Lawrence, Phys. Rev. **139**, B1227 (1965)

272. R. W. Hasse, Nucl. Phys. **A128**, 609 (1969)

273. S. Trentalange, S. E. Koonin, A. J. Sierk, Phys. Rev. C **22**, 1159 (1980)

274. A. J. Sierk, Phys. Rev. C **33**, 2039 (1986)

275. K. Pomorski, J. Bartel, Int. J. Mod. Phys. **E15**, 417 (2006)

276. J. Bartel, A. Dobrowolski, B. Nerlo-Pomorska, K. Pomorski, F. A. Ivanyuk, in *Proc. 4th Intern. Workshop on Nuclear Fission, Fission Product Spectroscopy, Cadarache, France, 2009* AIP Conf. Proc. **1175**, 231 (2009)

277. V. S. Stavinsky, N. S. Rabotnov, A. A. Seregin, Yad. Fiz. **7**, 1051 (1968), [Sov. J. Nucl. Phys. **7**, 631 (1968) 631]

278. V. V. Pashkevich, Nucl. Phys. **A169**, 275 (1971)

279. P. Moon, D. E. Spencer, *Field Theory Handbook* (Springer, Berlin, 1971) Sec. III, E2C

280. G.-T. Tillack, *PHD Thesis* (Technische Universität Dresden, 1992)

281. G.-T. Tillack, Phys. Lett. **278B**, 403 (1992)

282. G.-T. Tillack, R. Reif, A. Schülke, P. Fröbrich, H. J. Krappe, H. G. Reusch, Phys. Lett. **296B**, 296 (1992)

283. J. R. Nix, Nucl. Phys. **A130**, 241 (1969)

284. J. Błocki, W. J. Swiatecki, *Nuclear Deformation Energies* (report LBL 12811) (Lawrence Berkeley Laboratory, 1982)

285. W. J. Swiatecki, Nucl. Phys. **A428**, 199c (1984)

286. J. Błocki, J. de Physique Coll. **45**, C6-489 (1984)

287. J. Błocki, K. Grotowski, R. Płaneta, W. J. Swiatecki, Nucl. Phys. **A445**, 367 (1985)

288. J. Błocki, M. Dworzecka, F. Beck, H. Feldmeier, Phys. Lett. **99B**, 13 (1981)

289. H. J. Krappe, Nucl. Phys. **A269**, 493 (1976)

290. M. G. Mustafa, P. A. Baisden, H. Chandra, Phys. Rev. C **25**, 2524 (1982)

291. W. J. Swiatecki, in *Proc. Int. Conf. on Nuclear Reactions Induced by Heavy Ions, Heidelberg, 1969*, eds. R. Bock, W. R. Hering (North-Holland, Amsterdam, 1970), pp. 729

292. P. Möller, private communication

293. P. Möller, Nucl. Phys. **A192**, 529 (1972)

294. C. Gustafsson, P. Möller, S. G. Nilsson, Phys. Lett **34B**, 349 (1971)

295. R. Bengtson, I. Ragnarsson, S. Åberg, A. Gyurkovich, A. Sobiczewski, K. Pomorski, Nucl. Phys. **A473**, 77 (1987)

296. S. E. Larsson, P. Möller, S. G. Nilsson, Phys. Scripta **10A**, 53 (1974)

297. J. E. Lynn, Pramāṇa **33**, 33 (1989)

298. H. Weigmann, in *The Nuclear Fission Process*, ed. C. Wagemans (CRC, Boca Raton USA, 1991) pp. 7

299. B. Nerlo-Pomorska, K. Pomorski, M. Zwierzchowska, Int. J. Mod. Phys. **E16**, 474 (2007)

300. A. Sobiczewski, M. Kowal, Int. J. Mod. Phys. **E18**, 869 (2009)

301. J. Blons, Nucl. Phys. **A502**, 121c (1989)

302. A. Kraszahorkay, D. Habs, M. Hunyadi, D. Grassmann, M. Csatlós, Y. Eisermann, T. Faestermann, G. Graw, J. Gulyás, R. Hertenberger, H. J. Maier, Z. Máté, A. Metz, J. Ott, P. Thirolf, S. Y. van der Werft, Phys. Lett. **461B**, 15 (1999); Acta Phys. Polon. **B32**, 657 (2001)

303. L. Csige, M. Csatlós, T Faestermann, Z. Gácsi, J. Gulyás, D. Habs, R. Hertenberger, M. Hunyadi, A. Krasznahorkay, R. Lutter, H. J. Maier, P. G. Thirolf, H.-F. Wirth, Acta Phys. Polon. **B38**, 1503 (2007)

304. J. F. Berger, K. Pomorski, Phys. Rev. Lett. **85**, 30 (2000)

305. K. Pomorski, unpublished data

306. J. Schirmer, J. Gerl, D. Habs, D. Schwalm, Phys. Rev. Lett. **63**, 2196 (1989)

307. H. J. Specht, J. Weber, E. Konecny, D. Heunemann, Phys. Lett. **41B**, 43 (1972)

308. A. Sobiczewski, S. Bjørnholm, K. Pomorski, Nucl. Phys. **A202**, 274 (1973)
309. D. Habs, Nucl. Phys. **A502**, 105c (1989)
310. D. Pansegrau, P. Reiter, D. Schwalm, H. Bauer, J. Eberth, D. Gassmann, D. Habs, T. Härtlein, F. Köck, H. G. Thomas, Phys. Lett. **484B**, 1 (2000)
311. B. Nerlo-Pomorska, Nucl. Phys. **A259**, 481 (1976)
312. S. Prerez-Martin, L. M. Robledo, Int. J. Mod. Phys. **E18**, 788 (2009)
313. J. A. Wheeler, Physica **22**, 1103 (1956)
314. G. Andersson, S. E. Larsson, G. Leander, P. Möller, S. G. Nilsson, I. Ragnarsson, S. Åberg, R. Bengtsson, J. Dudek, B. Nerlo-Pomorska, K. Pomorski, Z. Szymański, Nucl. Phys. **A268**, 205 (1976)
315. J. L. Egido, L. M. Robledo, Phys. Rev. Lett. **85**, 1198 (2000)
316. K. Neergård, V. V. Pashkevich, S. Frauendorf, Nucl. Phys. **A262**, 61 (1976)
317. W. D. Myers, W. J. Swiatecki, Ark. Fys. **36**, 343 (1967)
318. D.R. Inglis, Phys. Rev. **96** (1954) 1059; Phys. Rev. **103**, 1786 (1956)
319. L. D. Landau, E. M. Lifshitz, *Statistical Physics, Part 1*, 3rd edition by E. M. Lifshitz, L. P. Pitaevskiĭ, (Pergamon, Oxford, 1980)
320. M. Hillman, J. R. Grover, Phys. Rev. **185**, 1303 (1969)
321. S. Hilaire, J. P. Delaroche, M. Girod, Eur. Phys. J. **A12**, 169 (2001)
322. J. Töke, W. J. Swiatecki, Nucl. Phys. **A372**, 141 (1981)
323. H. J. Krappe, Phys. Rev. C **59**, 2640 (1999)
324. H. A. Bethe, Phys. Rev. **50**, 332 (1936)
325. T. Ericson, Adv. Phys. **9**, 425 (1960)
326. C. Van Lier, G. E. Uhlenbeck, Physica **4**, 531 (1937)
327. H. J. Krappe, Int. J. Mod Phys. **E13**, 277 (2004)
328. G. H. Hardy, E. M. Wright, *An Introduction into the Theory of Numbers* (Clarendon, Oxford, 5th edition, 1979) Chap. XIX
329. F. C. Auluck, D. S. Kothari, Proc. Camb. Phil. Soc. **42**, 272 (1946)
330. G. H. Hardy, S. Ramanujan, Proc. London Math. Soc. **17**, 75 (1918)
331. C. A. Engelbrecht, J. R. Engelbrecht, Ann Phys. (N.Y.) **207**, 1 (1992)
332. A. V. Ignatyuk, V. P. Lunev, Y. N. Shubin, *Proc. Advisory Group Meeting, Beijing, 1987* (IAEA-TECDOC-483) pp. 122
333. A. M. Anzaldo-Meneses, in *Nuclear Data for Science, Technology, Proc. Int. Conf. Antwerp, 1982*, ed. K. H. Böckhoff (Reidel, Dordrecht, Holland, 1983) pp. 534
334. A. M. Anzaldo-Meneses, in *Methods for the calculation of neutron nuclear data for structural materials in fast, fusion reactors, Proc. of the final meeting of a Coordinated Research Programme of the IAEA Nuclear Data Section*, ed. D. W. Muir, report INDC(NDS)-247 (IAEA, Vienna, 1991) pp. 131
335. A. V. Ignatyuk, M. G. Itkis, V. N. Okolovich, G. N. Smirenkin, A. S. Tishin, Yad. Fiz. **21**, 1185 (1975), [Sov. J. Nucl. Phys. **21**, 612 (1975)]
336. F. C. Williams Jr., Nucl. Phys. **A166**, 231 (1971)
337. C. Y. Fu, Nucl. Sci. Eng. **109**, 18 (1991)
338. A. V. Ignatyuk, *Statistical Properties of Excited Atomic Nuclei (in Russian)* (Energoatomizdat, Moscow, 1983), [IAEA report INDC-233(L) (IAEA, Vienna, 1985)]
339. C. Bloch, Phys. Rev. **93**, 1094 (1954)
340. D. W. Lang, Nucl. Phys. **77**, 545 (1966)
341. M. G. Mustafa, M. Blann, A. V. Ignatyuk, S. M. Grimes, Phys. Rev. C **45**, 1078 (1992)

342. D. W. Lang, K. J. Le Couteur, Nucl. Phys. **14**, 21 (1959)

343. A. S. Iljinov, M. V. Mebel, N. Bianchi, E. de Sanctis, C. Guaraldo,V. Luccherini, V. Muccifora, E. Polli, A. R. Reolon, P. Rossi, Nucl. Phys. **A543**, 517 (1992)

344. R. H. Fowler, E. A. Guggenheim, *Statistical thermodynamics* (Cambr. University, Cambridge, 1956)

345. A Gilbert, A. G. W. Cameron, Can. J. Phys. **43**, 1446 (1965)

346. E. Gadioli, I. Iori, Nuovo Cim. **51B**, 100 (1967)

347. H. K. Vonach, A. A. Katsanos, J. R. Huizenga, Nucl. Phys. **A122**, 465 (1968)

348. H. Vonach, M. Hille, Nucl. Phys. **A127**, 289 (1969)

349. W. Dilg, W. Schantl, H. Vonach, M. Uhl, Nucl. Phys. **A217**, 269 (1973)

350. T. D. Newton, Can. J. Phys. **34**, 804 (1956)

351. F. Pühlhofer, Nucl. Phys. **A280**, 267 (1976)

352. A. V. Ignatyuk, G. N. Smirenkin, A. S. Tishin, Yad. Fiz. **21**, 485 (1975), [Sov. J. Nucl. Phys. **21**, 255 (1975)]

353. P. B. Kahn, R. Rosenzweig, Phys. Rev. **187**, 1193 (1969)

354. L. Moretto, Phys. Lett. **38B**, 393 (1972)

355. S. K. Kataria, V. S. Ramamurthy, S. S. Kapoor, Phys. Rev. C **18**, 549 (1977)

356. S. Goriely, Nucl. Phys. **A605**, 28 (1996)

357. S. Hilaire, Phys. Lett. **583B**, 264 (2004)

358. P. A. Gottschalk, T. Ledergerber, Nucl. Phys. **A278**, 16 (1977)

359. J. des Cloizeaux, in *Many-body physics, Les Houches, 1967*, eds. C. de Witt, R. Balian (Gordon, Breach, New York, 1968) pp. 1

360. K. Junker, J. Hadermann, N. C. Mukhopadhyay, in *Proc. 4th IAEA Symposium on the Physics, Chemistry of Fission, Jülich 1979* (IAEA, Vienna, 1980), vol. 1, pp. 445

361. A. I. Blokhin, A. V. Ignatyuk, Yu. N. Shubin, Yad. Fiz. **48**, 371 (1988), [Sov. J. Nucl. Phys. **48**, 232 (1988)]

362. S. Bjørnholm, A. Bohr, B. Mottelson, in *Proc. Third IAEA Symposium on the Physics, Chemistry of Fission, Rochester, 1973* (IAEA, Vienna, 1974), vol. **1**, pp. 367

363. D. M. Brink, Rep. Prog. Phys. **21**, 144 (1958)

364. G. Hansen, A. S. Jensen, Nucl. Phys. **A406**, 236 (1983)

365. A. V. Ignatyuk, Yad. Fiz. **21**, 20 (1975), [Sov. J. Nucl. Phys. **21**, 10 (1975)]

366. D. Vauterin, N. Vinh Mau, Phys. Lett. **120B**, 216 (1983); N. Vinh Mau, D. Vauterin, Nucl. Phys. **A445**, 245 (1985)

367. A. R. Junghans, M. de Jong, H.-G. Clerc, A. V. Ignatyuk, G. A. Kudyaev, K.-H. Schmidt, Nucl Phys. **A629**, 635 (1998)

368. V. A. Plujko, O. M. Gorbachenko, I. M. Kadenko, Int. J. Mod. Phys. **E16**, 372 (2007)

369. A. V. Ignatyuk, in *Proc. OECD Meeting on Nuclear Level Densities, Bologna, 1989*, eds. G. Reffo, M. Herman, G. Maino (World Scientific, Singapore, 1989) pp. 3

370. A. V. Ignatyuk, Yad. Fiz. **17**, 502 (1973), [Sov. J. Nucl. Phys. **17**, 258 (1973)]

371. L. G. Moretto, Nucl. Phys. **A185**, 145 (1972)

372. A. V. Ignatyuk, Yu. N. Shubin, Isv. Akad. Nauk SSSR, Ser. Fiz. **37**, 1947 (1973), [Bull. USSR Acad. Sci. Phys. Ser. **37**, 127 (1973)]

373. A. V. Ignatyuk, K. K. Istekov, G. N. Smirenkin, Yad. Fiz. **29**, 875 (1979), [Sov. J. Nucl. Phys. **29**, 450 (1979)]

374. International Atomic Energy Agency, *Handbook for Calculations of Nuclear Reaction Data*, (IAEA-TECDOC-1034, Vienna, 1998)

375. W. Reisdorf, Z. Phys. **A300**, 227 (1981)

376. C. Guet, E. Strumberger, M. Brack, Phys. Lett. **205B**, 427 (1988)

377. J. Bartel, K. Pomorski, B. Nerlo-Pomorska, Int. J. Mod. Phys. **E15**, 478 (2006)

378. J. M. Alexander, M. T. Magda, S. Landowne, Phys. Rev. C **42**, 1092 (1990)

379. T. Baer, W. L. Hase, *Unimolecular Reaction Dynamics* (Oxford University, Oxford, 1996) sec. 6.2

380. M. Polanyi, E. Wigner, Z. phys. Chemie **139**, 439 (1928)

381. R. J. Charity, Phys. Rev. C **53**, 512 (1996)

382. J. P. Lestone, Phys. Rev. C **59**, 1540 (1999)

383. V. S. Ramamurty, S. S. Kapoor, Phys. Rev. C **32**, 2182 (1985)

384. D. Vorkapić, B. Ivanišević, Phys. Rev. C **52**, 1980 (1995)

385. I. Halpern, V. M. Strutinsky, in *Proc. of the Second UN Intern. Conf. on the Peaceful Uses of Atomic Energy, Geneva 1958* (UN, Geneva, 1958) vol. **15**, pp. 408

386. L. G. Moretto, Nucl. Phys. **A247**, 211 (1975)

387. R. J. Charity, M. A. McMahan, G. J. Wozniak, R. J. McDonald, L. G. Moretto, D. G. Sarantites, L. G. Sobotka, G. Guarino, A. Pantaleo, L. Fiore, A. Gobbi, K. D. Hildenbrand, Nucl. Phys. **A483**, 371 (1988)

388. A. Bohr, in *Proc. of the Intern. Conf. on the Peaceful Uses of Atomic Energy, Geneva 1955* (UN, New York, 1956) vol. **2**, pp. 151

389. S. Bjørnholm, V. M. Strutinsky, Nucl. Phys. **A136**, 1 (1969)

390. E. V. Gai, A. V. Ignatyuk, N. S. Rabotnov, G. N. Smirenkin, Yad. Fiz. **10**, 542 (1969), [Sov. J. Nucl. Phys. **10**, 311 (1969)

391. S. Bjørnholm, J. E. Lynn, Rev. Mod. Phys. **52**, 725 (1980)

392. H. Weigmann, H.-H. Knitter, F.-J. Hambsch, Nucl. Phys. **A502**, 177c (1989)

393. S. S. Kapoor, V. S. Ramamurthy, Pramāṇa **33**, 161 (1989)

394. G. V. Danilyan, B. D. Vodennikov, V. P. Dronyaev, V. V. Novitskiĭ, V. S. Pavlov, S. P. Borovlev, Pis'ma Zh. Eksp. Teor. Fiz. **26**, 197 (1977), [JETP Lett. **26**, 186 (1977)]

395. G. V. Danilyan, B. D. Vodennikov, V. P. Dronyaev, V. V. Novitskiĭ, V. S. Pavlov, S. P. Borovlev, Yad. Fiz. **27**, 42 (1978), [Sov. J. Nucl. Phys. **27**, 21 (1978)]

396. B. D. Vodennikov, G. V. Danilyan, V. P. Dronyaev, V. V. Novitskiĭ, V. S. Pavlov, S. P. Borovlev, Pis'ma Zh. Eksp. Teor. Fiz. **27**, 68 (1978), [JETP Lett. **27**, 62 (1977)]

397. O. P. Sushkov, V. V. Flambaum, Phys. Lett. **94B**, 277 (1980)

398. O. P. Sushkov, V. V. Flambaum, Yad. Fiz. **33**, 59 (1981), [Sov. J. Nucl. Phys. **33**, 31 (1981)]

399. O. P. Sushkov, V. V. Flambaum, Usp. Fiz. Nauk **136**, 3 (1982), [Sov. Phys. Usp. **25**, 1 (1982)]

400. F. C. Michel, Phys. Rev. **133**, B329 (1964)

401. E. G. Adelberger, W. C. Haxton, Ann. Rev. Nucl. Part. Sci. **35**, 501 (1985)

402. D. A. Zaikin, Nucl. Phys. **86**, 638 (1966)

403. U. Fano, A. R. P. Rau, *Symmetries in Quantum Physics* (Academic, San Diego, 1996) Chap. 4.4

404. J. E. Lynn, *Theory of neutron resonance reactions* (Clarendon, Oxford, 1968)

405. A. L. Barabanov, W. I. Furman, in *Proc. Thirteenth Meeting on the Physics of Nuclear Fission in the Memory of G. N. Smirenkin, Obninsk, 1995*, ed. B. D. Kuzminov (State Scientific Center of the Russian Federation, Obninsk, Russia, 1995) pp. 52

406. A. Kötzle, P. Jesinger, F. Gönnenwein, G. A. Petrov, V. I. Petrova, A. M. Gagarsky, G. Danilyan, O. Zimmer, V. Nesvizhhevsky, Nucl. Instr. Meth. Sec. A **440**, 750 (2000)

407. U. Graf, F. Gönnenwein, P. Geltenbort, K. Schreckenbach, Z. Phys. **A351**, 281 (1995)

408. P. Jesinger, A. Kötzle, F. Gönnenwein, M. Mutterer, J. v. Kalben, G. V. Danilyan, V. S. Pavlov, G. A. Petrov, A. M. Gagarski, W. H. Trzaska, S. M. Soloviev, V. V. Nesvizhevski, O. Zimmer, Yad. Fiz. **65**, 662 (2002), [Phys. Atom. Nucl. **65**, 630 (2002)]

409. F. Gönnenwein, A. V. Belozerov, A. G. Beda, S. I. Burov, G. V. Danilyan, A. N. Martem'yanov, V. S. Pavlov, V. A. Shchenev, L. N. Bondarenko, Yu. A. Mostovoĭ, P. Geltenbort, J. Last, K. Schreckenbach, Nucl. Phys. **A567**, 303 (1994)

410. H. J. Krappe, K. Möhring, M. C. Nemes, H. Rossner, Z. Phys. **A314**, 23 (1983)

411. P. M. Morse, H. Feshbach, *Methods of Theoretical Physics* (McGraw-Hill, New York, 1953), § 9.3

412. C. Eckart, Phys. Rev. **35**, 1303 (1930)

413. L. D. Landau, E. M. Lifshitz, *Quantum Mechanics, Non-Relativistic Theory* (Pergamon, London, 1958)

414. D. Wilmore, P. E. Hodgson, Nucl. Phys. **55**, 673 (1964)

415. F. G. Perey, Phys. Rev. **131**, 745 (1963)

416. J. M. Lohr, W. Haeberli, Nucl. Phys. **A232**, 381 (1974)

417. F. D. Becchetti, Jr., G. W. Greenless, in *Polarization Phenomena in Nuclear Reactions*, eds. H. H. Barschall, W. Haeberli (Univ. Wisconsin, Madison, 1971) pp. 682

418. J. R. Huizenga, G. Igo, Nucl. Phys. **29**, 462 (1962)

419. R. Blendowske, T. Fliessbach, H. Walliser, Nucl. Phys. **A464**, 75 (1987)

420. G. H. Rawitscher, Nucl. Phys. **85**, 337 (1966)

421. P. Fröbrich, R. Lipperheide, *Theory of Nuclear Reactions* (Clarendon, Oxford, 1996) Chaps. 9, 10

422. I. Dostrovsky, Z. Fraenkel, G. Friedlaender, Phys. Rev. **116** 683 (1959)

423. V. Weisskopf, Phys. Rev. **52** 295 (1937)

424. A. A. McMahan, J. M. Alexander, Phys. Rev. C **21**, 1261 (1980)

425. J. R. Huizenga, A. N. Behkami, I. M. Govil, W. U. Schröder, J. Töke, Phys. Rev. C **40**, 664 (1989)

426. A. Iwamoto, R. Herrmann, Z. Phys. **A338**, 303 (1991)

427. K. Dietrich, K. Pomorski, J. Richert, Z. Phys. **A351**, 397 (1995)

428. J. O. Rasmussen, B. Segal, Phys. Rev. **103**, 1298 (1956)

429. P. Fröman, Mat. Fys. Skr. Dan. Vid. Selsk. **1** (3) (1957)

430. R. G. Stokstad, in *Treatise on heavy-ion science*, ed. D. A. Bromley, (Plenum, New York, 1984), vol. **3**

431. J. R. Grover, J. Gilat, Phys. Rev. **157**, 802 (1967)

432. M. A. Preston, R. K. Bhaduri, *Structure of the Nucleus* (Addison-Wesley, Reading, 1975) Appendix

433. M. Blann, Phys. Rev. C **21**, 1770 (1980)

434. M. Hillman, Y. Eyal, in *Proc. European Conference on Nuclear Physics with Heavy Ions, Caen, 1976*, eds. B. Fernandez et al., (Commissariat à l'Energie Atomique, 1976) vol. **1** pp. 109

435. A. Gavron, Phys. Rev. C **21**, 230 (1980)

436. J. P. Lestone, J. R. Leigh, O. Newton, D. J. Hinde, J. X. Wei, J. X. Chen, S. Elfström, M. Zielinska-Pfabé, Nucl. Phys. **A559**, 277 (1993)

437. J. J. Griffin, Phys. Rev. Lett. **17**, 478 (1966)

438. P. Obložinský, I. Ribanský, E. Běták, Nucl. Phys. **A226**, 347 (1974)

439. M. Blann, Ann. Rev. Nucl. Sci. **25**, 123 (1975)

440. K. Seidel, D. Seeliger, R. Reif, V. D. Toneev, Fiz. Elem. Chastits At. Yad. **7**, 499 (1976), [Sov. J. Part. Nucl. **7**, 192 (1976)]

441. E. Gadioli, P. E. Hodgson, *Preequilibrium Nuclear Reactions* (Clarendon, Oxford, 1992)

442. A. S. Iljinov, M. V. Kazarnovsky, E. Ya. Paryev, *Intermediate-Energy Nuclear Physics* (CRC, Boca Raton, 1994)

443. W. D. Myers, Nucl. Phys. **A145**, 387 (1970)

444. H. J. Krappe, Int. J. Mod. Phys. **E16**, 396 (2007)

445. H. J. Krappe, S. Fadeev, Nucl. Phys. **A690**, 431 (2001); *Proc. Workshop of the Nuclear Many-Body Problem, Brijuni (Croatia) 2001*, eds. W. Nazarewicz, D. Vretenar (Kluwer Academic, Amsterdam, 2002) pp. 195

446. S. Grossmann, H. J. Krappe, Phys. Rev. C **34**, 914 (1986)

447. M. Martinot, M. Gaudin, Rev. Roum. Phys. **22**, 17 (1977)

448. S. Großmann, A. Müller, Z. Phys. **B57**, 161 (1984)

449. P. Fong, Phys. Rev. **89** (1953) 332; Phys. Rev. **102**, 434 (1956)

450. P. Fong, *Statistical Theory of Nuclear Fission* (Gordon and Breach, New York, 1969)

451. A. V. Ignatyuk, Yad. Fis. **9** 357 (1969), [Sov. J. Nucl. Phys. **9**, 208 (1969)]

452. B. D. Wilkins, E. P. Steinberg, R. R. Chaseman, Phys. Rev. C **14**, 1832 (1976)

453. A. S. Jensen, J. Damgaard, Nucl. Phys. **A203**, 578 (1973)

454. L. G. Moretto, Phys. Lett. **40B**, 1 (1972)

455. A. C. Wahl, A. E. Norris, R. A. Rouse, J. C. Williams, *Proc. 2nd IAEA Symposium on the Physics, Chemistry of Fission, Vienna, 1969* (IAEA, Vienna, 1970) pp. 813

456. M. G. Itkis, V. V. Okolovich, G. N. Smirenkin, Nucl. Phys. **A502** 243c (1989)

457. J. P. Unik, J. E. Gindler, L. E. Glendenin, K. F. Flynn, A. Gorski, R. K. Sjoblom, *Proc. Third IAEA Symposium on the Physics, Chemistry of Fission, Rochester, 1973* (IAEA, Vienna, 1974) vol. **2**, pp. 19

458. W. John, E. K. Hulet, R. W. Lougheed, J. J. Wesolowski, Phys. Rev. Lett. 45 **27** (1971)

459. E. Konecny, H. W. Schmitt, Phys. Rev. **172**, 1213 (1968)

460. W. Reisdorf, J. P. Ulik, H. C. Griffin, L. E. Glendenin, Nucl. Phys. **177**, 337 (1971)

461. J. Terrell, Phys. Rev. **127**, 880 (1962)

462. A. Ruben, H. Märten, D. Seeliger, Z. Phys. **A338**, 67 (1991)

463. V. E. Viola, K. Kwiatkowski, M. Walker, Phys. Rev. C **31**, 1550 (1985)

464. M. G. Mustafa, U. Mosel, H. W. Schmitt, Phys. Rev. C **7** 1519 (1973)

465. P. Fong, Phys. Rev. C **17** 1731 (1978)

466. M. Prakash, V. S. Ramamurty, S. S. Kapoor, in *Proc. 4th IAEA Symposium on the Physics, Chemistry of Fission, Jülich, 1979* (IAEA, Vienna, 1980) vol. **2**, pp. 353

467. V. A. Rubchenya, S. G. Yavshits, Yad. Fis. **40**, 649 (1983), [Sov. J. Nucl. Phys. **40**, 416 (1984)]

468. L. Meitner, Nature **165**, 561 (1950)

469. A. Turkevich, J. B. Niday, Phys. Rev. **84**, 52 (1951)

470. E. Konecny, H. J. Specht, J. Weber, in *Proc. Third IAEA Symposium on the Physics, Chemistry of Fission, Rochester, 1973* (IAEA, Vienna, 1974) vol. **2**, pp. 3

471. H. C. Britt, H. E. Wegner, J. C. Gursky, Phys. Rev. **129**, 2239 (1963)

472. F. Gönnenwein, E. Pfeiffer, Z. Phys. **207**, 209 (1967)

473. E. K. Hulet, J. F. Wild, R. J. Dougan, R. W. Lougheed, J. H. Landrum, A. D. Dougan, M. Schädel, R. L. Hahn, P. A. Baisden, C. M. Henderson, R. J. Dupzyk, K. Sümmerer, G. R. Bethune, Phys. Rev. Lett. **56**, 313 (1986); Phys. Rev. C **40** 770 (1989)

474. S. Ćwiok, P. Rozmej, A. Sobiczewski, Z. Patyk, Nucl. Phys. **A491**, 281 (1989)

475. U. Brosa, H.-H. Knitter, in *Proc. 18th Intern. Symp. Nucl. Physics, Gaußig 1988*, eds. H. Märten, D. Seeliger, Rossendorf report ZfK-732, pp. 145 (unpublished)

476. V. V. Pashkevich, Nucl. Phys. **A477**, 1 (1988)

477. U. Brosa, S. Großmann, A. Müller, Z. Naturfor. **41a**, 1341 (1986)

478. S. I. Mulgin, V. N. Okolovich, S. V. Zhdanova, Phys. Lett. **462B**, 29 (1999)

479. S. Oberstedt, F. J. Hambsch, F. Vivés, Nucl. Phys. **A644**, 289 (1998)

480. A. Wolf, E. Cheifetz, Phys. Rev. C **13**, 1952 (1976)

481. L. G. Moretto, R. P. Schmitt, Phys. Rev. C **21**, 204 (1980)

482. J. R. Nix, W. S. Swiatecki, Nucl. Phys. **71**, 1 (1965)

483. R. P. Schmitt, A. J. Pacheco, Nucl. Phys. **A379**, 313 (1982)

484. L. G. Moretto, G. F. Peaslee, G. J. Wozniak, Nucl. Phys. **A502**, 453c (1989)

485. R. Freifelder, M. Prakash, J. M. Alexander, Phys. Rep. **133C**, 315 (1986)

486. J. O. Newton, Fiz. Elem. Chastits At. Yadra **21**, 821 (1990), [Sov. J. Part. Nucl. **21**, 349 (1990)]

487. R. P. Schmitt, G. Muchaty, D. R. Haenni, Nucl. Phys. **A427**, 614 (1984)

488. J. O. Rasmussen, W. Nörenberg, H. J. Mang, Nucl. Phys. **A136**, 465 (1969)

489. M. Zielińska-Pfabé, K. Dietrich, Phys. Lett. **49B**, 123 (1974)

490. M. Zielińska-Pfabé, Nukleonika **23**, 113 (1978)

491. F. Gönnenwein, I. Tsekhanovich, V. Rubchenya, Int. J. Mod. Phys. **E16**, 212 (2007)

492. J. Trochon, G. Simon, C. Signarbieux, in *Proc. Conf. "50 Years with Nuclear Fission", Gaithersburg 1989*, eds. J. W. Behrens, A. D. Carlson (Amer. Phys. Soc., La Grange Park, Ill. 1989) vol. **1**, pp. 313

493. P. Armbruster, Rep. Prog. Phys. **62**, 465 (1999)

494. C. Signarbieux, M. Montoya, M. Ribrag, C. Mazur, C. Guet, P. Perrin, M. Maurel, J. Physique Lett. **42**, L-437 (1981)

495. J. Kaufmann, W. Mollenkopf, F. Gönnenwein, P. Geltenbort, A. Oed, Z. Phys. **A341**, 319 (1992)

496. G. Simon, J. Trochon, F. Brisard, C. Signarbieux, Nucl. Inst. Meth. **A286**, 220 (1990)

497. F. Gönnenwein, B. Börsig, Nucl. Phys. **A530**, 27 (1991)

498. Y. Boneh, Z. Fraenkel, I. Nebenzahl, Phys. Rev. **156**, 1305 (1967)

499. C. Guet, H. Nifenecker, C. Signarbieux, M. Asghar, in *Proc. 4th IAEA Symposion on the Physics, Chemistry of Fission, Jülich, 1979* (IAEA, Vienna, 1980), vol. **2**, pp. 247

500. G. Pik-Pichak, Yad. Fiz. **40** 336 (1984), [Sov. J. Nucl. Phys. **40**, 215 (1984)]
501. A. S. Roshchin, V. A. Rubchenya, S. G. Yavshits, Yad. Fiz. **57** 974 (1994), [Phys. Atom. Nucl. **57** 914 (1994)]
502. Y. Gazit, A. Katase, G. Ben-David, R. Moreh, Phys. Rev. C **4**, 223 (1971)
503. W. Baum, *PHD Thesis* (Technische Hochschule Darmstadt, 1992)
504. I. S. Guseva, Yu. I. Gusev, in *Proc. of the XIV Intern. Seminar on Interactions of Neutrons with Nuclei, Dubna 2006*, pp. 101
505. I. Halpern, Annu. Rev. Nucl. Sci. **21**, 245 (1971)
506. M. Wöstheinrich, M. Hesse, F. Gönnenwein, in *Proc. of the Third Intern. Conference on Dynamical Aspects of Nuclear Fission, Častá-Papiernička, Slovac Republic, 1996*, eds. J. Kliman, B. Pustylnik (JINR Report E6,7-97-49, Dubna, 1997) pp. 231
507. F. Gönnenwein, private communication
508. L. M. Robledo, J. L. Egido, B. Nerlo-Pomorska, K. Pomorski, Phys. Lett. **B201**, 409 (1988)
509. H. Goutte, P. Casoli, J. F. Berger, Nucl. Phys. **A734**, 217 (2004)
510. H. A. Kramers, Physica **VII**, 284 (1940)
511. D. J. Jackson, M. Rhoades-Brown, Ann. Phys. (N.Y.) **105**, 151 (1977)
512. T. Fliessbach, Z. Phys. **A272**, 39 (1975)
513. T. Fliessbach, H. J. Mang, Nucl. Phys. **A263**, 75 (1976)
514. R. Blendowske, T. Fliessbach, H. Walliser, Z. Phys. **A339**, 121 (1990)
515. W. Greiner, M. Ivascu, D. N. Poenaru, A. Sandulescu, in *Treatise on Heavy-Ion Science, Vol. 8: Nuclei Far From Stability*, ed. D. A. Bromley (Plenum, New York, 1989) pp. 641
516. B. Podolsky, Phys. Rev. **32**, 812 (1928)
517. W. Pauli, in *Encyclopedia of Physics*, ed. S. Flügge (Springer, Berlin, 1958) Vol. **5** *Principles of Quantum Theory*, Part **1**, pp. 1
518. H. Hofmann, Z. Phys. **A250**, 14 (1972)
519. D. R. Bès, Mat. Fys. Medd. Dan. Vid. Selsk. **33** (2) (1961)
520. A. K. Kerman, Ann. Phys. (N.Y.) **12**, 300 (1961)
521. A. Sobiczewski, Z. Szymański, S. Wycech, S. G. Nilsson, J. R. Nix, X. F. Tsang, C. Gustafson, P. Möller, B. Nilsson, Nucl. Phys. **A131**, 67 (1969)
522. Z. Łojewski, A. Baran, Z. Phys. **A322**, 695 (1985)
523. L. G. Moretto, R. P. Babinet, Phys. Lett. **49B**, 147 (1974)
524. A. Staszczak, A. Baran, K. Pomorski, K. Böning, Phys. Lett. **161B**, 227 (1985)
525. A. Staszczak, S. Piłat, K. Pomorski, Nucl. Phys. **A504**, 589 (1989)
526. J. Griffin, J. Wheeler, Phys. Rev. **108**, 311 (1957)
527. D. M. Brink, A. Weiguny, Nucl. Phys. **A120**, 59 (1968)
528. J. Libert, M. Girod, J.-P. Delaroche, Phys. Rev. C **60**, 054301 (1999)
529. A. N. Tikhonov, V. Y. Arsenin, *Solution of Ill-posed Problems* (John Wiley, New York, 1981)
530. H. Hofmann, K. Dietrich, Nucl. Phys. **A165**, 1 (1971)
531. N. Onishi, T. Une, Prog. Theor. Phys. **53**, 504 (1975)
532. A. Lichnerowicz, *Elements of Tensor Calculus* (Methuen, London, 1962) Chap. 4
533. A. Bobyk, K. Pomorski, Z. Phys. **A339**, 11 (1991)
534. A. Kamlah, in *Proc. Int. Conf. on Nucl. Phys., Munich, 1973* eds. J. de Boer, H. J. Mang (North Holland, Amsterdam, 1973) vol. **1**, pp. 127
535. B. Giraud, B. Grammaticos, Nucl. Phys. **A223**, 373 (1974); **A255**, 141 (1975); **A330**, 40 (1979).

536. H. Goutte, J. F. Berger, P. Casoli, D. Gogny, Phys. Rev. C **71** 024316 (2005)
537. R. E. Peierls, D. J. Thouless, Nucl. Phys. **A38**, 154 (1962)
538. K. Goeke, P.-G. Reinhard, Ann. Phys. (N.Y.) **112**, 328 (1978); Ann. Phys. (N.Y.) **124**, 249 (1980)
539. K. Goeke, F. Grümmer, P.-G. Reinhard, Ann. Phys. (N.Y.) **150**, 504 (1983)
540. P.-G. Reinhard, F. Grümmer, K. Goeke, Z. Phys. **A317**, 339 (1984)
541. J. Dudek, Phys. Lett. **34B**, 181 (1971); Nucl. Phys. **A194**, 552 (1972); Nucl. Phys. **A203**, 121 (1973)
542. G. Gamov, Z. Phys. **51**, 204 (1928)
543. E. Fermi, *Nuclear Physics, a course given at the University of Chicago*. Notes compiled by J. Orear, A. H. Rosenfeld, R. A. Schluter, Revised Edition (Univ. of Chicago, Chicago, 1950)
544. N. Fröman, Ö. Dammert, Nucl. Phys. **A143**, 627 (1970)
545. A. Baran, Phys. Lett. **76B**, 8 (1978)
546. T. Ledergerber, H.-C. Pauli, Nucl. Phys. **A207**, 1 (1973)
547. A. Baran, K. Pomorski, A. Lukasiak, A. Sobiczewski, Nucl. Phys. **A361** 83 (1981)
548. A. Dobrowolski, K. Pomorski, J. Bartel, Phys. Rev. C **75**, 024613 (2007)
549. W. J. Swiatecki, Phys. Rev. **100**, 937 (1955)
550. A. E. S. Green, Phys. Rev. **95**, 1006 (1954)
551. NuDat 2.5, *Data Base of the Berkeley National Laboratory*, (http://www.nndc.bnl.gov/nudat2/)
552. P. Möller, J. R. Nix, W. J. Swiatecki, Nucl. Phys. **A469**, 1 (1981); Nucl. Phys. **492**, 349 (1989)
553. D. Hoffman, in *Proc. Conf. "50 Years with Nuclear Fission"* eds. J. W. Behrens, A. D. Carlson (Amer. Nucl. Society, La Grange Park, Ill. 1989) vol. **1**, pp. 83
554. D. Hoffman, Nucl. Phys. **A502**, 21c (1989)
555. J. Randrup, C. F. Tsang, P. Möller, S. G. Nilsson, S. E. Larsson, Nucl. Phys. **A217**, 221 (1973)
556. J. Randrup, S. E. Larsson, P. Möller, S. G. Nilsson, K. Pomorski, A. Sobiczewski, Phys. Rev. C **13**, 229 (1976)
557. E. O. Fiset, J. R. Nix, Nucl. Phys. **A193**, 647 (1972)
558. R. Smolańczuk, J. Skalski, A. Sobiczewski, Phys. Rev. C **52**, 1871 (1995)
559. R. Smolańczuk, H. V. Klapdor-Kleingrothaus, A. Sobiczewski, Acta Phys. Polon. **B24**, 685 (1993)
560. K. Pomorski, Nukleonika **23**, 125 (1978)
561. J. A. Maruhn, W. Greiner, Phys. Rev. C **13**, 2404 (1976)
562. S. K. Samaddar, D. Sperber, M. Zielińska-Pfabé, M. I. Sobel, Phys. Scripta **25**, 517 (1982)
563. A. Săndulescu, D. N. Poenaru, W. Greiner, Fiz. Elem. Chastits At. Yadra **11**, 1334 (1980), [Sov. J. Part. Nucl. **11**, 528 (1980)]
564. H. J. Rose, G. A. Jones, Nature **307**, 245 (1984)
565. A. Bonetti, A. Guglielmetti, Romanian Reports in Physics **59**, 301 (2007)
566. D. N. Poenaru, D. Schnabel, W. Greiner, D. Mazilu, R. Gherghescu, At. Data Nucl. Data Tab. **48** 231 (1991)
567. Yu. S. Zamyatnin, S. G. Kadmenskiĭ, S. D. Kurgalin, Yad. Fiz. **57**, 1981 (1994), [Phys. At. Nucl. **57**, 1905 (1994)]
568. G. A. Pik-Pichak, Yad. Fiz. **44**, 1421 (1986), [Sov. J. Nucl. Phys. **44**, 923 (1986)]

569. S. P. Tretyakova, A. A. Oglobin, G. A. Pik-Pichak, Yad. Fiz. **66**, 1665 (2003), [Phys. At. Nucl. **66**, 1618 (2003)]
570. W. Furman, S. Kadmensky, Yu. Tchuvilsky, Z. Phys. **A349**, 301 (1994)
571. A. Guglielmetti, R. Bonetti, G. Poli, R. Collatz, Z. Hu, R. Kirchner, E. Roeckl, N. Gunn, P. B. Price, B. A. Weaver, A. Westphal, Phys. Rev. C **56**, R2912 (1997)
572. B. G. Novatsky, A. A. Ogloblin, Vestnik Akad. Nauk. SSSR **1**, 81 (1988)
573. I. Perlman, A. Ghiorso, G. T. Seaborg, Phys. Rev. **77**, 26 (1950)
574. H. J. Mang, Phys. Rev. **119**, 1069 (1960); Ann. Rev. Nucl. Sci. **14**, 1 (1964)
575. S. G. Kadmenskiĭ, W. I. Furman, J. M. Chuvilskiĭ, Izv. Akad. Nauk SSSR, Ser. Fiz. **50**, 1786 (1986), [Bull. USSR Acad. Sci. Phys. Ser. **50** (9), 116 (1986)]
576. R. Blendowske, H. Walliser, Phys. Rev. Lett. **61**, 1930 (1986)
577. V. P. Bugrov, W. I. Furman, S. G. Kadmemsky, S. D. Kurgalin, Yu. M. Tchuvil'sky, in AIP Conf. Proc. **644** *Fourth Catania Relativistic Ion Studies, Exotic Clustering, 2001,* eds. S. Costa, A. Insolia, and C. Tuvè (Amer. Inst. Phys. 2002) pp. 142
578. A. A. Ogloblin, G. A. Pik-Pichak, S. P. Tretyakova, in *Proc. Int. Workshop on Fission Dynamics of Atomic Clusters and Nuclei, Luso, 2000,* eds. J. da Providência, D. M. Brink, F. Karpechine, and F. B. Malik (World Scientific, Singapore, London, New Jersey, 2001) pp. 143
579. A. A. Ogloblin, N. I. Venikov, S. K. Lisin, S. V. Pirozhkov, V. A. Pchelin, Yu. F. Rodionov, V. M. Semochkin, V. A. Shabrov, I. K. Shvetsov, V. M. Shubko, S. P. Tretyakova, V. L. Mikheev, Phys. Lett. **B235**, 35 (1990)
580. P. B. Price, Nucl. Phys. **A502**, 41c (1989)
581. Y.-J. Shi, W. J. Swiatecki, Nucl. Phys. **A464**, 205 (1987); Nucl. Phys. **A438**, 450 (1985)
582. D. N. Poenaru, J. A. Maruhn, W. Greiner, M. Ivaşcu, D. Mazilu, I. Ivaşcu, Z. Phys. **A333**, 291 (1989)
583. D. N. Poenaru, M. Ivaşcu, A. Săndulescu, W. Greiner, J. Phys. G **10**, L183 (1984)
584. D. N. Poenaru, M. Ivaşcu, A. Săndulescu, W. Greiner, Phys. Rev. C **32**, 572 (1985)
585. D. N. Poenaru, W. Greiner, M. Ivaşcu, D. Mazilu, I. Plonski, Z. Phys. **A325**, 435 (1986)
586. D. N. Poenaru, W. Greiner, Phys. Scr. **44**, 427 (1991)
587. D. N. Poenaru, W. Greiner, J. Phys. G **17**, S443 (1991)
588. D. N. Poenaru, R. A. Gherghescu, W. Greiner, Phys. Rev. C **73**, 014608 (2006)
589. D. N. Poenaru, D. Mazilu, M. Ivaşcu, J. Phys. G **5**, 1093 (1979)
590. H. Risken, *The Fokker-Planck Equation,* 2nd edition (Springer, Berlin, 1989)
591. I. I. Gontchar, P. Fröbrich, Nucl. Phys. **A551**, 495 (1993)
592. I. I. Gontchar, P. Fröbrich, N. I. Pischasov, Phys. Rev. c **47**, 2228 (1993)
593. P. Fröbrich, I. I. Gontchar, N. D. Mavlitov, Nucl. Phys. **A556**, 281 (1993).
594. P. Fröbrich, I. I. Gontchar, Phys. Rep. **292C**, 131 (1998)
595. A. E. Gettinger, I. I. Gontchar, J. Phys. G **26**, 347 (2000)
596. L. G. Kelly, *Handbook of Numerical Mathematics and Applications* (Addison-Wesley, Reading, Ma, 1969) Chap. 14.7
597. N. G. van Kampen, *Stochastic Processes in Physics and Chemistry* (North-Holland, Amsterdam, 1981)
598. E. Novak, *Deterministic, Stochastic Error Bounds in Numerical Analysis,* Lecture Notes in Mathematics, vol. **1349**, eds. A. Dold, B. Eckmann (Springer, Berlin, 1988)

599. Y. Abe, C. Grégoire, H. Delagrange, J. de Physique Coll. **47**, C4-329 (1986)
600. A. V. Karpov, P. N. Nadtochy, D. V. Vanin, G. D. Adeev, Phys. Rev. C **63**, 054610 (2001)
601. P. N. Nadtochy, G. D. Adeev, A. V. Karpov, Phys. Rev. C **65**, 064615 (2002)
602. T. Ichikawa, T. Asano, T. Wada, M. Ohta, J. Nucl. Radiochem. Sciences **3**, 67 (2002)
603. G. I. Kosenko, F. A. Ivanyuk, V. V. Pashkevich, J. Nucl. Radiochem. Sciences **3**, 71 (2002)
604. C. Schmitt, J. Bartel, K. Pomorski, A. Surowiec, Acta Phys. Polon. **B34**, 1651 (2003)
605. P. N. Nadtochy, G. D. Adeev, Phys. Rev. C **72**, 054608 (2005)
606. H. Delagrange, C. Grégoire, Y. Abe, N. Carjan, J. de Physique Coll. **47**, C4-305 (1986)
607. J. P. Lestone, Phys. Rev. Lett. **70**, 2245 (1993)
608. K. Pomorski, J. Bartel, J. Richert, K. Dietrich, Nucl. Phys. **A605**, 87 (1996)
609. H. Delagrange, C. Grégoire, F. Scheuter, Y. Abe, Z. Phys. **A323**, 437 (1986)
610. F. James, Comp. Phys. Comm. **60**, 329 (1990)
611. H. Niederreiter, Bull Am. Math. Soc. **84**, 957 (1978)
612. E. Holub, D. Hilscher, G. Ingold, U. Jahnke, H. Orf, and H. H. Rossner, Phys. Rev. C **28**, 252 (1983)
613. W. P. Zank, D. Hilscher, G. Ingold, U. Jahnke, M. Lehmann, H. H. Rossner, Phys. Rev. C **33**, 519 (1986)
614. D. Hilscher, D. J. Hinde, H. H. Rossner, in *Proc. of the Texas A&M Symposium on Hot Nuclei, College Station, Texas, 1987*, eds. S. Shlomo, R. P. Schmitt, J. B. Natowitz (World Scientific, Singapore, 1988) pp. 193
615. A. Gavron, A. Gayer, J. Boissevain, H. C. Britt, T. C. Awes, J. R. Beene, B. Cheynis, D. Drain, R. L. Ferguson, F. E. Obenshain, F. Plasil, G. R. Young, G. A. Petitt, C. Butler, Phys. Rev. C **35**, 579 (1987)
616. J. O. Newton, D. J. Hinde, R. J. Charity, J. R. Leigh, J. J. M. Bokhorst, A. Chatterjee, G. S. Foote, S. Ogaza, Nucl. Phys. **A483**, 126 (1988)
617. D. Hilscher, H. Rossner, Ann. Phys. (Paris) **17**, 471 (1992)
618. Y. Abe, in *Proc. Third IN2P3-Riken Symposium on Heavy Ion Collisions, Saitama, 1994*, eds. T. Motobayashi, N. Frascaria, M. Ishihara (World Scientific, Singapore, 1995) pp. 127
619. H. J. Krappe, in *Proc. 13th Meeting on the Physics of Nuclear Fission in Memory of Prof. G. N. Smirenkin, Obninsk 1995*, ed. B. D. Kuzminov (State Scientific Centre, Russian Federation, IPPE, Obninsk, 1995) pp. 134
620. Y. Abe, S. Ayik, P. G. Reinhard, E. Suraud, Phys. Rep. **275C**, 49 (1996)
621. G. D. Adeev, A. V. Karpov, P. N. Nadtochii, D. V. Vanin, Fiz. Elem. Chastits Atom. Yadra, **36**, 712 (2005), [Phys. Part. Nucl. **36**, 378 (2005)]
622. D. H. E. Gross, H. Kalinowski, Phys. Rep. **45**, 175 (1978)
623. U. Brosa, D. H. E. Gross, Z. Phys. **A298**, 91 (1980)
624. D. H. E. Gross, Phys. Lett. **68B**, 412 (1977)
625. S. Grossmann, Z. Phys. **A296**, 251 (1980)
626. P. Fröbrich, G.-R. Tillack, Nucl. Phys. **A540**, 353 (1992)
627. P. Grangé, H. A. Weidenmüller, Phys. Lett. **96B**, 26 (1980)
628. P. Grangé, Li Jun-Qing, H. A. Weidenmüller, Phys. Rev. C **27**, 2063 (1983)
629. Z. Lu, B. Chen, J. Zhang, Y. Zhuo, H. Han, Phys. Rev. C **42**, 707 (1990)
630. Y. Aritomo, J. Nucl. Radiochem. Sciences **3**, 17 (2002)

631. V. A. Drozdov, D. O. Eremenko, O. V. Fotina, G. Giardina, G. Malaguti, S. Yu. Platonov, O. A. Yuminov, Nucl. Phys. **A734**, 225 (2004)

632. A. V. Karpov, R. M. Hiryanov, A. V. Sagdeev, G. D. Adeev, J. Phys. G **34**, 255 (2007)

633. N. D. Mavlitov, P. Fröbrich, I. I. Gotchar, Z. Phys. **A342** 195 (1992)

634. H. A. Weidenmüller, Zhang Jing-Shang, J. Statist. Phys. **34**, 191 (1984)

635. D. J. Hinde, D. Hilscher, H. Rossner, Nucl. Phys. **A538**, 243c (1992)

636. D. J. Hinde, D. Hilscher, H. Rossner, B. Gebauer, M. Lehmann, M. Wilpert, Phys. Rev. C **45**, 1229 (1992)

637. G. D. Adeev, V. V. Pashkevich, Nucl. Phys. **A502**, 405c (1989)

638. G. I. Kosenko, I. G. Cagliari, G. D. Adeev, Yad. Fiz. **60**, 404 (1997), [Phys. Atom. Nucl. **60**, 334 (1997)]

639. G. G. Chubarian, M. G. Itkis, S. M. Lukyanov, V. N. Okolovich, Yu. E. Penionzkevich, A. Ya. Rusanov, V. S. Salamatin, G. N. Smirenkin, Yad. Fiz. **56**, 3 (1993), [Phys. Atom. Nucl. **56**, 286 (1993)]

640. V. I. Zagrebaev, Y. Aritomo M. G. Itkis, Yu. Ts. Oganessian, M. Ohta, Phys. Rev. C **65**, 014607 (2001)

641. Y. Abe, A. Marchix, C. Shen, B. Yilmas, G. Kosenko, D. Boilley, B. G. Giraud, Int. J. Mod. Phys. **E16**, 491 (2007)

642. A. V. Karpov, P. N. Nadtochy, E. G. Ryabov, G. D. Adeev, J. Phys. G **29**, 2365 (2003)

643. R. Graham, Z. Phys. **B26**, 397 (1977)

644. H. Grabert, R. Graham, M. S. Green, Phys. Rev. B **26**, 2136 (1980)

645. E. Strumberger, *PHD thesis* (Technische Universität München, 1990)

646. E. Strumberger, K. Dietrich, K. Pomorski, Nucl. Phys. **A529**, 522 (1991)

647. P. Fröbrich, S. Y. Xu, Nucl. Phys. **A477**, 143 (1988)

648. O. Edholm, O. Leimar, Physica **98A**, 313 (1979)

649. V. M. Strutinsky, Phys. Lett. **47B**, 121 (1973)

650. I. I. Gontchar, P. Fröbrich, Nucl. Phys. **A575**, 283 (1994)

651. S. Glasstone, K. J. Laidler, H. Eyring, *The Theory of Rate Processes* (McGraw-Hill, New York, 1941) Chap. 1

652. C. Ngô, H. Hofmann, Z. Phys. **A282**, 83 (1977)

653. S. Grossmann, H. J. Krappe, Z. Phys. **A298**, 41 (1980)

654. P. Fröbrich, B. Strack, M. Durand, Nucl. Phys. **A406**, 557 (1983)

655. K. Pomorski, H. Hofmann, J. de Physique **42**, 381 (1981)

656. G. D. Adeev, I. I. Gontchar, Z. Phys. **A320**, 451 (1985)

657. G. D. Adeev, I. I. Gontchar, V. V. Pashkevich, N. I. Pischasov, O. I. Serdyuk, Fiz. Elem. Chastits At. Yadra **19**, 1229 (1988), [Sov. J. Part. Nucl. **19**, 529 (1988)]

658. W. Schönauer, K. Reith, G. Glotz, Z. angew. Math. Mech. **62**, T352 (1982)

659. U. Brosa, Z. Phys. **A298**, 77 (1980)

660. U. Brosa, W. Cassing, Z. Phys. **A307**, 167 (1982)

661. F. Scheuter, H. Hofmann, Nucl. Phys. **A394**, 477 (1983)

662. H. Lamb, *Hydrodynamics*, 6th edition (Dover Publications, New York, 1945)

663. R. L. Hatch, A. J. Sierk, Nucl. Phys. **A341**, 513 (1980)

664. U. Brosa, H. J. Krappe, Quart. J. Mech. Appl. Math. **33**, 159 (1980)

665. A. A. Seregin, Yad. Fiz. **55**, 2639 (1992), [Sov. J. Nucl. Phys. **55**, 1473 (1992)]

666. F. G. Werner, J. A. Wheeler, *Superheavy Nuclei, the Kolesnikov-Larkin-Mottelson Effect in Sponaneous Fission, Appendix 3*, (unpublished manuscript, no date)

667. M. H. C. Knudsen, *The Kinetic Theory of Gases* (Wiley, New York, 1950)
668. D. H. E. Gross, Nucl. Phys. **A240**, 472 (1975)
669. J. Blocki, Y. Boneh, J. R. Nix, J. Randrup, M. Robel, A. J. Sierk, W. J. Swiatecki, Ann. Phys. (N.Y.) **113**, 330 (1978)
670. W. Nörenberg, Nucl. Phys. **A409**, 191c (1983)
671. H. Hofmann, F. A. Ivanyuk, S. Yamaji, Nucl. Phys. **A598**, 187 (1996)
672. S. E. Koonin, R. L. Hatch, J. Randrup, Nucl. Phys. **A283**, 87 (1977)
673. J. R. Nix, A. Sierk, Phys Rev. C **21**, 396 (1980)
674. H. Hofmann, *The Physics of Warm Nuclei with Analogies to Mesoscopic Systems* (Oxford University, Oxford, 2008) Chap. 7
675. S. Pal, T. Mukhopadhyay, Phys. Rev. C **54**, 1333 (1996)
676. G. E. Brown, M. Bolsterli, Phys. Rev. Lett. **3**, 472 (1959)
677. A. B. Migdal, *Theory of Finite Fermi Systems, Applications to Atomic Nuclei* (Wiley, New York, 1967)
678. J. R. Nix, A. J. Sierk, in *Proc. Sixth Adriatic Int. Conf. on Nucl. Physics, Dubrovnik, 1987*, eds. N. Cindro, R. Čaplar, W. Greiner (World Scientific, Singapore, 1990) pp. 333
679. D. Cha, G. F. Bertsch, Phys. Rev. C **46**, 306 (1992)
680. M. Thoennessen, J. R. Beene, F. E. Bertrand, C. Barktash, M. L. Halbert, D. J. Horen, D. C. Hensley, D. G. Sarantites, W. Sprang, D. W. Stracener, R. L. Varner, Phys. Lett. **282B**, 288 (1992)
681. R. Butsch, D. J. Hofman, C. P. Montoya, P. Paul, M. Thoennessen, Phys. Rev. C **44**, 1515 (1991)
682. I. Diószegi, D. J. Hofman, C. P. Montoya, S. Schadmand, P. Paul, Phys. Rev. C **46**, 627 (1992)
683. J. Błocki, R. Płaneta, J. Brzychczyk, K. Grotowski, Z. Phys. **A341**, 307 (1992)
684. T. Wada, Y. Abe, N. Carjan, Phys. Rev. Lett. **70**, 3538 (1993)
685. D. Hilscher, I. I. Gontchar, H. Rossner, Yad. Fiz. **57** 1255(1994), [Phys. Atom. Nucl. **57**, 1187 (1994)]
686. R. Hasse, Ann. Phys. (N.Y.) **93**, 68 (1975)
687. A. J. Sierk, J. R. Nix, Phys. Rev. C **21**, 982 (1980)
688. J. Randrup, W. J. Swiatecki, Ann. Phys. (N.Y.) **125**, 193 (1980)
689. R. Beck, D. H. E. Gross, Phys. Lett. **47B**, 143 (1973)
690. H. J. Krappe, Nucl. Phys. **A505**, 417 (1989)
691. W. Swiatecki, J. de Physique Coll. **33**, C5-45 (1972)
692. J. Randrup, Ann. Phys. (N.Y.) **112**, 356 (1978)
693. C. M. Ko, Phys. Rev. C **19**, 2417 (1979)
694. J. R. Nix, A. J. Sierk, Nucl. Phys. **A428**, 161c (1984)
695. H. Feldmeier, Rep. Progr. Phys. **50**, 915 (1987) 915
696. J. Blocki, J.-J. Shi, W. J. Swiatecki, Nucl. Phys. **A554**, 387 (1993)
697. J. Blocki, F. Brut, T. Srokowski, W. J. Swiatecki, Nucl. Phys. **A545**, 511c (1992)
698. G. Bennetin, L. Galgani, J. M. Strelcyn, Phys. Rev. A **14**, 2338 (1976)
699. G. Chaudhuri, S. Pal, Phys. Rev. C **63**, 064603 (2001)
700. A. J. Sierk, S. E. Koonin, J. R. Nix, Phys. Rev. C **17**, 646 (1978)
701. C. Yannouleas, M. Dworzecka, J. J. Griffin, Nucl. Phys. **A339**, 219 (1980)
702. L. D. Landau, E. M. Lifshitz, *Statistical Physics, Part 2 (Theory of the Condensed State)*, by E. M. Lifshitz, L. P. Pitaevskiĭ, (Pergamon, Oxford, 1981), §39-§40

703. F. Oberhettinger, L. Badii, *Table of Laplace Transforms* (Springer, Berlin, 1973)

704. P. J. Johansen, P. J. Siemens, A. S. Jensen, H. Hofmann, Nucl. Phys. **A288**, 152 (1977)

705. G. F. Bertsch, Z. Phys. **A289**, 103 (1978)

706. F. A. Ivanyuk, K. Pomorski, Phys. Rev. C **53**, 1861 (1996)

707. W. Nörenberg, Phys. Lett. **104B**, 107 (1981)

708. P. Morel, P. Nozières, Phys. Rev. **126**, 1909 (1962)

709. G. F. Bertsch, in *Les Houches Session XXX, 1977*, eds. R. Balian, M. Rho, G. Ripka (North Holland, Amsterdam, 1978) vol. **1**, pp. 175.

710. Y. Boneh, Z. Fraenkel, Phys. Rev. C **10**, 893 (1974)

711. T. Ledergerber, Z. Paltiel, H. C. Pauli, G. Schütte, Y. Yarif, Z. Fraenkel, Phys. Lett. **56B**, 417 (1975)

712. G. Schütte, L. Wilets, Phys. Rev. C **12**, 2100 (1975); Nucl. Phys. **A252**, 31 (1975); Z. Phys. **A286**, 313 (1978)

713. G. Schütte, P. Möller, J. R. Nix, A. J. Sierk, Z. Phys. **A297**, 289 (1980)

714. D. Glas, U. Mosel, Nucl. Phys. **A264**, 268 (1976)

715. S. Pal, D. H. E. Gross, Z. Phys. **A329**, 349 (1988)

716. J. B. Delos, Rev. Mod. Phys. **53**, 287 (1981)

717. S. Ayik, W. Nörenberg, Z. Phys. **A309**, 121 (1982)

718. L. Landau, Phys. Z. UdSSR **1**, 426 (1932); **2**, 46 (1932)

719. C. Zener, Proc. Roy. Soc. **A137**, 696 (1932)

720. E. C. G. Stückelberg, Helv. Phys. Acta **5**, 369 (1932)

721. K. Niita, W. Nörenberg, S. J. Wang, Z. Phys. **A326**, 69 (1987)

722. A. Lukasiak, W. Cassing, W. Nörenberg, Nucl. Phys. **A426**, 181 (1984)

723. P. Rozmej, W. Nörenberg, Phys. Lett. **177B**, 278 (1986)

724. A. Lukasiak, W. Nörenberg, Z. Phys. **A326**, 79 (1987)

725. D. Berdichevsky, A. Lukasiak, W. Nörenberg, P. Rozmej, Nucl. Phys. **A499**, 609 (1989); Nucl. Phys. **A502**, 395c (1989)

726. J. Błocki, J. Skalski, W. J. Swiatecki, Nucl. Phys. **A594**, 137 (1995)

Index

H. J. Krappe and K. Pomorski, *Theory of Nuclear Fission*,
Lecture Notes in Physics 838, DOI: 10.1007/978-3-642-23515-3,
© Springer-Verlag Berlin Heidelberg 2012